Twin Cities across Five Continents

This international collection provides a comprehensive overview of twin cities in different circumstances – from the emergent to the recently amalgamated, on 'soft' and 'hard' borders, with post-colonial heritage, in post-conflict environments and under strain.

With examples from Europe, the Middle East, Africa, Asia, South America, North America and the Caribbean, the volume sees twin cities as intense thermometers for developments in the wider urban world globally. It offers interdisciplinary perspectives that bridge history, politics, culture, economy, geography and other fields, applying these lenses to examples of twin cities in remote places. Providing a comparative approach and drawing on a range of methodologies, the book explores where and how twin cities arise; what twin cities can tell us about international borders; and the way in which some twin cities bear the spatial marks of their colonial past. The chapters explore the impact on twin-city relations of contemporary pressures, such as mass migration, the rise of populism, East-West tensions, international crime, surveillance, rebordering trends and epidemiological risks triggered by the COVID-19 pandemic. With case studies across the continents, this volume for the first time extends twin-city debates to fictional imaginings of twin cities.

Twin Cities across Five Continents is a valuable resource for researchers in the fields of anthropology, history, geography, urban studies, border studies, international relations and global development as well as for students in these disciplines.

Ekaterina Mikhailova is Postdoctoral Researcher at the Department of Geography and Environment, University of Geneva and Visiting Lecturer at the Graduate Institute of International and Development Studies (Switzerland). Ekaterina's work lies at the crossroads of urban studies, border studies and Russian studies.

John Garrard was Senior Lecturer in Politics and Contemporary History at Salford University (UK) until 2011. Although primarily a historian, his central teaching and research interests have bordered with political science.

Global Urban Studies

Providing cutting edge interdisciplinary research on spatial, political, cultural and economic processes and issues in urban areas across the US and the world, books in this series examine the global processes that impact and unite urban areas. The organizing theme of the book series is the reality that behavior within and between cities and urban regions must be understood in a larger domestic and international context. An explicitly comparative approach to understanding urban issues and problems allows scholars and students to consider and analyse new ways in which urban areas across different societies and within the same society interact with each other and address a common set of challenges or issues. Books in the series cover topics which are common to urban areas globally, yet illustrate the similarities and differences in conditions, approaches, and solutions across the world, such as environment/brownfields, sustainability, health, economic development, culture, governance and national security. In short, the Global Urban Studies book series takes an interdisciplinary approach to emergent urban issues using a global or comparative perspective.

The Millennial City
Trends, Implications, and Prospects for Urban Planning and Policy
Edited by Markus Moos, Deirdre Pfeiffer and Tara Vinodrai

Twin Cities
Urban Communities, Borders and Relationships over Time
Edited by John Garrard and Ekaterina Mikhailova

Disassembled Cities
Social and Spatial Strategies to Reassemble Communities
Edited by Elizabeth L. Sweet

Metropolis, Money and Markets
Brazilian Urban Financialization in Times of Re-emerging Global Finance
Jeroen Klink

Animals in the City
Edited by Laura A. Reese

Twin Cities across Five Continents
Interactions and Tensions on Urban Borders
Edited by Ekaterina Mikhailova and John Garrard

Twin Cities across Five Continents

Interactions and Tensions on Urban Borders

Edited by Ekaterina Mikhailova and
John Garrard

LONDON AND NEW YORK

First published 2022
by Routledge
2 Park Square, Milton Park, Abingdon, Oxon OX14 4RN

and by Routledge
605 Third Avenue, New York, NY 10158

Routledge is an imprint of the Taylor & Francis Group, an informa business

British Library Cataloguing-in-Publication Data
A catalogue record for this book is available from the British Library

Library of Congress Cataloging-in-Publication Data
Names: Mikhailova, Ekaterina (Postdoctoral researcher), editor. | Garrard, John, editor.
Title: Twin cities across five continents : interactions and tensions on urban borders / edited by Ekaterina Mikhailova and John Garrard.
Description: Abingdon, Oxon ; New York, NY : Routledge, 2022. | Series: Global urban studies | Includes bibliographical references and index.
Identifiers: LCCN 2021025683 (print) | LCCN 2021025684 (ebook) | ISBN 9780367609221 (hbk) | ISBN 9780367609245 (pbk) | ISBN 9781003102526 (ebk)
Subjects: LCSH: Cities and towns--Case studies. | Borderlands--Case studies. | Interregionalism--Case studies.
Classification: LCC HT151 .T89 2022 (print) | LCC HT151 (ebook) | DDC 307.76--dc23
LC record available at https://lccn.loc.gov/2021025683
LC ebook record available at https://lccn.loc.gov/2021025684

ISBN: 978-0-367-60922-1 (hbk)
ISBN: 978-0-367-60924-5 (pbk)
ISBN: 978-1-003-10252-6 (ebk)

DOI: 10.4324/9781003102526

Typeset in Goudy
by SPi Technologies India Pvt Ltd (Straive)

Contents

PART III
Twin Cities in Fiction and Editors' Dreams

Figures

Tables

Acknowledgements

The idea to producing the second volume of *Twin Cities* was approved by both editors in April 2019 when Ekaterina visited John and Eve Garrard in Manchester, UK. Hence the first acknowledgement goes to two editors' enthusiasm for studying twin cities and the Garrard family's hospitality.

This book's splendid author team comprises 33 scholars from 17 countries. They were invited to contribute their chapters by Ekaterina between July 2018 and December 2019 at various academic events from Incheon, South Korea, to Chandigarh, India, to Barcelona, Spain, and also through her professional network. Ekaterina wishes to thank the University of Eastern Finland, the Nordic Centre in India, Young European Research University Network and Northeast Asian Economic Forum for travel grants to undertake these trips and attend these academic events. In September 2020 Ekaterina started working at the University of Geneva as an Awardee of the Swiss Government Excellence Scholarship for Postdoctoral Researchers, providing much appreciated financial and institutional support during the final stages of the book preparation process. We hope all these places will feel we have produced something worthy of their financial help.

As individuals sentient of urban charm and fascinated with borders, the editors are thankful to places that have been their safety corners, creative playgrounds and sources of inspiration over time and more recently:

- medieval church-studded Pskov, Ekaterina's hometown in Russian North-West, 40 km away from the Russia-Estonia border;
- brick-laid, sky-scraping and deeply musical Manchester, the city where both *Twin Cities* edited volumes have been conceived, and from where John Garrard could contemplate its twin city of Salford, just across the river Irwell, indistinguishable from its vast neighbour but still determinedly separate;
- elegant, peaceful and liberal Geneva, Ekaterina's current home base next to the Swiss-French border;
- the decidedly un-urban village of Brora way up on the north-east coast of Scotland where John Garrard contemplated deserted beaches and thought about twin cities.

Finally, but importantly, we wish to thank the following people:

- for stimulating academic debates and encouragement: all our authors and Ekaterina's colleagues from the University of Geneva, notably Prof. Frédéric Giraut;
- for endless patience and great help: Nonita Saha, our Editorial Assistant at Routledge;
- for supplying the frontispiece that adorns this volume, just as it did volume 1: Eileen Dudley, artist and ex-mayor of the city of Twin City (Georgia, USA);
- for lifting Ekaterina's morale before challenging fieldtrips and for plentiful cultural influences including the discovery of the *Neverwhere*: Alexander Chernykh;
- for invaluable support, patience and sympathy during the long production process: Eve Garrard; Ekaterina Mikhailova's family – Zinaida, Lyudmila, Vladimir, Oksana and Dmitry; Leila Abbasova; Daria and Severyan Dyakonov; Simon Gunn; Michael Goldsmith; Theo Jepsen; Olga Leonova; Alexey Piven; Leyla Sayfutdinova; Lena Soboleva; Zina Vasilyeva and Harry, John Garrard's remarkable whippet-cross.

List of Contributors

Dr. Dorte Jagetic Andersen is Associate Professor at the Centre for Border Region Studies, Department of Political Science and Public Administration, University of Southern Denmark, with a background in European ethnology, European Continental philosophy and political science. Her main research interest concerns identity-formation in areas influenced by geopolitically drawn borders, focusing particularly on conflicts and their resolution and she has published widely on these issues in internationally recognized journals. Her recent work includes ethnographic studies in the border regions of Istria, Neum-Neretva and the Vukovar area in Slavonia, addressing issues like regional identity and belonging, war heritage as bordering practice and borderlands resilience.

Dr. Anna Anisimova is Senior Research Fellow in the Department of Medieval and Early Modern Studies at the Institute of World History, the Russian Academy of Sciences; and Associate Professor at the State Academic University for the Humanities. Anna holds an MA in history from Lomonosov Moscow State University, an MSt in medieval history from Oxford University and a PhD from the Institute of World History, the Russian Academy of Sciences. Anna is the author of 57 publications. Starting with her doctoral thesis, titled 'Town and Monastery in the Medieval England (on the material of South-Eastern Counties)', Anna has been researching the phenomenon of small towns, monastic towns and all aspects of their history including monastic participation in the process of urbanisation.

Dr. Anthony I. Asiwaju is Professor Emeritus of Comparative History and Borderlands Studies, University of Lagos; Fellow of Nigerian Academy of Letters, and Historical Society of Nigeria. Anthony was the First Commissioner of International Boundaries at Nigeria's National Boundary Commission (1988–1994) and Foundation Member of the African Union Border Programme Steering Committee (2005–2016). He has authored and edited several pertinent books.

Dr. Tamar Arieli heads the Governance and Politics Program in the Tel Hai College of Israel. Tamar's research combines the study of conflict and cooperation with spatial and social manifestations of borders in national and municipal contexts, focusing on local and national policies and practices regarding security and development.

Dr. Nick Baxter-Moore recently retired after 35 years teaching at Brock University, most recently as Associate Professor in the Department of Communication, Popular Culture and Film, and previously in the Department of Political Science. He also taught in the MA Program in Canadian-American Studies and the Graduate Program in Popular Culture. He is a former Associate Dean of Social Sciences at Brock and was founding President of the Popular Culture Association of Canada. His research interests include various aspects of popular music, local popular culture and Canadian–American relations. He recently co-edited *The Routledge Companion to Popular Music and Humor* (2019).

Dr. Michel S. Beaulieu is Professor of History, the Associate Vice-Provost Academic and Director of the Community Zone at Lakehead University. In addition, he is a Docent of Social Science History at the University of Helsinki, a Docent of Modern North American History at the University of Oulu and an Associate at the L.R. Wilson Institute for Canadian History at McMaster University. Michel's research and publications focus largely on the political, labour, and social history of Northern Ontario. He is also currently President of the Ontario Historical Society and President of the Champlain Society.

Dr. Paul Burton is Professor of Urban Management and Planning and Director of the Cities Research Institute at Griffith University. Before moving to Australia in 2007, he was Head of the School for Policy Studies at the University of Bristol. Paul's research interests include the theory and practice of public participation in planning, the emergence of new metropolitan governance and planning regimes, and the everyday professional lives of planners. He has led a research partnership on planning and growth management with the City of Gold Coast for over 12 years.

Dr. Aysin Dedekorkut-Howes is Senior Lecturer in Urban and Environmental Planning in the Griffith School of Engineering and Built Environment and a Member of the Cities Research Institute. Before moving to Australia, she taught and conducted research at Florida State University, USA and İzmir Institute of Technology, Turkey. Her research interests include climate-change adaptation and disaster resilience, water-resource management and urbanisation in subtropical areas and coastal cities. She also conducts historical research on the development history of the Gold Coast as well as current issues and problems the city is facing, such as climate change.

Dr. Munroe Eagles is Professor of Political Science and Chair of the Political Science Department at the University at Buffalo (UB) – State University of New York, where he has taught since 1989. While at UB he served almost a decade as an Associate Dean and as Director of the Canadian Studies Academic Program. He is Past-President of the Association of Canadian Studies in the

United States and is currently serving a two-year term as President of the International Council for Canadian Studies. His research interests include political geography, electoral politics, and Canadian-American relations.

Omri Elmaleh is a PhD candidate at Tel Aviv University. Omri is the recipient of Rotenstreich outstanding doctoral student fellowship, and his research has been supported by the S. Daniel Abraham Center for International and Regional Studies, the Sverdlin Institute for Latin American History and Culture and the Moshe Dayan Center for Middle Eastern and African Studies. He has published several articles on the Lebanese diaspora in the Triple Frontier region.

Dr. Hilda García-Pérez is a researcher and professor in El Colegio de la Frontera Norte (Mexico). She is a demographer and received a PhD in epidemiology from the University of Michigan. Her research and teaching focuses on the social determinants of health in the U.S.-Mexico borderlands through evidence-based, transcultural and community-based approaches. Her current research evaluates the efficacy of school-based programs to prevent substance-abuse among adolescents, as well as the emotional impact of immigration policies on high-school students in border cities. She has authored many articles in peer-reviewed publications and has served as a consultant in several border-wide projects.

John Garrard was Senior Lecturer in Politics and Contemporary History at Salford University (UK) until 2011. Although primarily a historian, his central teaching and research interests have bordered with political science. In 2019 John co-edited volume 1 of Twin Cities: Urban Communities, Borders and Relationships over Time (Routledge).

Cathrine Olea Johansen is a teacher of English, Norwegian and Social Science at an upper secondary school in Tromsø. She has a master's degree in English literature from UiT – The Arctic University of Norway. She wrote her thesis on *Border Theory: A New Point of Access into Literature* (2018), where she did a border-theoretical reading of China Miéville's *Un Lun Dun, The City & the City* and *Embassytown*.

Professor Simrit Kahlon is current Chair in the Department of Geography at Panjab University, Chandigarh (India). Simrit's doctoral thesis examined the state of housing in Indian cities. Subsequent research publications are located at the crossroads of Population, Environmental and Urban Geography. Her stint as Lecturer at the Centre of Women's Studies at the Guru Nanak Dev University, Amritsar, early in her career, exposed her to gender issues within geography and stimulated wider interest in the sub-discipline of cultural geography. She offers master's courses in urban, cultural and social geography alongside teaching a compulsory course on the philosophy of geography. Her current research interests focus on the shifting power dynamics and their material expression in space, particularly regarding urban spaces.

Jenna L. Kirker is a PhD candidate in history, working in the Wilson Institute for Canadian History at McMaster University. In 2019–2020 she was the Canadian Council for the Advancement of Education and TD Insurance Fellow and is currently the Advancement Coordinator at the Northern Ontario School of Medicine.

D.Sc. Vladimir Kolosov is Deputy Director of the Institute of Geography of the Russian Academy of Sciences and Head of its Laboratory of Geopolitical Studies. He is Past-President of International Geographical Union and Vice-President of Russian Geographical Society. He has been for five years Professor at the University of Toulouse (France) and Visiting Professor at a number of foreign universities, head of the teams participating in international projects funded by European Framework Programmes and other foreign and national foundations.

Dr. Verena La Mela is Junior Researcher at the Department of Social Sciences (Social Anthropology Unit) of the University of Fribourg, Switzerland. As a member of the research project 'Roadwork: An Anthropology of Infrastructure at China's Inner Asian Borders' she focuses on infrastructure, logistics and the Belt and Road Initiative. For her PhD she conducted long-term field research at the Sino–Kazakh border with a focus on women, trade and social networks. Her PhD was supported by the Max Planck Institute for Social Anthropology in Halle/Saale, Germany, and the University of Zurich, Switzerland. Verena is particularly drawn to border studies and the informal economy of the former socialist countries.

Dr. Francisco Lara-Valencia is Associate Professor and Director of the Transborder Policy Lab at Arizona State University. As a border scholar of space and territory, his scholarship is situated at the intersection of urban studies, community development, and cross-border cooperation. He is the author of several articles on cross-border cooperation, border environmental policy and urbanization, and cross-border governance and planning. His most recent writings engage the topics of identity and borders in North America, local responses to crisis and disruption, and the pedagogy of borders. He is also former President and current Executive Secretary of the Association for Borderlands Studies.

Dr. Clémence Léobal is a sociologist and a permanent member of the French National Center for Scientific Research. She belongs to the Lavue (Nanterre University). Her work on French Guiana began in 2009, when she was employed by the Saint-Laurent-du-Maroni town hall to prepare its Center for the Interpretation of Architecture and Heritage. In her PhD thesis, defended in 2017, she analyzed together the implementation of urban policies and their reappropriation by bushinenge inhabitants. Her current research focuses on the way in which the border takes on concrete consistency through the diversity of the administrative network and its uses on the Maroni.

Emeritus Professor Thomas Lundén is in Human Geography at Södertörn University and former director of its Centre for Baltic and East European

Studies, CBEES. His scholarship includes articles in political and social geography, border interaction and the history of geopolitics and Baltic relations.

Dr. Tony Michell is an experienced consultant on all Northeast Asia issues and managing director for Asia for EABC Ltd, responsible for local offices. Consultancies include the World Bank, UNDP, ILO and EU. Tony holds a Cambridge University Regional Development PhD. He has taught at Hull University, KDI School of Public Policy and Management, and has been visiting scholar at European, US and Asian universities. He publishes mostly on Northeast Asia business and economic development.

Dr. Ekaterina Mikhailova is Postdoctoral Researcher at the Department of Geography and Environment, University of Geneva and Visiting Lecturer at the Graduate Institute of International and Development Studies (Switzerland). Ekaterina's work lies at the crossroads of urban studies, border studies and Russian studies. She is an author of over 40 English and Russian publications on twin cities, cross-border communities and transfrontier cooperation, border tourism, governance and migration. Ekaterina has done fieldwork and analysed cross-border interactions along Russian borders with Norway, Finland, Estonia, Belarus, Ukraine and China, as well as at the Swedish-Norwegian and Spanish-French borders. She worked in several international research projects including the FP7 research project *EUBORDERREGIONS* (2012–2014) and *The Transformation of Soviet Republic Borders to International Borders* (2018–2019). In 2019 Ekaterina co-edited volume 1 of *Twin Cities: Urban Communities, Borders and Relationships over Time* (Routledge).

Dr. Ruben Moi is Associate Professor at UiT, Arctic University of Norway in Tromsø where he also belongs to the Border Aesthetics research group. He has published widely in English and Norwegian on borders and contemporary culture, and the imaginative arts of writers like Seamus Heaney, Derek Mahon, Ciaran Carson, T.S. Eliot, Samuel Beckett, Martin MacDonagh and Irvine Welsh. He is current President of the Norwegian Society for English Studies, treasurer of the Nordic Irish Studies Network, and member of the Norwegian Academic Council for English. He was chairman of Ordkalotten, Tromsø's International Literature Festival, for many years.

Dr. Dario Musolino is Researcher in Economic Geography and Lecturer in Economic Prospects at the Bocconi University (Milan, Italy), and at the Università della Valle d'Aosta (Aosta, Italy). He holds a PhD from the University of Groningen (the Netherlands). He has been a Board Member of the Italian Regional Science Association, and he is Editor and Co-Founder of *EyesReg*, the Italian online journal of regional science. Dario's academic interests include territorial competitiveness, development of peripheral and rural areas, agri-food value chain, industrial clusters and socio-economic impact of natural disasters. He is an author of over 120 national and international publications.

Dr. Paul Nugent is Professor of Comparative African History at the University of Edinburgh (UK). He holds a European Research Council Advanced Grant for the AFRIGOS project, comparing transport corridors, border towns and port cities across four regions of Africa. He has published *Boundaries, Communities and State-Making in West Africa* (CUP, 2019), comparing border dynamics in Ghana/Togo and Senegal/Gambia. He has also worked on border towns in Kenya/Uganda. Paul was a member of the Institute of Advanced Study in Princeton in 2015–2016. He is Founder/Chair of the African Borderlands Research Network (ABORNE). Paul is a member of the Africa-Europe Strategic Taskforce on transport and connectivity, supporting the collaboration of the European and African Unions. He is currently completing a history of South African wine.

Dr. Valerià Paül is Associate Professor at the Geography Department, University of Santiago de Compostela (Spain). Previously, he was Assistant Professor at the University of Western Australia (2014–2015). He holds a PhD in Geography from Barcelona University and has received its Special Awards both for Bachelor Studies (2002) and Doctoral Studies (2007). His research interests include regional planning and management focusing on open spaces, protected and mountain areas and development; historical and cultural geography of landscape; agriculture, food and rural studies; political geography; and tourism (specifically, cultural, natural and rural tourism). He has participated in numerous projects, usually working on inter-disciplinary rural and regional studies; these were funded by the European Union and the Australian, Spanish, Galician and Catalan governments.

Luigi Pellegrino is a policeman at the Police Headquarters in Aosta with 13-year-long and meritorious service. In 2003–2008 Luigi served as a Volunteer of the Italian Army participating in the UN peacekeeping missions in Lebanon. In 2020 he completed his master's degree in Economics and Politics of the Territory and Enterprise at the Università della Valle d'Aosta, where he successfully defended his master's thesis *'Towards a Bi-Regional city? An Application of the Delphi Policy to the Realization of the Integrated Area of the Strait of Messina'*. Since then he has continued researching the Messina Strait area.

Dr. Lena Poschet El Moudden is Head of Section in the Office for Spatial Development of the Swiss government. She holds a degree in Architecture and Logistics Management and completed her PhD on transformation of urban space in border towns using the case of Haiti and the Dominican Republic as an example. Lena participated in the Swiss National research program 'Mitigating Syndromes of Global Change' and led several research projects with focus on transport and urbanization at the Federal Polytechnic School of Lausanne.

Dr. Fabio Santos is Assistant Professor of Sociology at the Institute for Latin American Studies at Freie Universität Berlin (Germany). He previously taught as Guest Professor of International Development at the University of Vienna (Austria) and earned his PhD at the International Research

Training Group 'Between Spaces. Movements, Actors and Representations of Globalisation' (Freie Universität Berlin). From intersectional, post- and decolonial perspectives, his research focuses on global inequalities, entangled histories and contested borders. He is the author of *Bridging Fluid Borders. Entanglements in the French-Brazilian Borderland* (Routledge, 2022).

Dr. Gianlluca Simi is Creative Project Manager for global design agency Superside. He has a PhD in critical theory and cultural studies from the University of Nottingham (UK). Previously he worked as a teaching assistant at the University of Nottingham as well as a reporter and editor for independent publications both in Brazil and in the UK. His research focuses on the semiotics of everyday life, especially the construction of meaning and the uses of discourses on categories such as 'internationality' and 'cultural diversity' as they are expressed both through the consumption of material objects and the adoption of 'cosmopolitan' practices; as well as, at the same time, circumstantial re-bordering practices that feed into broader discourses on 'border control'.

Ganeshwari Singh is a PhD candidate in the Department of Geography at Panjab University, Chandigarh, India, and is currently Assistant Professor of Geography at the Department of Higher Education, Haryana. She was previously adjunct faculty at Department of Geography, Panjab University, Chandigarh, where she taught courses in disaster management, human geography and environmental geography. Her research interests span the fields of cultural geography, urban geography and environmental geography. She is currently working on discourse, socio-spatial dialectic, and production of space in the context of Chandigarh, the first planned city of independent India. Her research focuses on cultural landscapes, power relations and feminist geography. She is particularly interested in the dynamics of power relations and how these materialise in space.

Dr. Juan-Manuel Trillo-Santamaría is a Lecturer in the Department of Geography at the University of Santiago de Compostela (Spain) and currently a Co-Editor for Europe of the *Journal of Borderland Studies*. His research has benefited from research stays in international centers in France, the Netherlands and the UK. Juan-Manuel holds a PhD in humanities from the University Carlos III of Madrid and defines his research as multidisciplinary. His approach to border studies has emphasized the geographical tradition with a comprehensive spatial analysis of social, political, economic and cultural processes related to the production and reproduction of borders. Juan-Manuel's academic interests include cross-border cooperation and multilevel governance, historical and cultural geography, tourism and geopolitics.

Roberto Vila-Lage is a PhD candidate in the Department of Geography at the University of Santiago de Compostela (Spain). He holds a BA degree in economics from the University of A Coruña, a BA degree in geography and spatial planning and an MA in spatial planning, management and development from the University of Santiago de Compostela. Roberto's doctoral project focuses on the study of cooperation across external (Galicia-North of Portugal) and internal (Galicia-Asturias/Galicia-Castile and Leon) Spanish borders and is funded by a research grant from the Spanish Ministry of Science, Innovation and Universities obtained through a publicly competitive process.

Figure 00.a Frontispiece. Logo of Twin City, Georgia (USA)
Source: Eileen Dudley, by permission.

1 Introduction

Towards a global overview of Twin City Studies

Ekaterina Mikhailova

This is the second volume about twin cities around the world. Like its predecessor *Twin Cities: Urban Communities, Borders and Relationships over Time* (Garrard and Mikhailova 2019), it aims to advance the theoretical foundations for the cross-national comparative study of twin cities and to assemble empirical studies from around the globe. It seems natural to start by recapitulating and elaborating some of the key points made in the first volume, including the definition, classification and distinctive features of twin cities.

Twin cities are adjacent, mostly independently founded but closely interrelated urban pairings on either side of a border. They see themselves, and are seen by others, as especially adjacent, thereby having a *special* relationship if not necessarily a friendly or functional one. Twin citizens see this relationship with regret, pride or some blend of both, with the balance, mostly from the first to the second, changing over time.

Twin cities can be defined according to *where* and *how* they arise. They can emerge within one country and be separated by a municipal and/or federal-state border (hence called *internal* or *intranational* twin cities) or straddle international borders (consequently called *external* or *international* twin cities). Internal twin cities are often points where the implications of conurbanisation are first and most urgently observable. External twin cities are intense thermometers, even barometers, for what is happening in the wider urban world – globally, around and across international borders. They are trans-shipment points for legitimate and illegitimate trade, and the interface and interchange of cultures and peoples. More recently, they have become points of stress where the implications of, say, international mass migration and the political responses to/exploitations of this phenomenon are experienced with particular intensity (see more below).

Based on *how* twin cities arise, they can be classified into:

1. *twin cities that grow* ('crash') *into each other* (e.g., Minneapolis and St. Paul, USA; Garrard and Mikhailova 2019, chapter 1);
2. *twin cities* that arise on either side of a normally river-defined and/or internationally defined border and *expand outwards*, the growth of one often stimulated by the presence of the other (e.g., Manchester and Salford, UK; ibid., chapter 2);

DOI: 10.4324/9781003102526-1

3. *partitioned twin cities* originating from initially one city divided by a later imposed boundary (e.g., Görlitz, Germany/Zgorzelec, Poland; Chapter 7, this volume);
4. *duplicated twin cities*, one of which is the reason for the foundation of the other by duplicating it due to a change of the boundary and resettling of some inhabitants to a new place in order to keep their jurisdiction as before (e.g., Laredo, USA/Nuevo Laredo, Mexico; Garrard and Mikhailova 2019, chapter 11) or due to strategic factors such as defence or trade (e.g., Narva, Estonia/Ivangorod, Russia; ibid., chapter 16);
5. *twin cities of mixed origins* emerging as a result of *both partition and duplication* (e.g., Svetogorsk, Russia/Imatra, Finland; ibid., chapters 18–19);
6. *planned twin cities* – somewhat more distant entities that are perceived by central or federal-state governments (e.g., Ahmadabad-Gandhinagar, India; ibid., chapter 7), to be rapidly spreading towards each other and are legally declared twins in order to take planning control of the merging process – so as to reduce chaos and take economic advantage of it by ensuring complementarity;
7. *engineered twin cities* – again, more distant but still neighbouring urban places whose special relationship arises from the development of transport technology (like the Øresund Bridge that includes a bridge, an artificial island and a tunnel connecting Copenhagen, Denmark/Malmö, Sweden);
8. *nesting twin cities* – thus-far rare examples where one city for historical or other reasons comes to be located inside another (e.g., Taipei–New Taipei, Taiwan; see more in this book's conclusion).

This eightfold classification points to the great variety and globality of twin cities as a peculiar urban form spread around the world. We suggest approaching them as a *twin cities' family* that, despite the diversity of its members, have important resemblances defined as twin-city features. They are:

1. economic, social and political *interdependence from the start* of their mutual existence;
2. *tensions between inwardness* – orientation to the parent state and itself – *and openness* – orientation across the border, towards the other city, cross-border region, the neighbouring state or the broader outside world;
3. *mostly unequal relationships*, enhancing already-inbuilt inter-city conflict;
4. *ongoing formal and informal negotiations* due to the need for coordination, the overlap of jurisdictions and the high number of stakeholders with border-straddling interests (particularly in external twin cities) based in either twin city or in both, and representing various sectors from business to civil society to criminal gangs;
5. *persistence*, understood as endurance of pressures from the growing conurbations around them.

For a more detailed overview of these five essential and mutually reinforcing characteristics and their interrelation with previous attempts to characterise twin cities, see Garrard and Mikhailova (2019, 9–17).

This volume, like its predecessor, sees twin city relationships as dynamic and constantly changing, with ups and downs, periods of discord, even confrontation, followed by more peaceful and collaborative phases. Consequentially, the manifestations of each twin-city characteristic could alter over time, influenced by, or independent of, each other. The dynamism of twin-city relationships derives from twin cities situations of durable and intensive neighbourship.

To the best of our knowledge, there is no complete inventory of twin cities. While assembling and editing the two volumes, time and again we have been coming across new cases and case studies previously unknown to us. At the moment of writing (April 2021), we estimate that there are at least 250 twin-city examples, or 500 cities in twinned relationships across the world. Their number is increasing within individual states and on international borders. The reasons for ongoing twin city breeding can be roughly summarised as follows:

1. *Continuous boundary changes*, i.e. (re)drawing international and administrative boundaries that affects the relations of close-by urban settlements. All partitioned twins, and the majority of duplicated twin cities, have emerged due to large international boundary changes including the Mexican-American treaty of 1853, the 1918–1919 Versailles Settlement, the post-1945 peace settlement and others.

2. *Recurrent changes of border regimes*. Fluctuations in border porosity, known as re-bordering and de-bordering, often influence border twin cities. Following the end of the Cold War, Europe witnessed the increasing permeability of international borders stimulating more interactions between border twin cities at the boundaries of the Capitalist and Socialist blocks. Exploring Hungarian-Slovakian border twin cities, Tamaska calls this period a 'reintegration stage' in twin cities' history (Garrard and Mikhailova 2019, chapter 13). Meanwhile, the collapse of the Soviet Union and Yugoslavia in the early 1990s led to rebordering among their former republics. This resulted in the transformation of former *internal* twin cities, such as Valga, Estonia/Valka, Latvia (ibid., chapter 16), or Slavonski Brod, Croatia/Brod, Bosnia-Herzegovina (Chapter 9, this volume) into *external* ones.

3. *Institutional encouragement*. Twin cities could be seen as a symbolic tool to ameliorate binational relations and/or a pragmatic problem-solving tool to provide better living conditions for twin-city communities and their hinterland. No wonder the tendency towards twin city breeding in the 20th and 21st centuries has been originating at different administrative levels – municipal, regional, national, macro-regional and international. Manifestations of institutional encouragement for twin-city cooperation include the EU's ambition for European integration, related cross-border cooperation policies of EU member-states and states seeking EU membership, alongside the domestic policies of some states aiming to control rapid urbanisation in the modern world (as in India).

4. *Developments in fast-transit engineering*. The advance of transportation systems and technologies enables relatively proximate cities to become more and ever-faster connected. The shrinking of effective distances has already

given birth to some new cross-border regions, like the already-mentioned bridge-induced Øresund Region twins with 3,700,000 inhabitants. More regions are considering such investments, including Helsinki and Tallinn across the Gulf of Finland that consider exploring a Hyperloop – an experimental high-speed vacuum transportation technology where passengers and goods are propelled in magnetically levitated capsules through low-pressure underground tubes at speeds of around 1,000 km/h. In September 2017, the Estonian Prime Minister and representatives of the Hyperloop One company (later renamed Virgin Hyperloop) signed a letter of intent aimed at working together within the Tallinn-Helsinki tunnel project (Baltic Course 2018).

5. *Economic incentives and encouragement from global economic players.* Increasing global territorial competition stimulates nearby places to cooperate to obtain better visibility and improve their chances of attracting investment, new industries, jobs and residents. In 2020 Nova Gorica won the Slovenian national competition to become a European Capital of Culture in 2025 (outperforming Ljubljana, Slovenia's capital) thanks to teaming up with its Italian twin city, Gorizia. The award-winning bid used the slogan 'Go! Borderless' and listed transforming Nova Gorica and Gorizia into 'one cross-border European city' as its first priority (GO!2025 Team 2020, 10).

Due to increasing international trade, some infrastructural projects connecting border twin cities are funded by faraway state and non-state actors wanting to be primary beneficiaries of growing international flows. Thus, today the most debated issue regarding Helsinki and Tallinn's connectivity project across the Gulf of Finland (which has now returned to the original rail-tunnel possibility) is whether to involve the Chinese-backed Finnish venture proposing a $17 billion investment. This is due to strategic security concerns of outside (e.g., the USA and NATO) and inside actors (Eglitis 2021).

The other example of external investments stimulating twin-city interactions comes from Brest and Terespol, twin cities on the Belarus-Poland border. Since 2016, when Russia's blockade of rail transit via Ukraine started, most cargo flows between China and Europe run through Belarus and Poland and are unloaded from wide-track onto European standard-gauge trains in Brest and Malaszewicze trans-shipment terminals. Malaszewicze is located in Terespol administrative district, 9 km from Terespol, and is one of the largest 'dry' ports in Poland and the EU. The terminals' limited capacity has been repeatedly named among major bottlenecks on the prospective fast-train link planned for implementation within the Belt and Road Initiative (Jakobowski et al. 2018). To partially solve the problem, in 2015–2017 the Chinese government granted Belarus modern X-ray equipment for inspecting goods transported by rail to be installed in Brest. Simultaneously, the Malaszewicze terminal modernisation project was implemented with around 40% of funding coming from the EU (Railway Pro 2020).

6. *The mix of all or some of the aforementioned factors.* Often, the establishment of twin-city relationships between somewhat distant cities relies both on

institutional and economic incentives as well as new transport infrastructure enhancing inter-city interactions. If this occurs across a previously 'hard' border – one barely possible to cross due to its predominant barrier function in contrast to a 'soft', highly porous border primarily functioning as a point of contact – the border regime change becomes a necessary prerequisite. If one twin city is located in a highly centralized state, governmental support of twinning becomes a crucial factor. All these reasons facilitated the twinning of Nikel, Russia/Kirkenes, Norway (Garrard and Mikhailova 2019, chapter 17).

To conclude this section on definitions, it is worth touching upon the decades-long heated debate on the 'twin-city' term itself. Academic debate regarding its appropriateness has centred on external twin cities and the metaphor of 'twins' or the so-called 'Gemini complex' (Arreola 1996, 356). The criticism typically concentrates around three points: the gap (often significant) in the dates of cities' foundation; their lack of genetic inter-connections; and their stark visual and other contrasts, sometimes described as 'cartoonish' (Vanneph and Mouroz 1994, 13). Besides, the manifold disciplinary backgrounds of authors studying twin cities have fuelled the terminological disputes. Several alternative terms have been coined as more neutral and hence accurate descriptors – such as 'coupled settlements', 'paired border cities' (Arreola 1996), 'double cities', 'neighbouring cities', 'border-crossing cities', 'binational cities' (Buursink 2001), 'cross-border conurbation/agglomeration', 'transborder metropolis' (Herzog and Sohn 2014) and others. However, none has gained sufficient traction, and 'twin cities' remains the most widespread term to designate the phenomenon of nearby urban pairings on municipal and international borders. Interestingly, many languages have produced their own equivalents of the 'twin-city' term (see Table 1.1), suggesting a common need to name this phenomenon and differentiate it from other types of urban interrelations.

Further evidence of the twin-city term's deep roots is the fact that for over a hundred years there have been at least two places officially called twin cities. They are Minneapolis and St. Paul on the Mississippi River in Minnesota, and the city of Twin City in Georgia, USA. Given that the former is a far larger and populous urban area than the latter (the Minnesota Twin Cities metropolitan area is 28,422 sq km with a total population of around 4,000,000 inhabitants in 2019,[1] while Twin City, Georgia has an area of 9.3 sq km with 1,701 inhabitants in 2019), Minnesota's Twin Cities have naturally enjoyed greater visibility both nationally and internationally.

Minneapolis and St. Paul are the iconic example of internal twin cities. While maintaining a deeply conflictual relationship until the 1960s, they have been labelled (first colloquially, then officially) 'Twin Cities' since at least the 1890s. From early on, this label was used by and for outside audiences since the local identity was greatly reliant on the antagonism of the two cities. St. Paul was established in 1841, developed as a transport and trade hub on the Mississippi River and since 1849 has served as Minnesota's capital. Minneapolis was established in 1854 next to the St. Anthony Falls and soon became an important

Table 1.1 Linguistic adaptations of the term 'twin cities'

Language	Adaptation	Examples of cities that have been described with the adapted term in academic literature and periodicals
Chinese	Shuāng chéng	Heihe (China) – Blagoveshchensk (Russia)
Estonian	Kaksiklinnad	Tallinn (Estonia) – Helsinki (Finland), Narva (Estonia) – Ivangorod (Russia), Valga (Estonia) – Valka (Latvia)
Finnish	Kaksoiskaupunki	Tornio (Finland) – Haparanda (Sweden), Imatra (Finland) – Svetogorsk (Russia), Helsinki (Finland) – Tallinn (Estonia)
French	Villes jumelles	Comines (France) – Comines-Warneton (Belgium), Rosso (Senegal) – Rosso (Mauritanie)
German	Zwillingsstädte	Görlitz (Germany) – Zgorzelec (Poland), Guben (Germany) – Gubin (Poland)
Italian	Città gemellate	Gorizia (Italy) – Nova Gorica (Slovenia)
Norwegian	Tvillingbyer	Kirkenes (Norway) – Nikel (Russia)
Polish	Miasta bliźniacze	Cieszyn (Poland) – Český Těšín (Czech Republic), Zgorzelec (Poland) – Görlitz (Germany)
Portuguese	Cidades gêmeas	São Borja (Brazil) – Santo Tomé (Argentina), Uruguaiana (Brazil) – Paso de los Libres (Argentina), Ponta Porã (Brazil) – Pedro Juan Caballero (Paraguay)
Russian	a) Goroda-bliznetsy b) Dvoynoy gorod	a) Ivangorod (Russia) – Narva (Estonia), Blagoveshchensk (Russia) – Heihe (China), Nikel (Russia) – Kirkenes (Norway) b) Svetogorsk (Russia) – Imatra (Finland)
Spanish	Ciudades gemelas	Posadas (Argentina) – Encarnación (Paraguay), Rivera (Uruguay) – Santana do Livramento (Brazil)

centre of lumber production and flour-milling. By the 1880s, Minneapolis's population overtook St. Paul's; however, this has not affected the latter's administratively dominant position. St. Paul still remains Minnesota's state capital. The relocation of a major baseball league team from Washington, D.C., to Minneapolis and St. Paul and its renaming to become the Minnesota Twins in 1961 marked an important milestone on the way towards more harmonious twin-cities' coexistence and was one of the first local associations using the 'twin cities' label for local audiences.

The city of Twin City, Georgia is a historical example of twin cities. It was founded in 1921 after the merger of two urban settlements of Summit and Graymount trying to survive the effects of the Great Depression. The website of the city of Twin City (2021) asserts that 'although a touch of rivalry remains between "Summit" and "Graymont" neighborhoods, today's residents of Twin City are warm, friendly, and proud of their roots' – pointing at the persistence of the twin city breed. Recently for marketing purposes, Twin City produced a logo (that became the frontispiece of this volume and its predecessor) and a motto 'Twice as friendly, twice as nice'.

A trawl through digitalised academic sources (e.g., via Google Scholar), covering as far back as the mid-20th century, reveals the 'twin-city' term being applied to internal close-by urban settlements both in Europe and Asia. For instance, researchers writing on the Polish urban system used the term to describe Gdansk and Gdynya (Osborne 1959, 204), just as Lundén does in this book (Chapter 7). In the late 1950s, both American professor of political science Clifford Grant (1958) and British sinologist Victor Purcell (1959) called Saigon (today's Ho Chi Minh City) and Cholon twin cities. Cholon, initially a neighbouring independent Chinese settlement, expanded to Saigon by the 1930s, inducing its merger with the latter under the name 'Saigon-Cholon', two decades later shortened to 'Saigon'. Today, Cholon is one of Ho Chi Minh City's neighbourhoods and typically presented to tourists as its Chinatown.

Exploring the same digitalised academic sources, we come across the late-19th century book on the twin cities of Brownsville (Texas, USA) and Matamoros (Tamaulipas, Mexico) (see Chatfield 1893) and the mid-20th century doctoral dissertation on another twin-city pair – El Paso (Texas, USA) and Ciudad Juarez (Chihuahua, Mexico) (see D'Antonio 1958). The book was compiled by Lieutenant W.H. Chatfield, who uses the term 'twin cities' in the title – *The Twin Cities of the Border and the Country of the Lower Rio Grande* – and throughout the text. Chatfield writes that Brownsville and Matamoros 'became in fact as well as in name, "The Twin Cities of the Border"' during the 1861–1865 American Civil War. Then the two cities served as warehouses for exporting 'immense quantities of cotton and other accumulated products of the South' and importing 'munitions of war and the food staples' (Chatfield 1893, 2), other southern ports of the Confederacy being under blockade.

The doctoral dissertation on twin cities, by Italian-American sociologist William D'Antonio at Michigan State University, hardly features the term 'twin cities'. However, when justifying the case-study selection, D'Antonio refers to 'very high interaction rates' (D'Antonio 1958, 37) between the El Paso and Ciudad Juarez communities and gives multiple examples of constant formal and informal negotiations between the two, along with their inequality and interdependencies. For instance, in the mid-1950s, two of three Spanish-language newspapers serving El Paso were printed in Juarez (ibid., 61), and 10–15% of the Juarez labour force worked in El Paso (ibid., 63). Summarising, D'Antonio writes that 'while only a dry river-bed separated Juarez from its neighbor for most of the year' (ibid., 62), their urban growth, economy and socio-cultural characteristics differed considerably.

As the title of both edited volumes suggest, we find the 'twin-city' term credible and academically productive. Both volumes treat twin cities as a diverse phenomenon with various degrees of similarity and difference in how they look, evolve and behave. Overall, we hope this helps make earlier terminological debates obsolete.

*****

Notwithstanding twin cities' age-old existence – some are over seven centuries old (see Chapter 2) – they have been chronically under-researched in the constantly growing fields of urban studies and border studies. The shift from sporadic references to twin cities to examining them as a self-contained unit of research occurred in the 1960s. From then onwards, there has been a gradual increase in twin-city publications. However, for a long time, twin-city literature has largely involved applying the lens of exceptionalism, with attempts at theorisation confined to a few case studies. This particularism is evident in the numerous publications on particular twin cities and cross-border agglomerations. The first comparative studies addressed various aspects of twin cities on the US-Mexico border (e.g., Sloan and West 1977). In Europe, one of the first comparative studies of cross-border agglomerations and their theorisation focused on Lille and Geneva (Raffestin and Servin 1978). Efforts to overcome the field's fragmented state intensified from the 1990s, with multiple publications on cross-border urban areas in particular regions (see Lundén 2004, Jańczak 2013 for Europe; Herzog 1990, Vanneph and Mouroz 1994 for North America; Dilla et al. 2008 for South America; Nugent 2019 for West Africa).

The emergence of cross-regional comparisons marked an important milestone in the maturation of Twin City Studies. Such comparisons include thematic special issues (like *GeoJournal* 2001, *Geographica Helvetica* 2007 and *Journal of Borderlands Studies* 2017) and individual research studies (Herzog and Sohn 2014). The most recent indicator of where Twin City Studies are today has been *Twin Cities: Urban Communities, Borders and Relationships over Time* (Garrard and Mikhailova 2019) and this volume. The global authorship, broad case-study coverage and the novel approach involved in treating internal and external twin cities as one phenomenon, we think, make these volumes a significant breakthrough in the field.

It is unsurprising that the two enthusiasts of Twin City Studies – John Garrard and myself – have produced the two volumes on the subject. On the one hand, as it will be evident in the conclusion, the abundance of twin cities across the globe and the seriously fragmented state of Twin City Studies strongly suggests the necessity of at least two volumes wherein to present twin cities as a particular urban form in all its variety. On the other hand, for the two co-editors in their dedicated pursuit of paired settlements, it has been quite natural to have two volumes complementing each other.

We hope the two volumes from an assembled group of multi-disciplinary specialists demonstrate that enduring twin-city features, although at varying intensity, remain present regardless of their geographical location, genesis, morphology or duration of coexistence. This book, alongside its predecessor, showcases the diversity and constancy of twin-city characteristics with examples covering the entire globe – 70 cases of twin cities, or over 140 cities in twinned relationships in 40 countries (Figure 1.1): from the Baltic Sea region in the North (Prussian twin cities, Chapter 7, this volume) to Australia in the South (Gold Coast-Tweed Heads, Chapter 4, this volume).

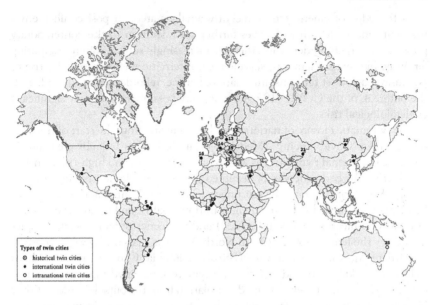

Figure 1.1 World map of twin cities presented in this volume.

The numbers on the map indicate: 1 – Thunder Bay (Canada), 2 – Niagara Falls (Canada and USA), 3 – Ambos Nogales (USA and Mexico), 4 – Ouanaminthe (Haiti) and Dajabón (Dominican Republic), 5 – Albina (Suriname) and Saint-Laurent (French Guiana, France), 6 – Saint-Georges (French Guiana, France) and Oiapoque (Brazil), 7 – Foz do Iguaçu (Brazil), Ciudad del Este (Paraguay) and Puerto Iguazu (Argentina), 8 – Chuí (Brazil) and Chuy (Uruguay), 9 – Stonar and Sandwich (UK), 10 – Fowey and Polruan (UK), 11 – Hamburg and Altona (Germany), 12 – Sovietsk (Russia) and Panemunė (Lithuania), 13 – Frankfurt an der Oder (Germany) and Słubice (Poland), 14 – Konstanz (Germany) and Kreutzlingen (Switzerland), 15 – Slavonsky Brod (Croatia) and Brod (Bosnia and Herzegovina), 16 – Chaves (Portugal) and Verín (Spain), 17 – Reggio Calabria and Messina (Italy), 18 – Aqaba (Jordan) and Eilat (Israel), 19 – Lomé (Togo) and Aflao (Ghana), 20 – Imeko (Nigeria) and Ketu (Bénin), 21 – Khorgos (Kazakhstan and China), 22 – Zabaikalsk (Russia) and Manzhouli (China), 23 – Chandigarh, Panchkula and Mohali (India), 24 – Dandong (China) and Sinuiju (North Korea), 25 – Gold Coast and Tweed Heads (Australia). Map by Marion Planque and Ekaterina Mikhailova.

Meanwhile, this volume fills some of the remaining gaps in Twin City Studies. First, it focuses on areas *unvisited* or *under-covered* in (Garrard and Mikhailova 2019). Notably, it sheds light on Latin-American twin cities. A quarter of the book (6 of 24 chapters) cover case studies on Brazil's borders with Argentina, Paraguay, Uruguay and French Guiana, as well as the borders of French Guiana-Suriname, Haiti–Dominican Republic and USA–Mexico. Furthermore, this volume develops the twin-city study of Africa and Asia, featuring examples on the borders of Ghana-Togo and Nigeria-Benin, plus examples on China's borders with Russia, Kazakhstan and North Korea. Finally, and importantly, this book for the first time extends twin-city debates to Australia and to a fictional imagining of a twin city. This volume also explores twin cities in different circumstances

– on the edge of emergence or recently amalgamated, in post-conflict environments and under stress. It gives further consideration to the contemporary pressures to which border twin cities are increasingly subject, often originating far away, like mass migration, surveillance, international terrorism, the rise of populism, East-West tensions, international crime, rebordering trends and, with the outbreak of the COVID-19 pandemic, very possibly increasingly significant epidemiological risks.

The systematic closure of national borders recurring from the start of the pandemic called 'Covid-fencing' (Medeiros et al. 2020) has brought great media coverage to external twin cities as arenas of civic protests and high-risk frontliners suffering the consequences of disrupted socio-economic ties. For instance, in Europe, Konstanz (Germany)/Kreuzlingen (Switzerland) often appeared in news headlines for repeatedly hosting demonstrations against and in favour of restrictive COVID measures (SwissInfo 2020) and for scenes of families and couples divided by the fence (BBC 2020). In North America, mass media were reporting on Hyder (Alaska, USA)/Stewart (British Columbia, Canada), tiny Northern twin cities 2 km apart and significantly remote from anything else. The two communities claimed they were used to isolation from the outside world but they 'don't want to be isolated from each other'. Instead, they sought permission to create 'their own bubble' (Chan 2020, The Economist 2020) – a special arrangement to avoid covid-fencing since Hyder, the smaller town, depends on Stewart for firewood and provisions to get through the winter. In Africa, Goma (Democratic Republic of Congo)/Gisenyi (Rwanda) were often presented as points of high risk as well as places where border closure reinforces inequality and hits vulnerable populations worst (e.g., Congo's disabled border-couriers reported by Reuters 2021).

*****

What follows is the summary of chapter content and commentaries emphasising the distinctive ongoing characteristics of twin cities that give unity to the two volumes we have assembled and edited with an eye to those distinctive commonalities.

This volume has three parts – first, on internal twin cities; second, a more extended section on external twins, and third, on twin cities in fiction and editors' dreams about real-world twins. As in Garrard and Mikhailova (2019), this volume begins with intranational city pairs. *Part I* with its five chapters comprises a considerable geographic variety (three continents – Eurasia, North America and Australia), time span (700 years) and evolutionary range – from the embryonic (but economically complementary) through the fully evolved and planned, to amalgamated twin cities. Their inter-relations are mostly dominant-subordinate and sometimes mutually resentful (even when, or perhaps because, they were planned). In two cases, relations are complicated by having to be conducted across federal-state as well as municipal borders.

In Chapter 2, Anna Anisimova explores English monastic twin towns (relatively small urban centres under monastic lordship) and examines how they emerged and coexisted during the 13th–15th centuries, finally tracing their fortunes into the 20th century. Anisimova concludes that each twin-town situation had different factors at play like origins, layouts, lordship, economic profile and connectivity. This impacted inter-twin interactions and degrees of conflict. She reminds us that, while dominant-subordinate relationships are distinctive features of her twin towns even in the Middle Ages, this should not be exaggerated: despite quarrels, twins cooperated for economic reasons and were riddled with kinship ties enhancing trends towards cooperation.

Michel Beaulieu and Jenna Kirker in Chapter 3 present the case of historical Canadian twin cities – Fort William and Port Arthur – that, after almost 80 years of rivalry, merged into the City of Thunder Bay in 1970. Their history, strongly imprinted by competition, at times, almost bankrupting both cities, showcases how thorny and protracted the path to harmonious coexistence could be. Besides, the chapter testifies to the fact that amalgamation does not neutralize rivalry; indeed, creating one new city with two historic cores is a tough undertaking, requiring multi-faceted planning and other efforts to unify twin-city residents. Even then, it can leave damaging residues: twin cities, even when amalgamated, take a long time to die.

In Chapter 4, Paul Burton and Aysin Dedekorkut-Howes examine Australian twin cities at the Queensland–New South Wales federal-state border. They critically review the impact of the two states' contrasting planning and local government regimes on the development of what has become Gold Coast/Tweed Heads twin cities. The chapter concludes that, apart from annoying anomalies like the seasonal time-difference and railway-gauge discrepancy, the conurban growth of this contiguous built-up area has not produced significant political conflict between the respective local governments, even while the two state governments have often been in conflict. Indeed, twin-city interactions have been favourably impacted by the 'faraway' character of one state capital, Sydney.

The other case study of internal twins, again divided by federal-state borders, comes from India. In Chapter 5, Ganeshwari Singh, Simrit Kahlon and I draw attention to Chandigarh (Union Territory), Panchkula (Haryana) and Mohali (Punjab), colloquially known as the Chandigarh tri-city. Despite being deeply divided by three jurisdictions, different ethno-religious and linguistic composition, the tri-cities are nevertheless very interdependent. The authors show that, while over time relationships within the tri-city have evolved from dysfunctional conflict to complementarity, with each city developing a distinctive character, their ability to cooperate on mutually beneficial joint planning is still limited, due to rivalry involving both cities and states, and remains a hindrance to cohesive development.

Dario Musolino and Luigi Pellegrino's Chapter 6 is inspired by the so-called 'twin-city model' and speculates on constructing twin cities across the Strait of Messina with the use of planning, like in India. Contemplating what they see as the 'embryonic twins' of Reggio Calabria and Messina, the authors seek complementarity instead of competition and the pooling of resources and economies of scale. They suggest approaching these two important South Italian cities as natural partners for close social and economic ties. They outline the potential for partnership and explore future integrative scenarios and strategies for implementing this ambitious project by using the Delphi technique – qualitative forecasting relying on opinions from an expert panel. The chapter concludes that applying the 'twin-city model' to Reggio Calabria and Messina requires improving transport networks and services in the Messina Strait to make it easy, safe, fast and efficient to move between the two urban areas.

Part II has far more chapters and covers five geographical subsections: (a) Europe, (b) Middle East and Africa, (c) Asia, (d) South America and (e) North America and the Caribbean. This part shows two contrasting things. First, cross-border inter-twin relationships can be more positive and fruitful than internal twins often are, due especially to the para-diplomatic roles assigned to cross-border twins by their respective nation states, and to the cross-border movements of people in pursuit of their own interests and agendas, sometimes quite independently of national centres. Second, and contrastingly, Part II (particularly Chapters 9, 10, 13 and 21) also underlines how cross-border twins can often become major points of stress due to developments originating elsewhere.

European subsection 'a' has three chapters, each following the tradition of comparative Twin City Studies. Thomas Lundén's Chapter 7 takes us back in time and provides over a dozen examples of Prussian border cities, some with a cross-border twin. He explores their changing fates from the 18th century to the present in the context of first Prussian and then German expansion and conquest, and then post 1945. Lundén proves that both internal and external twin cities are long-existent, and they endure and evolve, like the societies in and around which they originate.

Chapter 8, by Juan-Manuel Trillo-Santamaría, Valerià Paül and Roberto Vila-Lage, compares and contrasts cross-border inter-urban projects labelled 'eurocities' on the Spain-Portugal border that institutionalised their twinning in 2007 and 2018. The authors suggest these cross-border inter-urban projects come in two generations: the first based on bilateral agreements, and the second operating as a network of twin cities under the umbrella of joint governing structures composed of supra-municipal political bodies. Commonalities of related languages, daily interactions of peoples along the 'soft' Spanish-Portuguese boundary plus EU-funding availability for developing common projects all contributed to border inhabitants' empathy towards eurocities and their initiatives.

Dorte Andersen in Chapter 9 sets her discussion of twin-city development in post-conflict environments in former Yugoslavia, speculating about similarities and differences between, on the one hand, twin cities (like Slavonski Brod/Brod on the Croatia/Bosnia and Herzegovina border) and divided cities on the other – those that have undergone traumatic urban partition along ethnic lines following civil war (like Vukovar in Croatia). Andersen shows that twin and divided cities have significantly different ways of conceiving and constructing divisions, born both of history and of the cross-border twin-city trends we see elsewhere. Internally divided Vukovar is an inward-looking community maintaining asymmetries between ethnic groups in everyday life, while the twin city of Slavonski Brod/Brod, although also impacted by the civil war, nonetheless attempts to open up for interaction, realising their interdependence and the fragility of peaceful coexistence.

Tamar Arieli's Chapter 10 continues the theme of post-conflict twin-city relationships, also opening subsection 'b' on the Middle East and Africa. By examining Eilat (Israel)/Aqaba (Jordan) in a conflictual region, within the context of the 1994 Israel–Jordan Peace Treaty that set the background and expectations for cross-border cooperation (CBC), Arieli argues that cross-border twin-city development in conflict and post-conflict environments is 'a slow, subtle, and even reversible process rather than a permanent state'. This is due to its political sensitivity and how nationally originating priorities and politically driven necessities can so easily intrude on much-needed CBC. Deference by local authorities to national directives alongside social mistrust limit CBC to low-key, low-visibility problem-solving in non-controversial areas with not much spill-over into social cross-border ties outside of the important but ever-threatened commute of Jordanian workers into Eilat hotels.

Paul Nugent, in Chapter 11 on the twin cities of the decidedly dominant Lomé, capital of Togo, and the distinctly subordinate Aflao, the adjacent Ghanaian border town, vividly exemplifies how their interactions entail both cooperation and friction, and how they have been impacted by another variable common in several of the following chapters – the post-colonial legacy of rival European powers. By examining the twin cities' distinct historical and demographic trajectories and their close economic ties along this indistinguishable, fluid conurban border, Nugent shows how the twins' destinies have been intertwined since the 19th century through and beyond independence, and are reliant on mutual advantages arising from trade and associated livelihoods.

Anthony Asiwaju's Chapter 12 adds tribal identity to rival French and British colonial legacies in the cross-border twin-city mix of variables on the Benin-Nigerian border. He draws on the twin-city ties of Ketu and Imeko, two leading Yoruba ancestral cities, the former being the capital of the ancient Yoruba kingdom and the latter evidently prominent but of vassal status. Asiwaju points out that, despite the separating impact of rival European colonial rule and post-colonial nation-building traditions in keeping the cities apart, today's Ketu and Imeko remain closely interlinked by shared culture, dialect and jointly celebrated festivals, kinship ties and rapidly expanding cross-border businesses centring on the border-straddling city of Ilara.

Sub-section 'c' unpacks the peculiarities and continuities of twin-city dynamics in Asia. It comprises three chapters, each covering the twin cities of China and her neighbours – North Korea, Russia and Kazakhstan. All three show how, on the one hand, the central state and tremendous geopolitical changes have been determining factors in twin-city interactions. On the other hand, these chapters also evidence the prevalence of economic ties, with the Chinese twins typically deriving major benefits and dominating twin-city relationships, thereby causing resentment and security concerns in subordinate partner cities and their parent states.

In Chapter 13, Tony Michell uses Dandong (China)/Sinuiju (DPRK) across the Yalu River to illustrate how susceptible cross-border twins are to external pressures originating both nearby and faraway. Linked by a steel railway bridge since 1911, the twin cities' histories have been both shared and disparate as governments changed policies in both colonial and post-colonial times and as the international environment also changed with the most recent changes occasioned by UNSC- and US-imposed sanctions impacting both cities.

Another example of railway-connected twin cities is Zabaikalsk (Russia)/ Manzhouli (China), major trans-shipment bases of Russia-China trade flows, analysed by Vladimir Kolosov in Chapter 14. Drastic changes in the two countries' bilateral relations during the cities' century of existence recast the border regime over and over again, alongside the twin cities' function and fortune. Kolosov points to the extreme asymmetry in the twin-city relationship of Manzhouli and Zabaikalsk, with the former now becoming a successful entrepreneurial city whose urban growth and prosperity stems from the skilful use of the border rent and tourism for Chinese visitors wanting a gentle taste of Russia, and the latter stagnating as a 'forgotten station settlement with poor living standards'.

Verena La Mela's Chapter 15 focuses on Khorgos and the eponymous free-trade zone straddling the Sino-Kazakh border that became one of the main logistical hubs between Asia and Europe. The commercially busy and vertically built Chinese side contrasts strongly with the flat and sleepy Kazakh side. La Mela immerses the readers into the twin cities as she guides them through her visit to the free-trade zone pinpointing both the apparent and the hidden inequalities and (inter)dependencies, with dominance on the Chinese side reinforcing suspicion on the subordinate Kazakh side.

Opening subsection 'd' on South America, Clémence Léobal's Chapter 16, like La Mela in Chapter 15, uses anthropological approaches to acquaint readers with the twin cities of Albina (Suriname)/Saint-Laurent (French Guiana) on the Lower Maroni River. Again, the post-colonial legacy is clear, as is the very 'soft' character of the border, reinforced by cross-border tribal identity and kinship. So too is the fact that post-colonial Saint-Laurent has become incorporated into metropolitan France, adding a dimension to the inter-city relationship. The chapter tells the story of one large family straddling the border, examining the experiences of its female member facing the complex entanglement of state administration policies on houses, healthcare, social benefits and land access. Illustrating her narration, Léobal uses remarkable self-made ethnographic cartography resting on Ndyuka inhabitants' perceptions, conceiving places according to their river location – upstream or downstream.

Fabio Santos in Chapter 17 unpacks the conflictual negotiations preceding and succeeding the construction of the cross-border bridge on the Oyapock River, connecting and dividing South America and Europe, Brazil and France and the twin cities of Saint-Georges and Oiapoque in the Amazon rainforest. Santos sets his argument against the background of a post-colonial border that has been traditionally fluid with much crossing and recrossing by multiple borderlanders, and seeks to show how a decision taken faraway, with primarily geopolitical considerations in mind, has impacted back to enhance existing inequalities between the French and Brazilian sides, turning the Bridge into 'a one-way-street', limiting rather than enabling cross-border mobilities. He also shows how domination and subordination between cross-border twins relates not just to comparative size (Saint-Georges is much smaller than Oiapoque) but also to history and geopolitics, greatly advantaging Saint-Georges inhabitants.

Drawing on surveys with residents of the Brazil-Uruguay twin cities of Chuí and Chuy, and applying semiotic analysis, Gianluca Simi's Chapter 18 discusses the meanings and values attributed to the border in everyday life. Simi concludes that the Chuí/Chuy residents see the border in predominantly positive terms and as fluid. This fluidity, however, is ambivalent – both allowing different people and cultures to come together, and allowing undesired mixtures to take place, hence causing feelings of insecurity and thus ambivalence amongst borderlanders.

Omri Elmaleh's Chapter 19 addresses the multicultural human wonder of Ciudad del Este (Paraguay)/Foz do Iguaçu (Brazil)/Puerto Iguazú (Argentina) in the Triple Frontier. The former was reported to be 'the world's third largest trading city' in 1995 due to booming mass-shopping tourism stimulated by Iguaçu waterfalls tourism and available trade brands from across the globe. However, the twin cities' Golden Age of benefitting from shopping tourism came to an end with the advance of macroregional economic cooperation and coordination such as Mercosur, again a decision taken far away for geopolitical reasons, but impacting deleteriously on these cities' fortunes and relationships.

Opening subsection 'e', Nick Baxter-Moore and Munroe Eagles in Chapter 20 depict the cities of Niagara Falls – another example of twin cities next to breath-taking waterfalls. This transborder conurbation straddling the Canada–USA border is one of the world's most popular tourist destinations. To explore tensions between cooperation and competition, inwardness and outwardness of twin cities, the authors juxtapose interviews with senior political and administrative figures in each city. They show how significant impediments to twin-city cooperation can arise from the electoral cycle and consequent tendency for elected officials to become immersed in their own 'small political worlds'. Besides, the authors show that, as in Chapter 19, twin cities' fate can be deeply and damagingly impacted by faraway factors. For the Niagara cities, it was the post-9/11 securitisation, the extrapolation of fears arising from the US-Mexico border on to the US-Canada border, and the decisions of policymakers in distant capital cities getting in the way of cooperation.

Chapter 21 by Hilda García-Pérez and Francisco Lara-Valencia explores Mexican border cities as vital points of South-North and North-South migrant flows across the USA–Mexico border comprising asylum seekers from the Caribbean and Latin America, as well as Mexican migrants returning voluntarily or involuntarily from the USA. Examining the experience of the twin cities of Ambos Nogales, the authors conclude that cities along the USA–Mexico border are showing increasing agency in responding to mixed migration, in a short time becoming home to many formal and informal organisations providing aid and managing migration at the local level. The authors emphasise that such responses and learning processes are consistent with the traditional openness and solidarity of border communities toward migration.

In ways reminiscent of themes set in earlier chapters in Chapter 22, Lena Poschet El Moudden discusses the urban development of Ouanaminthe (Haiti)/ Dajabón (Dominican Republic) located on either side of the Massacre River on Hispaniola island, from the time of French and Spanish colonisation up to 2004. Despite these cities playing commonly found roles in cross-border economic networks, this particular border has maintained strong separation, with the wealth generated by largely unacknowledged binational exchanges disappearing into the pockets of an already wealthy minority. Poschet El Moudden summarises twin-city realities as a failure to develop a solid relationship that would enhance stability and equal exchange, due to a lack of resources and support from the two intermittently hostile national governments. Cross-border relations have mainly been conducted rather privately by the two sets of civil organisations and international solidarity networks – somewhat in the manner of Eilat/Aqaba in Chapter 10.

Chapter 23 by Cathrine Olea Johansen, Ruben Moi, John Garrard and myself, opens Part III of the volume and addresses representations of twin cities in art, notably literature and film, focusing primarily on the fictional twin cities created by contemporary British writer China Miéville in his crime-fiction novel *The City & the City*. Miéville vividly describes two city-states sharing the same spatio-temporal location: Besźel and Ul Qoma. The chapter puts Miéville's twins in the perspective of real-world twin cities pointing at their resemblances to the two specific subgroups of twin cities – internal twins and external twins straddling hard rather than permeable international borders.

Chapter 24 by John Garrard concludes the volume by casting light on twin cities that escaped our attention. He mainly covers: (1) more examples of internal twins that bucked the persistent twin-city trend by amalgamating; (2) twin cities of both internal and cross-border kinds that we see as important enough to warrant inclusion; (3) *nesting* cities – those located *inside* other cities rather than just adjacently; and (4) cross-border twins that were particularly affected by ongoing faraway events like ongoing changes in US policies towards incoming and outgoing migration flows, the COVID-19 pandemic and political unrest, i.e. the influence of the Chinese crack-down on protest and political autonomy of Hong Kong on Hong Kong/Shenzhen relationship.

*****

To conclude, we want to briefly contemplate the promising research avenues for Twin City Studies. First, we believe future research will continue contextualising twin cities within major ongoing global trends, including globalisation, urbanisation, ageing, digitalisation, technological advances and increasing environmental responsibility. These trends provide the background for urban development generally across the world. Necessarily, therefore, they are and will reshape internal and external twin cities and their centrality within regional and (inter)national urban systems. Consequentially, studies on transformation of twin cities' economic specialisation, population and employment structure, emerging businesses, directions of urban sprawl, new connectivity projects, twin cities' marketing campaigns and climate-change actions are particularly relevant in the near future.

Second, several chapters in this book, and the conclusion, touch on the impact of COVID-19 on border twin cities; they do so tentatively, since the pandemic is still ongoing and, for many, worsening. We feel there will soon be highly useful and reflective research to be done on how COVID-19 has impacted multiple twin cities across the world, and how it compares with their counterparts during the so-called Spanish Flu Pandemic.

Third, noting that local governments play prominent roles in establishing and maintaining cooperative twin-city relations, we expect more studies focusing on local government structures and their reform in twin cities, their prerequisites, reasons, mechanisms and outcomes, including the rare but possible scenario of twin-city amalgamation.

Fourth, we realise that some geographic regions, as well as particular case studies (some named in the conclusion) still lack their full conceptualisation within a twin-city framework. One such region is Africa, with its history of colonial competition, porous borders 'conceived (not always accurately) as "arbitrary"' (Coplan 2010, 2), strong contemporary cross-border kinship and trade ties. This continent is often presented as fertile ground for the future of border studies. Naturally, we see African border twin cities as a prominent direction of Twin City Studies too.

Fifth, particularly given the porosity characterising so many external twin-city borders, we also see great potential in examining twin cities as reconciliation platforms in conflict and post-conflict environments. Cross-country and cross-conflict comparisons from different macro-regions would valuably contribute to the debate started in Chapters 9 and 10 in this volume.

Besides, another promising research avenue lies in looking at twin cities as examples of city diplomacy practices. Namely, we should start exploring when, where, why and how twin cities (particularly external ones) have looked for experience-transfer between each other, individually and once-only, or collectively and on a regular basis, via coming together in (inter)national city networks, like the City Twins Association formed in 2006.

Finally, there is a need for comparing twin cities with twinned cities – distant urban centres in different countries that enter a twinning agreement to stimulate cross-cultural, people-to-people exchanges as well as strengthening city-to-city economic ties (Jayne et al. 2013). As homonymy of the terms suggests, they share extensive commonalities including the substantial role of residents and local government in promoting twinned relationships.

Notes

1 Population statistics and area data in the introduction were collected at *Thomas Brinkhoff: City Population*, http://www.citypopulation.de on 22.03.2021.

References

Arreola, D.D. (1996). Border-city idee fixe. *Geographical Review*, 86, 356–369.
Baltic Course (2018). Estonian econmin: Hyperloop must be considered in Tallinn-Helsinki tunnel project. Available: http://www.baltic-course.com/eng/transport/?-doc=145267 (accessed 18.03.21)

BBC (2020) Love in the time of closed borders. Available: http://www.bbc.com/travel/story/20200409-couples-meeting-at-closed-borders-during-coronavirus (accessed 18.03.21)

Buursink, J. (2001). The binational reality of border-crossing cities. *GeoJournal*, 54(1), 7–19.

Chan, C. (2020) 'We will deliver right to the border'. *CBC News*. Available: https://www.cbc.ca/news/canada/british-columbia/stewart-hyder-covid-19-border-closure-1.5513395 (accessed 17.03.2021)

Chatfield, W.H. (1893). *The Twin Cities (Brownsville, Texas; Matamoros, Mexico) of the Border and the Country of the Lower Rio Grande*. Harbert Davenport Memorial Fund, New Orleans.

City of Twin City (2021) Available: https://twincityga.com/ (accessed 18.03.21)

Coplan, D. (2010). Introduction: From empiricism to theory in African border studies. *Journal of Borderlands Studies*, 25(2), 1–5.

D'Antonio, W. V. (1958). *National Images of Business and Political Elites in Two Border Cities*. Michigan State University of Agriculture and Applied Science.

Dilla, H. et al. (2008). *Ciudades en la frontera, Santo Domingo*, 2008.

Eglitis, A. (2021) World's longest undersea rail tunnel back on agenda in Estonia. *Bloomberg News*. Available: https://www.bnnbloomberg.ca/world-s-longest-undersea-rail-tunnel-back-on-agenda-in-estonia-1.1553459 (accessed 17.03.2021)

Garrard, J., and E. Mikhailova (eds) (2019). *Twin Cities: Urban Communities, Borders and Relationships Over Time*, Routledge, London and New York.

Geographica Helvetica (2007). Border towns; city boundaries; boundaryless spaces. Vol. 62(1). Available: https://www.geogr-helv.net/62/issue1.html (accessed 17.03.2021)

GeoJournal, Vol. 54(1) (2001). Available: https://link.springer.com/journal/10708/54/1 (accessed 17.03.2021)

GO! 2025 Team(2020) European capital of culture candidacy. Available: https://www.go2025.eu/wp-content/uploads/2021/03/GO2025engCompressed-Screen-1.pdf (accessed 17.03.2021)

Grant, J.A.C. (1958). The Viet Nam Constitution of 1956. *The American Political Science Review*, 52(2), 437–462.

Herzog, L.A. (1990). *Where North Meets South: Cities, Space, and Politics on the United States-Mexico Border*. University of Texas Press.

Herzog, L.A., & Sohn, C. (2014). The cross-border metropolis in a global age: A conceptual model and empirical evidence from the US-Mexico and European border regions. *Global Society*, 28(4), 441–461.

Jakobowski, J., Popławski, K., Kaczmarski, M. (2018), The Silk Railroad. The EU-China rail connections: Background, actors, interests, OSW Studies No. 72, Centre for Eastern Studies, Warsaw.

Jańczak, J. (2013). *Border Twin Towns in Europe: Cross-border Cooperation at a Local Level*. Logos Verlag.

Jayne, M., Hubbard, P., Bell, D. (2013). Twin cities: Territorial and relational geographies of 'worldly' Manchester. *Urban Studies*, 50(2), 239–254.

Journal of Borderlands Studies, 32(4) (2017) Special issue: Theorizing town twinning: Towards a global perspective. Available: https://www.tandfonline.com/toc/rjbs20/32/4 (accessed 17.03.2021)

Lundén, T. (2004). *On the Boundary: About Humans at the End of Territory*. Södertörns högskola.

Medeiros, E., Guillermo Ramírez, M., Ocskay, G., Peyrony, J. (2020). Covidfencing effects on cross-border deterritorialism: The case of Europe. *European Planning Studies*, 1–21.

Nugent, P. (2019). *Boundaries, Communities and State-making in West Africa: The Centrality of the Margins*, Vol. 144, Cambridge University Press.

Osborne, R.H. (1959). Changes in the urban population of Poland. *Geography*, 44(3), 201–204.

Purcell, V. (1959) *From a Chinese city* by Gontran de Poncins, book review. *Pacific Affairs*, 32(1), 122.

Raffestin, C., Servin, R. (1978). Deux agglomérations transfrontalières: La 'Regio insulensis' et la région franco-genevoise. *Actes du 101e Congrès national des Sociétés Savantes, Section d'histoire moderne et contemporaine. Tome I: Frontières et limites de 1610 à nos jours*, 41–54.

Railway Pro (2020) Contracts awarded for Malaszewicze terminal expansion. Available: https://www.railwaypro.com/wp/contracts-awarded-for-malaszewicze-terminal-expansion/ (accessed 29.03.2021)

Reuters (2021) Pandemic derails trade for Congo's disabled border couriers. Available: https://www.reuters.com/article/uk-health-coronavirus-congo-idUSKBN2A80T6 (accessed 29.03.2021)

Sloan, J.W., West, J.P. (1977). The role of informal policy making in US-Mexico border cities. *Social Science Quarterly*, 58(2), 270–282.

SwissInfo (2020) Covid sceptics form chain on Swiss-German border. Available: https://www.swissinfo.ch/eng/covid-sceptics-form-chain-on-swiss-german-border/46074680 (accessed 29.03.2021)

The Economist (2020) The closure of Canada's border with Alaska has split twin towns. Available: https://www.economist.com/the-americas/2020/09/05/the-closure-of-canadas-border-with-alaska-has-split-twin-towns (accessed 29.03.2021)

Vanneph, A., Mouroz, J.R. (1994). Ciudades fronterizas México-Estados Unidos. *Estudios Fronterizos*, 33, 9–35.

Part I
Intranational Twin Cities

2 Twin cities in medieval England
The case of small towns' development

Anna Anisimova

Many factors contributed to the process and form of urbanisation in the medieval period. Small towns, populated by fewer than 2000, characterised Europe's medieval urban landscape generally and England's particularly. These included several twin towns, whose coexistence in long-term close mutual proximity raises several questions about their origins, parallel development, separate existence or future amalgamation. All these factors help us understand the nature of the twin-city arrangement and whether it was a long-enduring step to broader urban unification or just a brief stopping place.

This chapter aims to explore the twin-towns phenomenon, primarily using the example of English monastic towns, i.e., relatively small urban settlements under monastic lordship. Several examples of long-existing pairs from different parts of the country are considered. The main focus is medieval, when these towns arose and developed, but later developments are also considered.

The towns' location depended on many factors, including natural factors (availability of essential resources, e.g. water) and the strategies and policies of urban lords. Several requirements needed to be met by a successfully developing prosperous town, and its environment is very important: its location's economic potential, and its distance from other settlements, especially towns. There are many instances of rivalry between towns (and their lords) because of various rights – market, fishery, tolls, etc. Sometimes they were not confined to legal proceedings concerning rights infringements, and they even produced violence.

From Henry II's reign, there was official regulation and restriction of market proliferation, supposedly protecting the interests of existing settlements. *The Gazetteer of Markets and Fairs in England and Wales to 1516* is full of notes about appeals concerning damage to already-existing markets caused by new ones being created (Letters 2013; Masschaele 1994, 264). Such complaints often induced the prohibition and closing of a new market or fair, especially during the later 13th century, with most markets already being existent and continuing their economic growth. There was also a rule that there must be at least $6^2/_3$ miles between markets (Swanson 1999, 47–48). This should have prevented or, at least, complicated the possibility of two towns being in mutual proximity.

The presence of twin cities indicated that a location could support both, by having the essential resources and economic potential. Some places, like estuaries, major road junctions, river crossings, were especially attractive, and rival

DOI: 10.4324/9781003102526-3

urban foundations could result (Table 2.1). Such cases show lordly aspirations and attempts to take advantage of a particular situation. It could also relate to the general logic of manor boundaries typically following roads and natural barriers like rivers. Curiously, advantageous locations like significant road intersections, usually attractive to settlements, do not characterize twin towns. But perhaps roads did not provide the sort of distancing that rivers presumably did.

The evidence on markets is based on: Letters 2013.

There are two essential factors underpinning twin-town development. The first is when they appeared. The towns under consideration were founded or emerged mostly during the 13th century, i.e., the period of rapid urban growth characterised by the development of small towns and a denser town network (Astill 2000, 46–49). The period also sees very active town foundations when different lords were particularly active participants in urbanisation. The only exception is Stonar and Sandwich in Kent, which can be explained by previous port development in this region due to intense interactions with the Continent.

Table 2.1 Several examples of medieval twin towns, located in the south of England.

Towns	County	Markets	Lordship	Location	Eventual disposition
Stonar	Kent	1189–1199	Monastic	Across river, no bridge, ferry	Separate
Sandwich		1086	Monastic		
Portsmouth	Hampshire	1194	Secular	Across harbour, no bridge, ferry	Separate
Gosport		1284	Bishopric		
Kingsbridge	Devon	1220	Monastic	Across river (?), bridge, ford	Unified in 1893
Dodbrooke		1257	Secular		
Newton Abbot	Devon	1220	Monastic	Across river, no bridge, ford	Unified in 1900s
Newton Bushel		1246	Secular		
East Teignmouth	Devon	1220	Monastic	Across river, no bridge	United in 1795–1805
West Teignmouth		1268/9	Bishopric		
Fowey	Cornwall	1225	Monastic	Across river, no bridge, ferry	Separate
Polruan		1292	Secular		
Weymouth	Dorset	1248	Monastic	Across estuary, no bridge	Unified in 1571
Melcombe Regis		1268	Secular		
West Looe	Cornwall	1227–1243	Secular	Across estuary, bridge (c. 1400)	Unified in 1883
East Looe		1237	Secular		

Many of the towns apparently emerged pretty much simultaneously. There is difficulty dating precisely when a town appeared. Market-granting is often considered an indicator if there is no foundation charter. When one town considerably predated the other, it could oppose and discourage another developing nearby. By contrast, concurrent development allows us to talk about parallel rather than suburban growth.

Another important factor underpinning twin-town development is their lordship. Different ownership is quite typical, indicating two lords trying to profit from the advantageous location promoting their settlements. It helped determine the towns' location. Two different lords meant they belonged to different parishes and manors and were situated near manorial borders. The remarkable presence of monastic towns in a twin-town development also indicates that monasteries came later in the urbanisation process and so had to share the advantageous places. Different lordship also promoted the fact that twin towns were two administratively separated places growing in more or less the same circumstances.

All the towns are situated on both communication and water routes, across the river from each other. They could have been considered so-called bridge towns, except that there was no bridge between most of them during the Middle Ages.

The difficulty of studying small medieval towns is often connected with the scarcity of evidence, and its disparity as conflicts usually left more traces than harmonious relationships.

To see how these towns originated, developed and coexisted, we will look in detail at Weymouth and Melcombe Regis, twin towns with a conflictual relationship, as well as examples of peaceful coexistence (East and West Teignmouth, Newton Abbot and Newton Bushel), and another confrontational relationship (East and West Looe) (Figure 2.1).

Weymouth and Melcombe Regis

These two boroughs were established across the Wey River (Figure 2.1a), although their early history is mostly unknown. In 1248, St Swithun's Priory, Winchester, was granted a weekly market and a fair at Weymouth in their manor of Wyke Regis. Then, in 1252, they founded there a borough ('liber burgus'), establishing its limits (Ballard and Tait 1923, 4; Moule 1883, 15–19). The charter mentions a port with full rights, modelling them after those of Southampton and Portsmouth.

Apparently, there was already some settlement there. The manor of Wyke was granted to the priory in 1042. Henry I's precept (1106?) mentions the ports of Weymouth and Melcombe among the monastery's possessions (Johnson and Cronne 1956, 52, No. 745). Soon after the borough's foundation in 1258, Richard de Clares, earl of Gloucester and Hertford, gained control over Weymouth, alongside the manors of Portland and Wyke, which he exchanged with the prior and convent for the manor of Mapledenham (Altschul 1965, 77, 209).

Across the river, another town emerged pretty much simultaneously. The borough of Melcombe was established by royal charter in 1280, soon after the manor

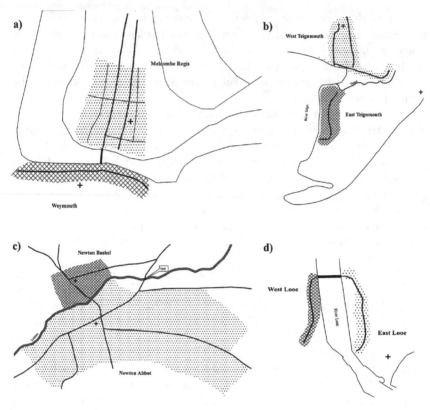

Figure 2.1 Medieval town plans: (a) Weymouth and Melcombe Regis, (b) West and East
Teignmouth, (c) Newton Abbot and Newton Bushel, (d) West and East Looe.
Map by Anna Anisimova.

was transferred from the abbey of Cerne to the king (Public Record Office [here-
after PRO] 1901, 365). A new borough's privileges were extensive and copied
the London charter of 1268 (Ballard and Tait 1923, 8, 24), although there is no
special mention of the rights of Melcombe to the harbour or of Weymouth. However, Melcombe had already existed as a settlement and port before this charter;
it is mentioned in some documents (e.g. PRO 1932, 304).

Both towns began as dependencies of neighbouring centres (Wyke Regis and
Radipole) and consequently lacked parish churches, only possessing chapels in
the Middle Ages. The towns, their inhabitants, and owners relied on the harbour's profits: tolls, customs, trade with France, etc. Weymouth was older, originally bigger, and dominated for a while. It was an important port for collecting
the king's tolls, and it sent ships on the king's service at least from 1254. Close
rolls and Patent rolls contain many documents concerning the summons to the
king's naval service. However, Melcombe [Regis] gained strength and became an
important port. In 1315 its burgesses petitioned Parliament to restore to them
the cocket of wool (i.e., right to export wool through their port), as they alleged

to have had it from the foundation (Phillips 2005, No. 152 (122)). In 1318 they were a port with the cocket (PRO 1903, 161). In 1365 it was decided to set up a wool staple in Melcombe, which was confirmed in 1371 (Ormrod 2005a, No. 30; Ormrod 2005b, No. 7).

The towns' layouts were influenced by the local landscape – a single street parallel to the southern side of the harbour in Weymouth, and a grid with four principal streets for Melcombe. While the former was constrained by the local relief, the latter could expand.

The towns' close proximity quickly produced conflict. The main controversy was about boundaries, especially on the water. Already in 1284, there was a suit in court concerning the moiety of Wey between Weymouth and Melcombe (Oliver 1934, 32). In 1320 Melcombe burgesses accused Weymouth's inhabitants of seizing the entire river (National Archives, SC8/193/9646). In turn, around 1332, Elizabeth de Bourg, Lady of Clare, complained that Melcombe's burgesses did not observe Weymouth's established bounds, trying to seize tolls from the boats arriving in the port and also building houses within these bounds (National Archives SC8/35/1724). On 14 August 1332, an inquisition ascertained that the port and water between Weymouth and Melcombe belonged entirely to Weymouth which received dock dues and other customs duties (PRO 1916, 338, No. 1383; Moule 1883, 22). Melcombe's burgesses accused Lady Clare of trying to obstruct the legal proceedings concerning this matter (National Archives SC8/240/1957), also complaining that coercion had been used against them (National Archives, SC8/127/6334). Conflicts concerning the water of Wey, customs collection and port functions, as well as territories between the towns, continued, with another outburst in 1367 (PRO 1937, 246–247, no. 658).

The wars with France impacted enormously on both towns. Their ships and trade suffered, and both were attacked during coastal raiding. Melcombe was burnt in 1377 and 1380 (Bettey 1986, 113). However, around 1378, Parliament refused a petition to licence the town's enclosure (National Archives SC8/236/11784). Because of the destruction, Melcombe's inhabitants requested and were repeatedly granted payment exemption or reduction from 1388 until at least 1512 (Given-Wilson 2005, item 147; PRO 1907, 74, etc.). The inquisition of 1408 found that, of 120 burgesses living there during Edward III's reign, only 20 remained by 1394, and 8 remained in 1408 itself (PRO 1968, 196–197, no. 368).

Meanwhile, in 1427, there was a Commons' request in Parliament to make Melcombe a staple (designated market) for the export of wool, etc., endorsed with the king's agreement (National Archives SC8/25/1225; Curry 2005a, item 44). However, its reduced circumstances led to the transfer of port rights from Melcombe to Poole in 1434, 'havynge consideration to the feblesse and non-sufficeante of youre porte of Melcombe, nouht enhabited' (Curry 2005b, item 38, 39).

Despite these difficulties, the Weymouth-Melcombe Regis animosity continued into the 16th century. By now, they were both royal possessions, since Weymouth had come under the ownership of the dukes of York and then the Queen (Horrox 2005a, item 5; Horrox 2005b, item 2). Constant conflicts meant the 1560s were filled of proceedings concerning their respective rights (National Archives STAC5/M7/26; Moule 1883, 9, 12–13, 22–23). As John Coker described it,

> between whom [Melcombe] and Weymouth arose great controversy, both enjoying like privileges, and both challenging the particular immunities of the haven, which lyeth in the very bosom of them; each of them taken the overthrow of the other: but not resting by that, continually commenced new suits.
>
> (Ellis 1829, 11)

Finally, in 1570 a commission was formed to enquire into the disputes (Moule 1883, 23, No. II, 10), which recommended the union of two towns. 'Whereunto they have consented upon this condition that they, and every of them may have and enjoy all their annciente Libties and privelege which they have enjoyed tyme oute of mynde' (Moule 1883, 23, no. II,11). In 1571 a Parliament act incorporated and united Weymouth and Melcombe Regis under a mayor, aldermen, bailiffs, burgesses and community (Weinbaum 2010, 33). The new corporation was to succeed to the two previously separate liberties in all respects. However, problems remained and quarrels between the two formally united settlements continued, as attested by a Quo Warranto enquiry in 1572 and a petition to the Privy Council (Moule 1883, 25, no. II, 13, 15). A justices' commission reported in 1575 that, though governmentally united, the towns continued separately in private actions. The commission recommended building a bridge to aid agreement (Moule 1993, 25, no. II, 17). In 1580 both sides filed complaints against each other (Moule 1883, 5–6, no. I,12; National Archives, SP46/32/fo104, SP46/18/fo30). Weymouth continued collecting customs independently, and its bailiffs acted independently of the mayor. Things continued thus, and the next decade was occupied by intertown proceedings, enquiries and disputes (Moule 1883, 6, no. I, 13, 14; National Archives E134/28and29Eliz/Mich25). The bridge was built during 1593–1597, and the Queen donated income for its upkeep (Moule 1883, 8, no. I, 17). In 1598, incorporation was legally confirmed, with some changes (Weinbaum 2010, 33). A 1616 charter reinforced urban union, but they continued with separate Courts Leet even into the 19th century (Drinkwater 1832, 155–157).

Thus, Weymouth and Melcombe Regis were amalgamated but remained in long-term steady dispute; even formal unification hardly induced harmony, still less genuine merger. During their competing coexistence, both were quite important ports influenced by changing circumstances – French destruction, also competition from other ports – Poole and Lyme Regis. The fact that they eventually came under the same lordship facilitated unification, although they retained considerable autonomy.

East and West Teignmouth

The towns of East and West Teignmouth (county of Devon, Figure 2.1b), belonging respectively to the canons and bishop of Exeter, followed a very different course. They were established during the 13th century on opposite sides of the

Tame rivulet flowing into the Teign estuary. In 1220, King Henry III granted the canons of Exeter Cathedral a market at its manor of Teignmouth (Lake 1904, 106). In 1256 a market charter was issued for West Teignmouth, later confirmed in 1270 (Griffiths 2001, 24). During the inquisition of Hundred Rolls in 1274, East Teignmouth inhabitants complained about the unauthorized Saturday market in West Teignmouth (Illingworth 1812, 85). Thus the settlements appeared pretty much simultaneously. They quickly developed; by the 14th century, there are references to the burgage tenements and a bishop's borough (Beresford and Finberg 1973, 91; Letters 2013; Lake 1904, 103–112). Curiously, their topography is somewhat similar to the previous pair. While East Teignmouth was a single street, West Teignmouth's layout consisted of three main thoroughfares – Exeter Street, Bitton Street, and Teign Street. The latter expanded with the development of land communications along the Teign and possessed more space for expansion southwards.

In the later Middle Ages, they are difficult to separate, since records do not necessarily distinguish them. Of the two, West Teignmouth was bigger, possibly due to its more advantageous position in relation to the Teign estuary. They continued their parallel and separate existence, separated by the stream and surrounding marsh until the early 19th century when, led by Robert Jordan, a local banker, the burgesses channelled the stream into a canal, reclaiming the surrounding land (Griffiths 2001, 68). That brought the two towns together between 1795 and 1805, although they continued as parts of different manors with different lords – the Courtenays and the Cliffords (Griffiths 2001, 83). Their rights were preserved by including them in local commissions. One town was much bigger than the other. According to the 1801 census, West Teignmouth had 350 houses and 1528 people, while East Teignmouth possessed 100 houses and 484 people (Smart 2016, 21). Amalgamated, they developed into a fashionable resort, while continuing as an active port.

There was apparently no inter-town rivalry after 1274, probably due to their small size and different locations – one was more advantageously situated for shipbuilding and other things. Apparently, long before the two towns amalgamated at the turn of the 18th century, they were perceived as one, probably because their main joint interest lay with the port of Teignmouth.

Newton Abbot and Newton Bushel

Not far from the Teignmouths, another 13th-century urban pairing emerged: Newton Abbot and Newton Bushel (Figure 2.1c), on opposite banks of the river Lemon, a Teign tributary, near the point where it was crossed by the Exeter-Totnes road. Newton Abbot was founded by Torre Abbey in Wolborough manor shortly after the monastery's foundation and endowment by this manor in 1196 (Seymour 2000, 51). The first mention of a 'new town' comes around 1200. It was situated at the edge of the manor (Wolborough) and parish (St Mary's). It is unclear when the abbey gained a market charter for its *nova villa*. A Letter Close of 1220 grants a market and a fair to the abbot of Torre (Letters 2013), but the Hundred Rolls of 1274 claimed there was a charter of King John with such a grant (Illingworth 1812, 72). The new town's focus was at St Leonard's chapel, and the market situated along Wolborough Street.

Newton Bushel originated under a lay lord in the neighbouring manor. In 1237 the manor of Teignwick was granted to Theobald de Englishville, and in 1246 he received a charter for a Tuesday market, which emerged at a triangular spot near St Mary's chapel (Beavis 1985, 24).

The towns coexisted for a long time, and there is no indication of hostility from either side. On the contrary, there is evidence of familial and business connections between inhabitants. Curiously, there was no bridge over the Lemon in the Middle Ages: the possible crossing was a ford at Back Road, with a bridge being built only in the 1800s. The markets continued coexisting until 1633, when Newton Bushel's was abandoned (Beavis 1985, 13). Eventually, Newton Abbot absorbed Newton Bushel in 1901, although the two manor jurisdictions still exist.

West Looe and East Looe

East Looe and West Looe (Figure 2.1d) in the county of Cornwall appeared simultaneously across the Looe-River estuary, both first mentioned in the 1201 assize roll. They got their market and borough charters from their secular lords – the Bodrigans and the Treverbyns. Both towns belonged to separate manors and parishes (Beresford 1967, 406). Topographically, they each comprised a single street along the shore. They are reputed to have had an antagonistic relationship. However, unlike other twin towns explored here, a bridge was built between them around 1400. It linked the route from Plymouth to Fowey and west Cornwall and was of regional importance. Originally of wood and burnt in 1405, it was replaced by a stone construction in 1411–1436, with 15 arches along with a chapel (Camp and Birchwood Harper 2008, 12, 17–18). This created additional inter-town friction over responsibilities for repairs. However, in the 17th century it was decided to be a county responsibility (Browne 1904, 9). The towns were both incorporated in the later 16th century; West Looe in 1574, East Looe in 1587 (Weinbaum 2010, 16). They became one parish in 1845 and were co-joined by the 1883 Municipal Corporations Act, although West Looe inhabitants opposed what they saw as an East Looe takeover (Browne 1904, 28). The latter was apparently always larger and, already in the 16th century, was considered the principal settlement. The 1801 Census lists 467 people in East Looe and 376 in West Looe (Bond 1823 146). Nevertheless, for a long time there was tradition of municipal cooperation, as they invited the members of the other town to corporate meetings (Browne 1904, 69).

Twin port towns

From the four examples of twin towns discussed here, it is possible to point out some characteristic features that can be put into a wider perspective of regional ports' development and some other examples of concurrent settlement development.

All these ports developed and thrived on the same economic location-defined factors on England's southern coast – trade with France, Spain and other homeports, usually wine but also exports of wool, tin and hides. Fishing and shipbuilding were also significant for them. Teignmouth had salt works. Notwithstanding restrictions, each town had its own market. Quite often, their origins related to lordly ambitions. Their fortunes were closely connected to their proximity to other ports, especially larger ones, like Plymouth, Dartmouth, later Falmouth, etc. From the 16th century, they became involved in the Newfoundland trade. Smuggling remained a significant revenue and activity source for centuries, alongside piracy and privateering, especially during war times. By the early 19th century, many had become resorts. All, even small places like Polruan, sent ships and sailors for king's service from the later 13th and early 14th centuries. The bigger ones had customs posts and controlled exit and entry to and from the country. All were also vulnerable to enemy attack; many were burnt and had their ships plundered during the Hundred Years War, disturbances during Charles II's quest for power (East and West Teignmouth in 1690), and the Napoleonic Wars.

All these towns, facing each other across a river or harbour, show the potential of such locations for supporting two settlements and the importance of trade for urban growth, relying heavily on commence and transit (tolls/customs). Some failed developments next to, or across from, larger ports, only confirm this. (For example, Newton and Poole, county of Dorset [Beresford 1967, 428], where in 1286 Edward I ordered the founding of a new borough on Poole Harbour's south side, granting the new town the same liberties as Lyme Regis and Melcombe Regis had. However, nothing was apparently built.) Sometimes, a new foundation was quickly incorporated into an existing settlement, as with Lostwithiel, a borough from the 1190s, and Penknight appearing as a separate borough by the mid-13th century next door and divided by a stream, but amalgamated with Lostwithiel in 1268 ('quod burgi nostril predicti de L. et P. unus et liber burgus si[n]t') (Ballard and Tait 1923, 5; Beresford 1967, 406–407). Amalgamation was probably rendered likely by their situation on the same side of the river.

Different lordship seems a significant factor for continuing separation, although Weymouth and Melcombe Regis were not united when they both ended in the king's hands. The importance of seigneurial support can be illustrated by Fowey and Polruan in Cornwall. Situated across the river Fowey, the two settlements appeared pretty much simultaneously. The priory of Tywardreath made Fowey a free borough around 1250 (Keast 1987, 137–138). In 1316, the prior got it a market charter, and during the 14th century, Fowey increased in prosperity. While Polruan also became a borough, it never reached Fowey's level (Henderson 1963, 30).

Different factors influenced whether both ports survived, and how prosperous and developed they became. Thus, Fowey and Polruan were initially influenced by the proximity of the port of Lostwithiel, as well as Plymouth's harbour. During the 14th century, Fowey replaced Lostwithiel as the estuary's principal port. Lostwithiel, at the upper reaches of the navigable river, suffered from silting caused by inland tin-streaming from as early as 1357, so trade transferred to Fowey at the more accessible estuary mouth (Henderson 1963, 33–34). Silting, regionally prevalent because of tin-working (Fox 2001, 5), and devastating storms played important roles in the ports' fate.

Meanwhile, Gosport's development opposite Portsmouth, in Hampshire, shows rather different dynamics. It was a humble early-13th century bishopric foundation which had possibly declined by the later Middle Ages, but then gained additional developmental impulse with Portsmouth becoming a naval base in the 16th century. Gosport controlled the ferry (National Archives E178/2054, E134/42and43Eliz/Mich22). Also important was Gosport's prominence for defending Portsmouth Harbour. Portsmouth briefly extended its limits to include Gosport via the 1682 charter of Charles II, later annulled (Murrell and East 1884, 385). This caused Gosport's inhabitants to complain that Portsmouth's authorities had infringed their liberties. Eventually, Gosport remained as a part of the bishop's estate (with the manor of Alverstoke), but it was closely connected with Portsmouth's dock, and its needs. It thrived on its naval connections, barracks and forts being built there; they even got a market and fair in 1717 (Page 1908).

Topographically, twin towns had two possible layouts – a single street plan or a grid, both oriented on the waterfront. Town shapes were usually determined by landscape peculiarities, and as they were situated on the different sides of a river, there is no evidence that towns specifically influenced each other in that respect. Some had space to expand in favourable economic circumstance, with one normally becoming dominant (Melcombe Regis, East Looe); some had no such opportunity.

One of the enduring features of twin towns is their dominant-subordinate relationship – a factor often enhancing inter-twin conflict (e.g., Melcombe Regis subdued Weymouth, and East Looe subdued West Looe). This could be determined by a more advantageous location, or some other external circumstances. The relationship could also change over the time.

Nevertheless, twin-town conflict should not be exaggerated, as there is evidence of close connections between local families and people with holdings in both towns at the same time, e.g., in Newton Abbot (Jenkins 2010, 15, 189, 238). In 1573–1576, John Rashleigh, a merchant and important landowner in Fowey, purchased the manor of Polruan (Archives and Cornish Studies Service R/1901/1, 2; R/1905/1, 2). The corporations of East Looe and West Looe invited the other's representatives to their meetings and had dinner parties where they toasted 'two Looes'. Even Weymouth and Melcombe Regis's quarrelling inhabitants sometimes plundered ships together. For the outside world, the twins were quite often perceived as a whole (just like many internal twin cities to this day), especially when they had similar names and were considered by their combined potentials. They shared the same trade and economic activities and, in a way, functioned as limbs of a single economic entity.

Thus, most twin towns survived the Middle Ages as separate units and were amalgamated much later or never. The reason for their continued separate existences could be because they had different lords who preferred to keep their own interests, and the conflict, if there was one, enhanced local attachment. Towns' relatively small size was also conductive to their continuing coexistence.

References

Altschul, M. (1965). *Baronial Family in Medieval England: The Clares, 1217–1314*. Baltimore: Johns Hopkins University Press.

Astill, G. (2000). General Survey 600–1300. In: D.M. Palliser, ed., *The Cambridge Urban History. Volume 1: 600–1540*. Cambridge: Cambridge University Press, 25–50. doi:10.1017/CHOL9780521444613.004.

Ballard, A. and Tait, J. eds., (1923). *British Borough Charters 1216–1307*. Cambridge: Cambridge University Press.

Beavis, D. (1985). *Newton Abbot: The Story of the Town's Past*. Buckingham: Barracuda.

Beresford, M. (1967). *New Towns of the Middle Ages. Town Plantation in England, Wales, and Gascony*, New York, Washington: Praeger.

Beresford, M. and Finberg, H.P.R. (1973). *English Medieval Boroughs: A Hand-list*. Newton Abbot: David & Charles.

Bettey, J.H. (1986). *Wessex from AD 1000*. London and New York: Longman Inc.

Bond, T. (1823). *Topographical and Historical Sketchers of the Boroughs of East and West Looe, in the County of Cornwall*. London: J. Nichols and Son.

Browne, A.L. (1904). *Corporation Chronicles*. Plymouth: John Smith Plymouth Limited, Printers, Stationers, &c.

Camp, M. and Birchwood Harper, B. (2008). *The Book of Looe: Tourism, Trawlers and Trade*. Wellington: Halsgrove.

Curry, A. (2005a). Henry VI: 1427, October. Text and Translation. In: C. Given-Wilson, ed., *The Parliament Rolls of Medieval England, CD-ROM*. Leicester: Scholarly Digital Editions.

Curry, A. (2005b). Henry VI: 1433, July. Text and Translation. In: C. Given-Wilson and et al., eds., *The Parliament Rolls of Medieval England, CD-ROM*. Leicester: Scholarly Digital Editions.

Drinkwater, J.E. (1832). Report on the Borough of Weymouth and Melcombe Regis. *Parliament Papers*, 28(2), 155–157.

Ellis, G.A. (1829). *The History and Antiquities of the Borough and Town of Weymouth and Melcombe Regis*. Weymouth, London: Messrs. Baldwin and Cradock.

Fox, H. (2001). *The Evolution of the Fishing Village: Landscape and Society along the South Devon Coast, 1086–1550*. Oxford: Leopard's Head Press.

Given-Wilson, C. (2005). Henry IV: 1399, October. Text and Translation. Part 1. In: C. Given-Wilson and et al., eds., *The Parliament Rolls of Medieval England, CD-ROM*. Leicester: Scholarly Digital Editions.

Griffiths, G. (2001). *History of Teignmouth*. Bradford-on-Avon: ELSP.

Henderson, C. (1963). *Essays in Cornish History*. Truro: D. Bradford Barton LTD.

Horrox, R. (2005a). Henry VII: 1495, October. Text and Translation. In: C. Given-Wilson et al., eds., *The Parliament Rolls of Medieval England, CD-ROM*. Scholarly Digital Editions.

Horrox, R. (2005b). Henry VII: 1497, January. Text and Translation. In: C. Given-Wilson et al., eds., *The Parliament Rolls of Medieval England, CD-ROM*. Leicester: Scholarly Digital Editions.

Illingworth, W. (1812). *Rotuli Hundredorum temp. Hen. III & Edw. I in Turr' Lond' et in Curia Receptae Scaccarii Westm. Asservati*. Vol. 1. London: S.N.

Jenkins, J.C. (2010). *Torre Abbey: Locality, Community, and Society in Medieval Devon*. Ph.D. Thesis. Oxford University.

Johnson, C. and Cronne, H.A. eds., (1956). *Regesta Regum Anglo-Normannorum/2, Regesta Henrici Primi 1100–1135*. Oxford: Clarendon Press.

Keast, J. (1987). *The Story of Fowey (Cornwall)*. Redruth: Dyllansow Truran.

Lake, W.C. (1904). Ancient Teignmouth. *Transactions of the Devonshire Association*, 36, 103–112.

Letters, S. (2013). *Gazetteer of Markets and Fairs in England and Wales to 1516*. [online] archives.history.ac.uk. Available at: https://archives.history.ac.uk/gazetteer/gazweb2.html [Accessed 5.5.20].

Masschaele, J. (1994). The multiplicity of medieval markets reconsidered. *Journal of Historical Geography*, 20(3), pp. 255–271. DOI: 10.1006/jhge.1994.1020.

Moule, H.J. (1883). *Descriptive Catalogue of the Charters, Minute Books and Other Documents of the Borough of Weymouth and Melcombe Regis. AD 1252–1800*. Weymouth: Sherren & Son.

Murrell, R.J. and East, R. (1884). *Extracts from Records in the Possession of the Municipal Corporation of the Borough of Portsmouth: and Other Documents Relating Thereto*. Portsmouth: H. Lewis.

Oliver, V.L. (1934). King Edward I & Melcombe Regis. *Proceedings of the Dorset Natural History and Archaeological Society*, 55, 32–40.

Ormrod, M. (2005a). Edward III: 1365, January. Text and Translation. In: C. Given-Wilson et al., eds., *The Parliament Rolls of Medieval England*, CD-ROM. Leicester: Scholarly Digital Editions.

Ormrod, M. (2005b). Edward III: 1371, February. Text and Translation. In: C. Given-Wilson et al., eds., *The Parliament Rolls of Medieval England*, CD-ROM. Leicester: Scholarly Digital Editions.

Page, W. (1908). The Liberty of Alverstoke with Gosport. In: W. Page, ed., *A History of the County of Hampshire: Volume 3*. [online] London: Victoria County History, pp. 202–208. Available at: http://www.british-history.ac.uk/vch/hants/vol3/pp202-208 [Accessed May 5 2020].

Phillips, S. (2005). Edward II: 1315, January. Text and Translation. In: C. Given-Wilson et al., eds., *The Parliament Rolls of Medieval England*, CD-ROM. Leicester: Scholarly Digital Editions.

Public Record Office (hereafter PRO). (1901). *Calendar of the Patent Rolls. Edward I. 1272–1281*. London: H.M.S.O.

PRO (1903). *Calendar of the Patent Rolls. Edward II. 1317–1321*. London: H.M.S.O.

PRO (1907). *Calendar of the Patent Rolls. Henry VI. Vol. 3. 1436–1441*. London: H.M.S.O.

PRO (1916). *Calendar of Inquisitions Miscellaneous (Chancery). Vol. 2*. London: H.M.S.O.

PRO (1932). *Close Rolls of the Reign of Henry III. 1256–1259*. London: H.M.S.O.

PRO (1937). *Calendar of Inquisitions Miscellaneous (Chancery). Vol. 3*. London: H.M.S.O.

PRO (1968). *Calendar of Inquisitions Miscellaneous (Chancery). Vol. 7*. London: H.M.S.O.

Seymour, D. ed., (2000). *The Exchequer Cartulary of Torre Abbey : (P.R.O. 164/19)*. Torquay: Friends of Torre Abbey.

Smart, H. (2016). *Devon Historic Coastal and Market Towns Survey: Teignmouth*. Exeter: Devon County Council. https://doi.org/10.5284/1041583.

Swanson, H. (1999). *Medieval British Towns*. Basingstoke: Macmillan.

Weinbaum, M. ed., (2010). *British Borough Charters 1307–1660*. Cambridge: Cambridge University Press. https://doi.org/10.1017/CBO9780511707704.

3 The continued evolution of Fort William and Port Arthur into the City of Thunder Bay, Ontario, Canada

Michel S. Beaulieu and Jenna L. Kirker

When Sir Arthur Conan Doyle and his wife arrived at the Canadian Northern Railway Station in Port Arthur, Ontario, Canada in 1907, they entered a vast and magnificent land between the nation's eastern portion and its rapidly expanding West. The author of the Sherlock Holmes stories was drawn to the twin cities of Fort William and Port Arthur, collectively known as the Lakehead, by fascination about their intertwined history and potential as the next 'Chicago of the North'. As the *Port Arthur News Chronicle* proclaimed in March 1906, the region's assets,

> are stupendous, the country reeks with underdeveloped riches, agricultural soil, minerals, water power, navigable lakes and rivers, a healthy invigorating climate … everything that makes a country great, waiting only for capital and energy … to develop it.

For much of its history, this concept has dominated the evolution of the twin cities, now collectively called the City of Thunder Bay. It both spurred their separate but intertwined desire for economic development central to their evolution and underpinned their 20th-century mutual competitiveness. As Thunder Bay's only comprehensive history argues, it 'does not fit into any of the historical models, Western Canadian or Ontarian' (Tronrud and Epp 1995, vii). Fort William and Port Arthur's history is complex, often shaping both persistent rivalry and a sense of unity.

Indigenous peoples inhabited the Lakehead for thousands of years before Europeans arrived in the late 17th century. What began as a launching point for French and English explorers hoping to maximize their lucrative fur trade eventually became Fort William, named after the trading post therein. After the Hudson's Bay Company and the North West Company merged in 1821, fur-trade redundancies shifted the economic makeup during the 19th century's latter half, eventually transforming the Lakehead to a resource-based industrial centre (Morrison 2001).

The region's potential for colonial-resource exploitation led first to treaties with the region's Indigenous peoples. The nations around Lake Superior,

DOI: 10.4324/9781003102526-4

noticing colonial government activities encroaching on lands elsewhere, had also long recognized the need for treaties. These calls, however, had been largely ignored by governments until an influx of prospectors and surveyors in the 1840s, desiring to exploit the region's mineral wealth, forced the issue to the forefront. Alongside a one-time payment and guaranteed yearly payments, the Robinson-Superior Treaty of 1850 included setting aside reserve lands for each Indigenous signing group and gave 'full and free privilege to hunt ... the territory now ceded by them and to fish in the waters ... as they have ... been in the habit of doing'. Little attention was given to matters of mining, resource development or settlement (Hele 2016; Robinson-Superior Treaty 1850) (Figure 3.1).

Signatories were persuaded by government negotiator W. B. Robinson that, since large-scale settlement of their lands by non-Indigenous peoples was unlikely, they could continue under the treaty as previously, with little interference. However, first surveyors, then prospectors, then miners and eventually settlers flooded Lake Superior's northern shore. Boundaries along the Kaministiquia River, for townships like Westfort (later absorbed into what became Fort William) and Prince Arthur's Landing (Port Arthur's eventual location, Figure 3.1) were established (Arthur 1986). However, the land that comprised part of the reserve of the Fort William First Nation covered part of what would become the modern City of Thunder Bay a century later.

With steady land absorption and abundant space, natural-resource expansion became central over the next two decades. Helped by Indigenous guides and labourers, silver mining in what soon became Silver Islet produced the region's largest in-migration after the fur trade's decline 20 years previously (Barr 1988; Newell 1985). Mining attracted many businesses, investors and immigrants, profoundly affecting the region's future. It also produced the region's first recorded

Figure 3.1 Prince Arthur's Landing, artists unknown, 1870.

Source: Thunder Bay Historical Museum Society, by permission.

urban Indigenous community, as many from what is now known as Fort William First Nation and other shore communities became wage-labourers, establishing in Silver Islet what non-Indigenous members of the community referred to as a 'ghetto' (Dunk 2008).

The settlers' first official municipality was Shuniah in 1873. 'For eight years', Scollie has argued, 'settlements at Prince Arthur's Landing and Fort William were united in this early form of regional government' (Scollie 2020, 27). However, controlling the municipality and growing competition over the location of railway construction induced increasing conflict between settlers at the Landing and those at Fort William Town Plot (Arthur 1975, 59). Constant pressure from Town Plot people induced Shuniah's disintegration and a new Neebing Municipality comprising its southern components, including the Town Plot (later Westfort). The silver boom also included the growth of Fort William's twin city, Prince Arthur's Landing, and infrastructural development like town halls, churches and schools. However, until the early 1880s, the two communities' populations still largely comprised Indigenous peoples and French-Canadian trader descendants (Epp 1995).

Rivalry between the future Fort William (founded in 1892) and Port Arthur (founded in 1884) is evident as railways expanded into the region. In contention were Fort William, Prince Arthur's Landing (becoming Port Arthur in 1884) and Westfort. The decision to build in Westfort meant expenditure on a competing line from Port Arthur. Railway competition ensured two nascent communities both surviving in an increasingly connected country ('Rivalry Forced Unique Railroad in Port Arthur' n.d.). The CPR's goal, besides fulfilling Prime Minister John A. Macdonald's promise to link British Columbia by rail to the rest of Canada, was to encourage European immigration to the West and stimulate East-West trade. For the Lakehead, the construction in 1875 of CPR's terminus in Fort William ensured the region's survival and success in the post-fur trade era (Muirhead 1995, 78). Although Ontario's government had designated Port Arthur as the region's administrative centre in 1871, the CPR's continued preference for Fort William shaped the latter's development (Scollie 1985).

Both communities thereby underwent an economic boom, partially based on speculation with Port Arthur's natural harbour spurring anticipatory commercial waterfront development (Arthur 1985, 22). The money spent attracting economic activity also hurt both communities. Fort William officials actively courted the CPR, offering $120,000 in bonuses. Illegal and unaffordable, they were still provided alongside a 25-year taxation respite (Muirhead 1995, 80). Meanwhile, the Grand Trunk Pacific Railway's (GTPR) expansion occurred on land expropriated by the federal government that had comprised part of the reserve of Fort William First Nation. Fort William also provided

> $300,000, a $50,000 contribution toward the cost of a bridge over the river, river frontage valued at $45,000, and a tax exemption for fifteen years. In total, the town paid and gave gifts in kind to the GTPR a total of $893,500 before any work was actually done.
>
> (Muirhead 1995, 82)

A referendum approved them all, and the first sod was turned during a Prime Ministerial visit, notwithstanding loud complaints from Port Arthur and Fort William First Nation.

A new century

Despite declining after the closure of the silver mines in the early 1890s, Port Arthur and Fort William's population rose from a few hundred to over 3,000 by 1885 (Arthur 1975, 60). The region evolved rapidly after 1900, becoming dependent on resource-extraction and transnational shipping, built on a leg-acy of trade and commerce (Dagostar et al. 1992; Ontario Royal Commission on the Northern Environment 1985). For much of their history, success for the Lakehead's two settler-communities has been measured by the same criteria used by turn-of-the-20th-century boosterism: the number of rail lines, wheat shipped from ports, numbers of lumber mills, land values, population size, manufacturing and trade value (Tronrud 1993).

Yet, industrial progress grew on the hardworking backs of regional Indige-nous peoples, plus first-, second- and third-generation Canadians from around the world, with men and women involved in all aspects of northern expansion. Aside from limited use as wage-labourers, members of Fort William First Nation benefitted little. Provincial policy in the north was inherently colonial, con-cerned only with natural-resource exploitation to ensure continuing expansion for southern industrial manufacturing. When policy moved beyond this, as some advocated, systems were imposed, and agricultural communities laid out in ways possibly working for southern Ontario but wholly unsuited to the north. This was often done at Indigenous peoples' expense, treaties notwithstanding (Beaul-ieu and Southcott 2010; High 2008; Innis 1967; Nelles 1974).

While a few officials and politicians fought to protect and expand Indige-nous rights, many in 'New Ontario', as all Northern Ontario was then known, viewed creating reserves as impediments to settlement and economic develop-ment (Arthur 1986, 22–25). Treaties, and federal and provincial government attempts to honour them, were viewed as 'withholding valuable land from set-tlers' (Rasporich and Tronrud 1995, 204). One Port Arthur councillor's 1891 election platform even advocated they be totally ignored, and colonization roads built through them (*Weekly Herald and Algoma Miner*, 28 February 1891). Unsur-prisingly, little engagement with the Fort William First Nation occurred in the 20th century's first half, aside from land expropriation, especially during the 1905 relocation to build the CPR's grain terminus and the establishment of a military rifle range (Lackenbauer 2006).

Following the first grain elevator's construction in 1883, others emerged in both communities; by the late 1920s, 32 elevators made the Lakehead the world's largest grain port (Di Matteo 1990, 4). As one economist has argued, 'much of the economic infrastructure of the modern city of Thunder Bay has its roots in the seminal years between 1900 and 1914 – the wheat boom era' (Di Matteo 1991, 7). As a major export epicentre for Canadian goods, the Lakehead presented unique labour opportunities for workers and settlers, with varied manufacturing, natural-resource extraction and transport industries providing jobs for thousands of skilled and unskilled labourers across the region, both annual and seasonal.

The railways' arrival in the early 1880s joined Fort William and Port Arthur to the western Canadian wheat economy (Di Matteo 1990). They employed a combined 4,000 men in the region by 1912, thereby becoming the single largest employer during its push to modernity. Rail transport, combined with the Great Lakes system, allowed the Lakehead to experience rapid industrial and population growth as the century turned. From 1891 to 1901, it grew from 1,965 to over 7,000 (Southcott 1987, 17–20). Thereby, the region transformed from transient trading post into two booming cities in the 80 years following the fur trade's decline.

Both towns viewed themselves as the next 'Chicago of the North' or the 'Geneva of Canada,' in the words of the *Port Arthur Illustrated* (1883). This publication predicted Port Arthur would become 'the greatest manufacturing centre in the West'. However, railway battles left politicians in both towns rather embittered. Coupled with Port Arthur's attempts to annex portions of the Fort William Plot, including one suggestion that the two communities unite as 'Europa', the southern community's politicians had little appetite for amalgamation despite many in both communities raising it (Scollie 1979, 24; Morrison 1981, 23).

Inter-urban rivalry defined the decades around 1900 (Black and High 1995; Tronrud 1990). Historians contend this created an atmosphere 'conducive to risk-taking and innovation' (Black and High 1995, 1). By the mid-1880s, civic leaders believed the region's private enterprise possessed insufficient capital to allow continued expansion by both communities, a narrative continuing into the early 20th century. Far from sources of finance capital and aware of each other's expansion attempts, both towns used municipal ownership to spur economic growth by private enterprise.

Thus, the formation of municipal street railways in both cities (North America's first), electrical development and municipal telephone systems, for example, owed much to inter-urban rivalry (Chochla 1989; Scollie 1990). Both towns were incorporated as cities in 1907 and continued competitively going to 'great lengths to advertise their locations, improve their port facilities and … civic infrastructures in an effort to be the progressive, modern industrial centres they advertised themselves as' (Di Matteo 1991, 11). Both also competed to construct hospitals, schools, churches and parks.

By 1912, the region's economic and social growth had produced significant changes. The two towns became national leaders in municipal ownership of services and infrastructure like sewers, safe water, electricity, telephone and streetcars. The Lakehead had also become the Great Lakes' premier rail and shipbuilding centre. It was a period of rampant boosterism (Tronrud 1993). While all this produced rapid economic development, inter-urban rivalry induced unnecessary expenditures. Both towns competed for industry to their combined detriment. While either city's growth would benefit the region, unwillingness to cooperate on many economic and social issues produced duplication and wasted resources.

During this hyper-competition for development, Industrial Commissioners in both cities ran competing national advertisements (Di Matteo 1991, 28, fn. 48). In the case of what became Canadian Car and Foundry (a leading long-term regional industrial manufacturer), both towns rancorously lobbied the company. After Fort William leaders slandered Port Arthur in attempting to lure the plant, Port Arthur countered with competing industry bonuses worth $666,000 (Tronrud 1995, 108–109). Neither town could really afford offers like this. Sensing Port Arthur losing the bid, a local MP even suggested amalgamation and using the intervening area. Even after Fort William received the plant, Port Arthur continued with a scheme for a competing railcar factory which never material-ized (Tronrud 1991, 1–7; Di Matteo 1990, 13).

The World Wars and aftermath

Although World War I heavily impacted the region, economic activity by 1918 'reached a zenith' in starting new plants and operations (Di Matteo 1991, 14). By 1921, the economy had reached new heights of diversification, inducing labour and capital shortages. Manufacturing's employment share would not be repli-cated until the early 1980s. Fort William and Port Arthur's population between 1911 and 1921 grew by 304.8% (Southcott 1987, 17–20). Post-war growth, however, halted during the late 1920s. Forest-related industries aside, the entire economy declined (Di Matteo 1991, 14). Economic collapse in 1928 and the Great Depression also widely affected the twin cities. In April 1931 alone, 200 Fort William families (1,000 people) were receiving goods in kind (Beaulieu 2011, 153–201). As Tronrud (1991, 8) writes, by year's end, 'almost one of every two male wage-earners found themselves out of work and most took over six months to regain employment'.

By 1933 things had slowly begun improving, but hardships and inability to provide some services incited further amalgamation talk. However, Port Arthur's *News-Chronicle* declared:

> Unnatural amalgamation would multiply [troubles] many times … over the location of every office, over the homes of every appointee, over the place of doing every piece of business and over every appropriation of money for municipal government, the north [Fort William] and the south [Port Arthur] constantly pitted against one another.
>
> (7 March 1936)

Only in 1933 did the cities begin seeing minor improvements, and not until World War II did twin-city development recommence. During 1940–1945, thou-sands of jobs were created, and, by war's end, the cities' combined population totalled over 55,000: collectively, Ontario's fifth largest city.

The decades after 1945 witnessed profound change. One of the most influ-ential ministers in the federal governments between 1935 and 1957 was Port Arthur MP Clarence Decatur Howe ensured the cities benefited from both war-time economic opportunities and post-war changes undertaken by successive

Liberal governments. Despite ongoing inter-city competition, both gained from expanding industrial manufacturing tied to regional natural-resource abundance. As the Wightmans (Wightman and Wightman 1997) have established, the 25 years following 1950 witnessed unprecedented advance in living standards and social-service levels across Canada. The twin cities were no different, directly benefitting from provincial and regional-development initiatives. Population increased in size and diversity due to an influx of immigrant workers to both cities with combined population climbing dramatically from 65,000 in 1951, to 90,000 in 1961, reaching its historic peak of 121,000 by 1981 (Southcott 2006, 24–27).

Together at last?

Both cities benefitted, albeit differently, from expanding highways, air services, hydroelectricity and oil pipelines, alongside national and global natural-resources demand in the post-war era (Wightman and Wightman 1993). The jewel of St Lawrence Seaway, Keefer Terminal, positioned Fort William to take advantage of western Canadian economic growth. The Port of Thunder Bay benefitted greatly from the province of Alberta's developing oil fields. Coupled with forest-industry revitalization and post-war immigration, both cities experienced substantial population growth; new cultures mingled with existing ones, bringing innovative ideas and community-building. However, although amalgamation was briefly discussed in 1920, producing a failed plebiscite, it largely became a topic for periodic editorial discussion (often derisive) by politicians and businessmen.

After 1945, amalgamation rose up the agenda, with changes to municipal structure coming under serious discussion (Fort William *Daily Times-Journal*, 17 July 1952). This produced an Intercity Development Association. Significantly, the proponents for joint municipal ventures were businessmen who had established themselves in the region after the war. With little support from either municipality, in the mid-1950s, the Association began calling for amalgamation. The 1958 municipal election contained a plebiscite, amalgamation losing in Fort William and winning in Port Arthur. Meanwhile many organizations ignored the failed plebiscite and began merging independently. The first major organizations were the Chambers of Commerce and Labour Councils. Both cities began coordinating the delivery of services such as telephones and participating in larger regional organizations like the Lakehead Harbour Commission, Lakehead Planning Board, Lakehead Board of Education, Thunder Bay District Health Unit and others. Other non-governmental and economic organizations, while still using their originating city names, began operating and being governed by citizens from both cities.

With growing communality, alongside mutual geographic expansion to encompass intervening industrial and commercial spaces, the cities became evermore entwined and indistinguishable. However, both continued competing, often to their own detriment. Thus, Fort William (Figure 3.2) failed to utilize federal urban renewal programs available in the early 1960s, instead waiting to see what projects Port Arthur would undertake. The latter renewed some of its

Figure 3.2 Aerial photograph of Fort William looking north, 1969.
Source: Thunder Bay Historical Museum Society, by permission.

downtown core, but the program was cancelled before Fort William could utilize it. One eventual result was the continued post-amalgamation decline of its Central Business District (CBD) (MacPhail 2001, 38–39). Competition notwithstanding, amalgamation increasingly gained steam in the 1960s.

Amalgamation was finally tabled in the late 1960s but remained controversial. Contrary to popular myth, the Ontario government's formal enquiry into municipal reorganization was requested by the mayors and reeves (township leaders), and fully supported by the Lakehead Planning Board. Led by Eric Hardy, *The Lakehead Government Review* ('the Hardy Review') was completed in 1968 and the Government of Ontario introduced Bill 118 in 1969. Opponents cried foul because no plebiscite was held this time around. Controversy also followed the decision to decide the new city's name by plebiscite. 'Thunder Bay' received the most votes, but the other options, 'The Lakehead' and 'Lakehead', so similar in wording, divided the overall majority between them. Despite decades of stubbornness, questionable ballots and intercity strife, on 1 January 1970, residents awoke to find themselves Thunder Bay citizens (Raffo 2020).

The years following amalgamation were hardly smooth. Both former cities had operated in very different political climates. Post-amalgamation planning remained contentious; pre-existing rivalries continued with many of those opposed being continually elected, particularly in former Fort William wards. The first council's activities, for example, were overshadowed by conflict-of-interest charges laid by former Fort William citizens against Port Arthur councillors (Raffo 1998). However, amalgamation did induce renewed efforts to redevelop both cities' downtown cores. City officials sought to unify their populace and diversify their economy by developing the 'inter-city' space long dividing the two. This area was to profoundly influence city development. Originally swampy, it had become the focus of the Lakehead Local Government Review leading to amalgamation and figured prominently in the 1972 official city plan. That plan, while committed to maintaining the two historic business centres, also encouraged their development.

In the late 1970s, the City of Thunder Bay took advantage of the Ontario Downtown Revitalization Program to try revitalizing the old Fort William's central business district. Signs of decay, especially in the former Fort William, were already apparent pre-amalgamation, having lost access to the urban renewal program a decade before. Residential expansion elsewhere, a movement of white-collar jobs to the inter-city area and business development outside the CBD exacerbated the pre-amalgamation situation. Another factor was how Thunder Bay's planning decisions were often guided by 'reconciling the tensions that arose from creating one new city, while still maintaining two historic cores' (MacPhail 2001, 39). The resultant folly remains evident in the economic challenges facing those areas to this day. When Port Arthur's urban renewal commenced in 1974 before Fort William's in 1978, old resentments resurfaced in the media and council meetings. As Kosny (1995) has established, the new city during the 1970s and 1980s remained prisoner to the past. MacPhail (2001, 35) notes how planning considerations post-unification 'were motivated by the inherited jealousies of two cities and two CBDs'.

The overall situation's current embodiment is the Victoriaville Centre. Opened in May 1980, this covered mall in the Fort Willian CBD quickly became known as 'Dusty's Ditch' after Thunder Bay Mayor Dusty Miller. Less than a year after construction, its anchor store closed, like many others. The construction of a new shopping mall in the Intercity Area (Intercity Mall) under two years later meant Victoriaville never yielded the city any dividend, remaining heavily subsidized to this day. In late 2020, Thunder Bay's City Council, after years of discussion and study, voted to demolish Victoriaville.

Changing global trade patterns, economic recession, and the energy crisis in the late 1970s and early 1980s profoundly impacted all of Northern Ontario and particularly Thunder Bay (Beaulieu 2013). While the City continued growing and developing (its population reaching 112,486), by the 1980s the pace slowed, hard hit by job losses in the grain industry, railways, mining and particularly the pulp and paper industry (Stafford 1995, 52–53). By the early 1990s, Thunder Bay's labour force and economy were shifting. As Southcott (1993) writes,

> there was increasing awareness that the forest and mining industries, the region's economic base, were no longer major producers of prosperity but instead ... concentrating on labour-saving technologies to lower production costs in attempting to complete in an increasingly globalized marketplace.

(17)

While forestry and mining remain important, and many yearn to return to the 'good old days', changes from the 1990s have entailed the city transforming itself from a 'workingman's town' into a post-industrial service centre.

The economic shift has not been easy. Nevertheless, Thunder Bay has produced new industry and spotlighted historic relationships, signified most meaningfully by increasing engagement with Fort William First Nation and other Indigenous nations, enabling them to take 'advantage of their much greater role in the economy of the region' (Conteh 2013, 51). Central to the future prosperity of settler communities like Thunder Bay, this directly contrasts with their actions a century ago, embracing truth and reconciliation with Indigenous peoples (Abele 2009; Madahbee 2013). In many respects, the region has come full circle, becoming more dependent on communities ignored for much of the 20th century. Fort William and Port Arthur's intertwined, yet distinctive, historical relationship reveals a history both defying existing historical models and lending itself to a notion of 21st-century Canada that Prime Minister Trudeau noted as 'a work in progress' (Canadian Broadcasting Corporation 2017).

Yet, all this notwithstanding, and having been amalgamated for over 50 years, the identities, even the historical rivalry, of Fort William and Port Arthur continue in some respects. Debates over where infrastructure is spent, ongoing attempts to maintain the two former downtown cores, and how residents identify spatially where they live in Thunder Bay persists (just a few of many examples). One joke is that only politicians and those who have lived in the city for less than a year will reference 'north side' and 'south side' rather than Port Arthur and Fort William respectively.

References

Abele, F 2009, *Policy Research in the North: A Discussion Paper*, Gordon Foundation, Ottawa.

Arthur, E 1975, 'Inter-urban Rivalry in Port Arthur and Fort William,' AW Rasporich (ed), *Western Canada Past and Present*, University of Calgary and McClelland and Stewart West, Calgary, 58–68.

Arthur, E 1985, 'Willian Van Horne, the CPR, and the Kaministiquia Property: A Selection of Letters,' *Thunder Bay Historical Museum Society Papers & Records*, vol. 13, 20–27,

Arthur, E 1986, *Simon J. Dawson, C.E.*, Thunder Bay Historical Museum Society, Thunder Bay.

Barr, E 1988, *Silver Islet: Striking it Rich in Lake Superior*, Natural Heritage Books, Toronto.

Beaulieu, M 2011, *Labour at the Lakehead: Ethnicity, Socialism, and Politics, 1900–35*, University of British Columbia Press, Vancouver.

Beaulieu, M 2013, 'A Historic Overview of Policies Effecting Non-Aboriginal Resource Development in Northwestern Ontario, 1900–1990,' in C Conteh & B Segsworth (eds), *Governance in Northern Ontario: Economic Development and Policy Making*, University of Toronto Press, Toronto, 94–114.

Beaulieu, M & Southcott C 2010, *North of Superior: An Illustrated History of Northwestern Ontario*, James Lorimer and Company, Toronto.

Black, DL & High, SC 1995, *Power for the People: Inter-Urban Rivalry over Electricity at the Lakehead, 1884–1917*, Research Report #39, Lakehead University Centre for Northern Studies, Thunder Bay.

Canadian Broadcasting Corporation, 2017, 'Canada Is 'Work in Progress,' Justin Trudeau Tells UN General Assembly,' 21 September (Accessed: 3 March 2021), https://www.cbc.ca/news/politics/justin-trudeau-un-speech-general-assembly-1.4300602

Chochla, M 1989, 'Sabbatarians and Sunday Street Cars,' *Thunder Bay Historical Museum Society Papers & Records*, vol. 17, 25–36.

Conteh, C 2013, 'Administering Regional Development Policy,' in C Conteh & B Segsworth (eds), *Governance in Northern Ontario: Economic Development and Policy Making*, University of Toronto Press, Toronto, 43–57.

Dagostar, B, Jankowski, WB, & Moazzami, B 1992, *The Economy of Northwestern Ontario: Structure, Performance, and Future Challenges*, Research Report #31, Lakehead University Centre for Northern Studies, Thunder Bay.

Di Matteo, L 1990, *The Impact of the Wheat Boom on the Canadian Lakehead, 1901–1911*, Research Report #14, Lakehead University Centre for Northern Studies, Thunder Bay.

Di Matteo, L., 1991, *The Economic Development of the Canadian Lakehead during the Wheat Boom Era: 1900–1914*, Research Report #29, Lakehead University Centre for Northern Studies, Thunder Bay.

Dunk, T 2008, 'Aboriginal Participation in the Industrial Economy on the North Shore of Lake Superior, 1869–1940,' M Beaulieu (ed), *Essays in Northwestern Ontario Working Class History: Thunder Bay and its Environs*, Lakehead University Centre for Northern Studies, Thunder Bay, 199–212.

Epp, AE 1995, 'The Achievement of Community,' in TJ Tronrud & AE Epp (eds), *Thunder Bay: From Rivalry to Unity*, Thunder Bay Historical Museum Society, Thunder Bay, 180–203.

Hele, K (ed) 2016, *This is Indian Land: The 1850 Robinson Treaties*. Aboriginal Issues Press, Winnipeg.

High, S 2008, 'Responding to White Encroachment: The Robinson-Superior Ojibwa and the Capitalist Labour Economy: 1880–1914,' in M Beaulieu (ed), *Essays in Northwestern Ontario Working Class History*, Lakehead University Centre for Northern Studies, Thunder Bay, 213–230.

Innis, HA 1967, 'An Introduction to the Economic History of Ontario from Outpost to Empire,' in EG Firth, *Profiles of a Province: Studies in the History of Ontario*, The Ontario Historical Society, Toronto, 45–155.

Kosny, ME 1995, 'Thunder Bay after a Century,' in TJ Tronrud & AE Epp (eds), *Thunder Bay: From Rivalry to Unity*, Thunder Bay Historical Museum Society, Thunder Bay, 227–242,

Lackenbauer, PW 2006, 'Of Practically No Use to Anyone': Situating a Rifle Range on the Fort William Indian Reserve, 1905–1914,' *Thunder Bay Historical Museum Society Papers & Records*, vol. 34, 3–28.

MacPhail, JA 2001, 'Downtown Revitalization and Victoriaville Centre,' *Thunder Bay Historical Museum Society Papers & Records*, vol. 24, 35–54.

Madahbee, D 2013, 'First Nations Inclusion: A Key Requirement to Building the Northern Ontario Economy,' in C Conteh & B Segsworth (eds), *Governance in Northern Ontario: Economic Development and Policy Making*, University of Toronto Press, Toronto, 76–93.

Morrison, J 2001, *Superior Rendezvous Place: Fort William in the Canadian Fur Trade*, Natural Heritage Books, Toronto.

Morrison, KL 1981, 'The Intercity Development Association and the Making of the City of Thunder Bay,' *Thunder Bay Historical Museum Society Papers & Records*, vol. 9, 22–30.

Muirhead, B, 1995, 'The Evolution of the Lakehead's Commercial and Transportation Infrastructure,' in TJ Tronrud & AE Epp (eds), *Thunder Bay: From Rivalry to Unity*, Thunder Bay Historical Museum Society, Thunder Bay, 6–98.

Nelles, HV 1974, *The Politics of Development: Forests, Mines, and Hydro-Electric Power in Ontario, 1849–1941*, Macmillan of Canada, Toronto.

Newell, D 1985, 'Silver Mining in the Thunder Bay District, 1865–1885,' *Thunder Bay Historical Museum Society Papers & Records*, vol. 13, 28–45.

Ontario Royal Commission on the Northern Environment 1985, *Final Report*, Ministry of the Attorney General, Toronto.

Port Arthur Daily News, 20 March 1906.

Port Arthur Illustrated, with a description of its Products, Resources and Attractions (1883), n.p., Winnipeg.

Port Arthur News-Chronicle, 7 March 1936.

Raffo, P 1998, 'Municipal Political Culture and Conflict of Interest at the Lakehead, 1969–72,' *Thunder Bay Historical Museum Society Papers & Records*, vol. 26, 26–45.

Raffo, P 2020, 'Saul Laskin and the Making of Thunder Bay,' *Thunder Bay Historical Museum Society Papers & Records*, vol. 48, 5–39.

Rasporich, AW & Tronrud, TJ 1995, 'Class, Ethnicity and Urban Competition,' in TJ Tronrud & AE Epp (eds), *Thunder Bay: From Rivalry to Unity*, Thunder Bay Historical Museum Society, 204–226.

'Rivalry Forced Unique Railroad in Port Arthur,' Research File, Held at: Thunder Bay Historical Museum Society Archives.

Robinson-Superior Treaty (1850), [Manuscript-online] Available: Government of Canada, https://www.rcaanc-cirnac.gc.ca/eng/1360945974712/1544619909155#rt

Scollie, FB 1979, 'The Population of Thunder Bay, 1884–1901,' *Thunder Bay Historical Museum Society Papers & Records*, vol. 7, 21–29.

Scollie, FB 1985, 'Falling into Line: How Prince Arthur's Landing Became Port Arthur,' *Thunder Bay Historical Museum Society Papers & Records*, vol. 13, 8–19.

Scollie, FB 1990, 'The Creation of the Port Arthur Street Railway, 1890–95," *Thunder Bay Historical Museum Society Papers & Records*, vol. 18, 40–58.

Scollie, FB 2020, *Biographical Dictionary and History of Victorian Thunder Bay (1850–1901)*, Thunder Bay Historical Museum Society, Thunder Bay.

Southcott, C 1987, 'Ethnicity and Community in Thunder Bay,' *Polyphony*, vol. 9, no. 2, 10–20.

Southcott, C (ed) 1993, *Provinical Hinterland: A Social Inequality in Northwestern Ontario*, Fernwood, Halifax.

Southcott, C 2006, *The North in Numbers: A Demographic Analysis of Social and Economic Change in Northern Ontario*, Lakehead University Centre for Northernn Studies, Thunder Bay.

Stafford, J 1995, 'A Century of Growth at the Lakehead,' in TJ Tronrud & AE Epp (eds), *Thunder Bay: From Rivalry to Unity*, Thunder Bay Historical Museum Society, Thunder Bay, 38–55.

Tronrud, TJ 1990, 'Buying Prosperity: The Bonusing of Factories at the Lakehead, 1885–1914,' *Urban History Review*, vol. 19, no. 1, 1–13.

Tronrud, TJ 1991, *The Search for Factories in a Staple Economy: Thunder Bay's Manufacturing Industries, 1880–1980*, Research Report #28, Lakehead University Centre for Northern Studies, Thunder Bay.

Tronrud, TJ 1993, *Guardians of Progress: Boosters & Boosterism in Thunder Bay, 1870–1914*. Thunder Bay Historical Museum Society, Thunder Bay.

Tronrud, TJ 1995, 'Building the Industrial City,' in TJ Tronrud & AE Epp (eds), *Thunder Bay: From Rivalry to Unity*, Thunder Bay Historical Museum Society, Thunder Bay, 99–119.

Tronrud, TJ & Epp, AE 1995, 'Introduction,' in TJ Tronrud & AE Epp (eds), *Thunder Bay: From Rivalry to Unity*, Thunder Bay Historical Museum Society, Thunder Bay, vii–ix.

Weekly Herald and Algoma Miner, 28 February 1891.

Wightman, N & and Wightman, WR 1993, 'Beyond rail and road: The Red Lake-Pickle Lake gold fields, 1925–1954,' *Thunder Bay Historical Museum Society Papers & Records*, vol. 21, 19–34.

Wightman, WR & Wightman, N 1997, *The Land Between: Northwestern Ontario Resources Development*, University of Toronto Press, Toronto.

4 Banana-benders and cockroaches[1]

Cross-border planning for Gold Coast-Tweed Heads

Paul Burton and Aysin Dedekorkut-Howes

Introduction

Flying into Coolangatta airport reveals the southern end of what is now Australia's sixth largest city, Queensland's Gold Coast. However, the visible built-up area also includes the town of Tweed Heads across the state border in New South Wales (NSW) (Figure 4.1). This whole area represents an amalgamation of small coastal towns emergent during the 20th century, driven mainly by burgeoning tourism but also new settlers looking for what is known locally as a 'sea-change'. Gold Coast-Tweed Heads epitomises the type of twin city where 'nearby but administratively separate urban places within one country … [are] either indistinguishable from the start or rapidly become so' (Garrard and Mikhailova 2019, 7). It also shows their interdependence, but without the conflict and dysfunction sometimes associated with inter-twin relationships.

This chapter traces urban growth on either side of the border, reviewing the influence and significance of Queensland's and NSW's different planning and local-government regimes, including explicit attempts to foster cross-border collaboration. Apart from inconveniences like seasonal time difference and a railway-gauge discrepancy, this built-up area's growth and amalgamation, including its *de facto* twin-town status, has not produced significant political conflict at either state or local-government levels.

We also explore ongoing debates about Australia's three-tier federal system's efficiency and effectiveness and occasional calls to redraw state boundaries and reorganise local governments into larger regional or metropolitan-scale entities. These inevitably raise cultural and identity issues alongside economic and administrative efficiency – especially pertinent here where Tweed Heads, indeed broader Tweed Shire, is closer geographically to Queensland's capital Brisbane than their own state capital, Sydney. An important question is whether this geographical proximity is matched by equivalent economic connectivity or socio-cultural affinity.

Steele et al. (2013) have written of planning's apparent obsession with borders, citing Parker et al.'s (2009) reference to their 'seductive charm', to advocate greater attention to politics, governance and mobility opportunities across borders, whether tangible or conceptual and whether relating to administrative or ecological/cultural entities (Aberley 1993). We focus on the material impacts

DOI: 10.4324/9781003102526-5

Figure 4.1 Coolangatta-Tweed Heads on the Queensland-New South Wales border, where Boundary Street starting from Point Danger (bottom left) separates Coolangatta in the foreground from Tweed Heads.

Source: Skyepics.com.au, by permission.

and practical consequences of having a state-government border bisecting a major Australian urban area.

Since federation in 1901, Australia has been characterised by interesting 'boundary issues' between states and territories, notably the three different railway gauges characterising some areas until the 1990s. More recently, metropolitan-scale planning's increasing importance has forced places like Gold Coast/Tweed to consider more integrated planning of a wider range of infrastructure, like water, waste and energy systems to better reflect on-ground realities rather than administrative arbitrariness and historical accident.

Gold Coast is not only Australia's sixth largest city; it is its second largest local-government entity, with an annual budget of approximately $1.7 billion serving over 600,000 residents. Tweed Shire's equivalent totals $192,000,000 serving 94,000, implying a classically dominant-subordinate twin-city relationship (Garrard and Mikhailova 2019). Bearing these differences in mind, we also consider some practical challenges of enhanced integration in managing and planning Gold Coast/Tweed, including the consequences of political-power imbalances and the attractiveness of greater integration to both places.

Australia's Bureau of Statistics defines this border-straddling place as a Significant Urban Area, indicating it comprises over 10,000 people. But this hides various political, socio-cultural and administrative differences unapparent until leaving the airport. Alongside the seasonal time difference, various road-user regulations and land-use-planning regimes change upon traversing the border, and while it does not actually follow the Tweed River, for many, crossing it marks interstate transition.

From colonies to states and the Commonwealth of Australia

Despite occupation for 50,000–60,000 years by people from over 200 of Australia's 700 known Aboriginal nations, Queensland's recorded history mostly comprises the 250 years since colonisation by the British and then a wider range of non-Indigenous people, initially from mainland Europe but increasingly from China and elsewhere in the Asia-Pacific region.

Continental sub-division proceeded inexorably through the 19th century, with the term 'Australia' unused in official correspondence about the whole until the late 1820s. South and Western Australia achieved some autonomy from the NSW founding colony by 1836, but Queensland was not self-governing until 1859, with NSW governor Dennison resisting separation and fighting for the border to be as far north as possible. As Queensland's Government (2018) now notes,

> the Secretary of State for the Colonies announced … their intention to create a new colony whose southern border-line would run not far … south of 30-degrees-south latitude but would be accommodated to suit the natural features of the country.

Thus, the border follows mountain-range tops and rivers until reaching flatter country almost 700 km west of the coast when latitude line 29° becomes the border until South Australia's eastern edge. Eschewing cotemporary conventions about rivers as borders, the border left the Tweed watershed and the Tweed Valley's lucrative cedar trade to NSW, forming an unusual boundary conforming to more recent concepts of bioregional planning (Dedekorkut-Howes 2014). As Peter Spearritt observes, in the 1850s the small settlements south of the Tweed were concerned also about losing connection with and political support from Sydney, capital of Australia's richest colony at the time (McCutcheon 2020). Meanwhile, Brisbane became the new colony's capital, annoying northerly towns like Gladstone and Rockhampton that felt more important economically. Many still feel the state remains too focused on Brisbane and its south-east corner, with periodic calls for greater self-rule, even secession.

Here we should note a highly significant aspect of Australian federalism: states retain responsibility for land-use planning. Each state and territory passes its own legislation setting out its overall planning intentions and how they and local governments, the implementing bodies, will achieve them.

Indigenous perspectives

Western colonial-border concepts conflict with Indigenous ways of life and custom. The Yugambeh-language people are traditional custodians of the land encompassing the Gold Coast/Tweed regions. They called the Gold Coast region Umbigumbi ('the place of the ant'), congregating there for special occasions with tribes travelling from as far north as Brisbane, as far west as the Great Dividing Range and across the border from today's northern NSW (Holthouse 1982). The colonial border ignored Indigenous customs, traditions and connections with the land, and numbers declined rapidly in the late 19th century, faced by systemic attacks and neglectful policies.

Little now remains of the area's Indigenous history besides some place-names and ceremonial areas. It was called 'Gulun' in local Bundjalung language (MALCC 2020), meaning blue fig (Sharpe 1998), a regional rainforest tree. Little is recorded about this name, besides an attempt to call the town of Tweed 'Cooloon' when surveyed in 1886 (Longhurst 1978): it quickly reverted to Tweed Heads. Across the border, Coolangatta ('pleasant outlook' in Nowra dialect) reflects an Indigenous name (GNB of NSW 2020), acquired when a schooner of that name sank off its coast in 1846 (Vader and Lang 1980).

NSW Surveyor-General Mitchell's instruction to retain Aboriginal place-names explains the remaining names (Longhurst 1994). Unfortunately, indigenous peoples fared much worse than the places they named. Cedar loggers arrived in 1844, proving disastrous to local Nganduwal people: settler interactions severely damaged their social fabric, alongside European diseases, systematic killings and interbreeding. These original inhabitants gradually disappeared: by 1933, no trace remained (Longhurst 1994).

Settlement and urban history

Longhurst (1978) defines the Gold Coast region more widely than the city's municipal boundaries, reporting wide acceptance that it includes the built-up area extending from the Coomera River in the north to Kingscliff in the south; references to the Gold Coast region almost always include Tweed Heads (Dedekorkut-Howes and Bosman 2015). Since their emergence the twin-town relationship has been symbiotic, from early Tweed residents' reliance on cross-border beef to the 1980s when many Queenslanders crossed to access NSW clubs' gaming machines, then forbidden in Queensland (Small 1981).

The earliest recorded European settlement was a short-lived military post (1829–1830) at Point Danger (on the current state border) located to capture escapees from Moreton Bay's penal settlement. In 1842 the Tweed River was opened to cedar-cutters. Considered unsuitable for grazing, the area remained economically reliant on timber until the early 1860s, when cotton, sugar and maize production emerged. This prompted a township nucleus at Tweed River Heads to be established: the first buildings were the pilot's house (1870), followed soon after by the *Customs House*, prompting Tweed residents to object and begin a long campaign to incorporate their town into Queensland (Longhurst 1978). This was hardly new: Clarence-Richmond settlers had campaigned for northern NSW's incorporation into the new colony before Queensland's foundation due to disenchantment with faraway Sydney governance (Vader and Lang 1980).

Settlement began around the border customs post, but Brisbane's inaccessibility and developments in nearby Murwillumbah and Tumbulgum hampered development. Queensland's side was surveyed in 1883 and land parcels auctioned in 1884. By late 1884, the double-storey *Commercial Hotel* became Coolangatta's first building, followed by the first private residence. The year 1884 also saw a boarding house built on the Tweed side, but NSW's side remained unsurveyed. Road construction from Tallebudgera to the Tweed started in 1885 with a daily mail-coach operating from 1886, and Brisbane holidaymakers appearing along the beach, initiating early tourism, eventually making the area famous and enhancing long-term growth. These developments prompted NSW authorities to survey and auction Tweed Heads in 1886–1887. The two-storey Hotel Pacifique emerged in 1888, facing the Tweed's mouth sandbar, by which time a passenger steamboat was operating from Murwillumbah to the Heads. Sandbar-induced navigation dangers, faraway Sydney's unresponsiveness to public-works demands, protective tariffs, import duties, increasing contact with Queensland and the prospective Southport-Nerang railway extension all increased calls for Tweed's incorporation into Queensland (Longhurst 1978).

Tweed Heads was originally larger than Coolangatta due to water transport's importance in the early settlement years, with Tweed Valley timber, seafood and agricultural produce shipped from its wharves (O'Hare and Burke 2016). However, only the Nerang-to-Tweed railway's opening in 1903 significantly accelerated growth: Tweed Heads became Brisbane's most accessible open-surf beach, while interstate agreement brought the Queensland railway, buildings and trains several hundred meters into NSW (Longhurst 1978). NSW, however, did not reciprocate this federalist gesture, leaving a never-filled gap from Murwillumbah where its northern railway terminates, fearing loss of economic benefits to Brisbane. However, rather than funnelling NSW produce to Brisbane, the railway somewhat unexpectedly brought Brisbane holidaymakers to the border (Longhurst 1978). O'Hare and Burke (2016) point to this final stage of the South Coast Railway as the first of three significant transport-infrastructure projects traversing the border and helping shape Gold Coast settlement patterns.

Around 1900, Coolangatta's population was just 125, while Southport, which it would soon rival as a resort, had 1538 (Green 1982), causing trains to bypass a place with just one hotel (Russell, 1995). But soon after the first train stopped, Coolangatta's first and most famous guest house, the Greenmount, heralded its tourist golden era. Major contributors were the opportune coincidence of mass-transport availability and surfing's rising popularity (Russell, 1995). With excellent surf beaches, Coolangatta boomed, eventually overtaking Tweed Heads' population (O'Hare and Burke 2016). By 1913 it was a sizeable and growing town, boasting three hotels, eight boarding houses, a cordial factory, real-estate agent, fruiterer, butcher, hairdresser and baker. This pressurised its sanitation services, strengthening Coolangatta's bid for independence from the mostly rural Nerang Shire Council which rather neglected its coast (Longhurst 1994). In 1914, it gained legal recognition as a town, albeit encompassing just 17 square kilometres (Jones 1988).

The transport revolution continued affecting regional development via the automobile. After, World War I, Southport benefited from proximity to Brisbane, better road access (Holthouse 1982) and more upmarket accommodation provision. Coolangatta, meanwhile, continued attracting working-class holidaymakers travelling by train to its more downmarket guesthouses and camping areas with up to 18 daily trains bringing passengers during the Depression era (Russell and Faulkner 1998).

More rail links ended the riverboat era, removing Tweed Heads' *raison d'être* as the river's entry point. By 1935, Coolangatta had surpassed it partly because its municipal facilities could not compete given its steadfast commitment to keeping council rates low. It gradually became just a service town for Coolangatta and local agriculture (Longhurst 1978) while Coolangatta came to rival Southport as Brisbane's main South Coast destination. In the early 1930s, Southport and the twin towns, located at the Gold Coast region's southern and northern ends, were attracting visitors from southern Queensland and other states, unlike newer intervening resorts which mostly attracted locals (Vader and Lang 1980). The next transport development, Coolangatta's airport (later renamed Gold Coast Airport), serving the Gold Coast region, evolved from an emergency landing strip created in 1936 for planes flying the Sydney–Brisbane route (O'Hare and Burke 2016). The airport's steady development is the second significant transport-infrastructure project straddling the border.

Many reasons seemingly underpin Coolangatta and Tweed Heads' differential development and Coolangatta surpassing its older twin. Coolangatta in the 1880s had no roads and shops, whereas Tweed Heads was already a busy seaport and commercial centre (McRobbie 1966). Once surfing became fashionable, Coolangatta's natural advantages stimulated resort development, whereas Tweed Heads lacked attractive, safe and accessible surf beaches, mainly becoming a minor port and distribution point for coastal shipping (Pigram 1977). The state capital Sydney's neglect also hampered Tweed Heads' attraction to tourists.

After World War II, Coolangatta's popularity peaked, with a record 50,000 attending the Australian Surf Lifesaving Championships on Easter Sunday in 1950 (Russell, 1995). However, the town had reached its physical limits given the NSW border, the river to the south, hills and swamps to the west, the Coolangatta airport to the north and the Pacific to the east. Discussions about boundary changes for South Coast local authorities began in late 1945. Yet again, calls emerged for Tweed Heads to be ceded to Queensland and amalgamated with Coolangatta, this time from Coolangatta Council (Longhurst 1994). In 1949, Coolangatta and Southport councils amalgamated to form the South Coast Shire (later renamed City of Gold Coast), for the first time, bringing all southern Queensland's coastal resorts under a single administration (Jones 1988).

In the 20th century's second half, improved interstate motorways and rising car ownership changed the dominant travel mode, and Coolangatta and Tweed's popularity waned. At opposite ends of the coast, both Southport and Coolangatta/Tweed started declining as Surfers Paradise began dominating from the mid-1960s (Scott et al. 2016). Railway closure in 1961 marked the end of this era, although former railway land provided profitable redevelopment opportunities, especially on Queensland's side of the border.

Over the following decades, the Gold Coast's development in the north outstripped Coolangatta and Tweed Heads (Russell and Faulkner 1998). While removing the border's tick fence in 1958 and rail closure promoted some unification, 'the sense of the border' (Allom Lovell Marquis-Kyle et al. 1997, 73) remains with the aptly named Boundary Street running along the headland between Queensland and NSW. The earlier tendency towards town-centre functions duplication has been replaced with Coolangatta specializing in recreational business and Tweed Heads becoming a more conventional central business district for both towns (Pigram 1977). In 1994 the City of Gold Coast serving the coastal areas was amalgamated with the hinterland jurisdiction of Albert Shire forming today's Gold Coast City Council. Tweed Heads' original twin, Coolangatta, is now a suburb in the city of Gold Coast.

The Pacific Highway's Togun Bypass opened in 2008, the last major interstate infrastructure project (O'Hare and Burke 2016). However, despite most of the seven-kilometre route being in NSW, its government refused any contribution, leaving Queensland and Commonwealth governments to pay the entirety. While built to ease traffic congestion in the Gold Coast's southern beachside suburbs, it also stimulated coastal suburban development south of the border.

The 2001 Tweed-River Entrance Sand Bypassing Project is another example of cross-border collaboration intended to maintain a safe and navigable entrance to the Tweed River and restore and maintain naturally occurring coastal sand drift to southern Gold Coast beaches (Queensland Government 2020). Two monuments mark both the border and interstate/inter-city cooperation: the Captain Cook Memorial Lighthouse on Point Danger opened in 1970 to commemorate the 200th anniversary of Cook's voyage along Australia's east coast, and the State-Border Marker erected in 2001 commemorating the Australia Federation's centenary. These cross-border projects and initiatives embody crucial common characteristics of twin cities: ongoing formal and informal negotiations (Garrard and Mikhailova 2019).

Cross-border planning

Cross-border planning, especially spatial or land-use planning, takes two forms: formal and relatively informal. Formal planning is generally associated with regional planning, insofar as each state's regional plans and strategies recognise the Gold Coast/Tweed as a conjoined place facing similar opportunities and challenges. Thus, the second iteration of the statutory strategic plan for South-East Queensland (SEQ) 2009 notes,

> a strong relationship between the Gold Coast and ... adjoining urban areas in Tweed Shire and northern NSW. Major planning and infrastructure issues require cross-border cooperation with the NSW government and relevant local government authorities.

Tweed Shire faced

> many of the same growth-management issues as SEQ: rapid population growth; high tourism visitations; development pressures on natural areas, the coastal zone and agricultural lands; and a requirement to invest in

additional infrastructure and community services. A positive partnership has been established between the Office of Urban Management and the North Coast Regional Office of the NSW Department of Infrastructure, Planning and Natural Resources. The aim ... is to share information and promote consistent planning.

(DIP 2009)

The plan's most recent manifestation (*Shaping SEQ*) observes,

SEQ has close relationships across regional boundaries with the surrounding areas of ... the Tweed Coast and northern NSW. These areas have unique social and economic linkages and can leverage opportunities provided by SEQ's continued population growth and diversified employment market. Maintaining and enhancing extensive infrastructure networks that connect these regions will support mutual social and economic benefits by providing access to employment and recreation. They will also enable the efficient movement of commodities, services and skills.

(DILGP 2017)

The NSW government has prepared similar planning documents, including its *Far North Coast Regional Strategy* in 2007 and *The North Coast Regional Plan 2036*, approved in 2016. This too acknowledged the importance of stronger ties with SEQ but focused more on economic benefits their region can gain from significant development north of the border.

Other cooperation areas are also evident. Since 2011 the two state governments have had a cross-border collaboration agreement, reiterated in 2016 with a new statement of principles wherein they

acknowledge the importance of each other in the border region and ... of working closely and collaboratively to improve service delivery, build local communities and further develop the region economically. Sustained population growth, interdependent economic relationships and the complex issues of service delivery require a coordinated approach to support the region's future.

(QLD/NSW 2016)

However, O'Hare and Burke (2016) note the 'out of state, out of mind' attitude of local planning scheme maps which, until the early 2000s, showed Boundary Street's 'other side' completely blank. Regional-scale planning in South-East Queensland was guided by voluntary, collaborative *Regional Frameworks for Growth Management* from 1990 to 2005. From 1995 these frameworks started acknowledging Gold Coast-Tweed Heads linkages and maps extended across the border. The 1998 and 2000 frameworks went one step further, including Tweed Shire Council completely in the planning area, the shire being part of the Southern Regional Organisation of Councils, a subregional organisation the plan area covered. However, as Tweed Shire is outside Queensland's jurisdiction and without formal collaborative-planning agreements, once plans became statutory instruments in 2005, regional plan maps reverted to depicting a 'terra nullius' south of the border (O'Hare and Burke 2016). Howes (2016) reports this discouraged cooperation and encouraged distrust during the water-supply crisis associated with the 2001–2009 severe drought. Major infrastructure projects proposed in the SEQ Regional Plan, like extending the heavy-rail and constructing a light-rail system, continue stopping at the state border in Coolangatta (Dedekorkut-Howes 2014).

In 2012 NSW's government created a Cross-Border Commissioner to work with and for border-community residents, businesses and other organisations. The Commissioner collaborates with governmental representatives from Victoria, the Australian Capital Territory and Queensland on various issues requiring cross-border cooperation and coordination and helping build significantly better working relations in the Gold Coast/Tweed region.

There are also more localised and less formal collaboration and cooperation channels, led more by the two local governments. The most significant situate around urban and regional planning, focusing on transport infrastructure, especially master planning Coolangatta airport and the Gold Coast Light Rail southern extension to and across the border. Since the 1990s, the Pacific Motorway extension and upgrade has been the most significant investment in cross-border infrastructure, improving tourist experience driving north into Queensland; also increasing the effective travel-to-work area for northern NSW residents, although both were disrupted during the COVID-induced border-crossing restrictions in 2020.

Coolangatta's Gold Coast airport presents one of the greatest challenges for cross-border coordination because the state border bisects the runway and terminal buildings. However, major airport planning lies with the Commonwealth government under the 1996 Airports Act. The current Gold Coast Airport Masterplan 2017 was approved, therefore, by the Federal Infrastructure and Transport Minister. However, relevant local-government planners participate in joint-planning exercises with the airport's owner, Queensland Airports Limited, to ensure planning consistency, especially in coordinating development of new public-transport infrastructure including light and heavy rail.

Finally, Australia has adopted several UK-pioneered urban-policy measures. The Commonwealth government recently has enthusiastically embraced the City-Deals concept. While City-Deals principles and practices in Australia have not always followed the UK experience, not least because Australia's three-tier federal system presents different challenges and opportunities (Burton 2016), they remain significant in federal urban policy, even extending to regional areas. Another Australian twin city, Albury-Wodonga, straddling the Victoria-NSW border, has recently been confirmed as the location of a new Regional Deal. This gathers all three governmental levels to promote local economic development and population growth outside the major urban areas that are more typically the focus of City Deals. Tweed/Gold Coast planners and politicians have noted this with interest, seeing potential for similar collaboration involving their twin city.

Conflicts and border closures

This twin city's history indicates that, despite small steps towards collaboration induced by ever-spreading urbanization, in some areas little has changed in over a century regarding cross-state-border issues. Early jocular accounts of the Custom House pre-echo the airport's condition today, highlighting current border issues accurately:

> The Customs officer stationed here has his house on one side and his kitchen on the other. He cooks in Queensland and sleeps in New South Wales – a sort of neutral inhabitant of both colonies.
>
> (*Brisbane Courier*, 15 May 1880, 3)

Longhurst (1978, 141) reports that, even in 1914, Coolangatta and Tweed Heads were regarded as 'a natural conurbation divided by petty officialdom in the way of a border fence'. The double border fence divided by a 'no man's land' emerged in 1898 to prevent tick-infected Queensland cattle from entering NSW, Coolangatta's turnstile allowing people through but not animals (Longhurst 1994).

In late March 2020, Queensland introduced state-entry restrictions to slow the spreading of COVID-19. Unless classified as 'exempt persons', people not normally Queensland-resident had to quarantine for 14 days if entering. However, there were substantial exemptions including essential-service providers like health professionals, emergency workers, transport workers and construction-sector workers. These reflect the border bisecting an urban area where many lives straddle the border in residence, employment or education terms. Freight could cross, although logistics workers were not expected to handle goods on both sides. These measures were relaxed on 10 July 2020, restricting entry only to hotspots, but remained under constant review. When Victoria experienced surging infections in mid-2020, on 21 July NSW introduced similar restrictions similarly impacting another cross-border twin city, Albury-Wodonga.

Until December 2020, those entering Queensland needed a pass establishing eligibility and entitlement at any border checkpoint, including the Tweed. This is relatively straightforward, and displaying them in vehicles usually allowed unimpeded crossing. While this regime applied at all Queensland's border crossings, the Brisbane-Sydney Pacific Motorway remains the most significant route, and the Tweed crossing the most significant checkpoint.

Unsurprisingly, there is more extensive quarantining and prohibition for transporting produce interstate in line with national biosecurity strategy: some animals and soil cannot enter Queensland from NSW, while plants like grapevines and banana-planting material cannot enter NSW from Queensland. There is little political debate or public concern about these long-standing biosecurity restrictions, but recent COVID-19 travel restrictions have aroused controversy. Indeed, several Australian High Court cases have questioned the constitutionality of Queensland's border closure. This is not the first time the Tweed River has had great significance during a pandemic. Australian troops returning from

Europe post-World War I brought Spanish flu with them, October 1918 seeing the first cases. Although quarantine measures were instituted and infection suspects inoculated, the disease spread, and in November, Commonwealth, state and territory governments agreed to manage the outbreak through border controls and other measures.

In ways strongly paralleling contemporary debates, some states quickly closed borders while others were more reluctant, worrying about disrupting everyday life, trade and post-war celebrations and commemorations. Queensland closed its border, allowing NSW travellers across only at Wallangarra (almost 300 km inland from Tweed Heads), where they were quarantined. At the Tweed River, closure severely disrupted many lives: local museum curator Erika Taylor observed, 'people were literally stuck. [They] had gone out to see a friend, come back [and] police … were not letting you through' (Young and Kinsella 2019). At that time, Coolangatta on Queensland's side had virtually no facilities: shops and schools were across the river and alternative facilities well northwards. Practical problems included Coolangatta children enrolled at the Tweed Heads Public School being refused access for months, and Coolangatta residents staying in Tweed Heads on closure night travelling hundreds of kilometres to return home. Masked police had to push cashboxes across the border stile with long poles to maintain Coolangatta accounts in the district's only bank located in Tweed Heads (Figure 4.2). Two decades later a newspaper called the border fence a 'hindrance to intercourse as a single community', bemoaning the barrier it posed 'to the development and progress of what is practically one town' (*Courier Mail*, 17 February 1938, 12).

A few years after the Spanish flu-induced problems, further challenges presented by the border and separate jurisdictions were revealed when fire engines

Figure 4.2 Men exchanging money over the Queensland-New South Wales border during the 1919 flu epidemic. Photographer unidentified.

Source: Image courtesy of Gold Coast Libraries Local Studies Collection.

from Tweed Heads could not cross the border to attend a fire at the nearby Digger's Theatre and found their hose water falling 150 feet short of target (Longhurst 1978)!

Like today, border closure produced considerable local hostility, not least among Queenslanders holidaying in northern NSW and unable to return, but, again like today, Queensland's government remained steadfast in its commitment to keeping the 'flu out'.

The two states' variable adoption of daylight savings time is another enduring source of dispute, profoundly affecting Gold Coast/Tweed twin-city life. Every October, when Tweed Heads residents move to Australian Daylight Saving Time (ADST), they find themselves in a different time zone to their Coolangatta near-neighbours until harmony resumes the following April. ADST prevails throughout Australia, excepting Queensland, Western Australia and the Northern Territory. A Queensland Chamber of Commerce and Industry survey (CCIQ 2010) found most favouring ADST's introduction, although regional and northern Queensland registered considerable opposition. A proposal to introduce it only in SEQ was unpopular across the rest of the state. While profoundly annoying residents and workers on either side of the Coolangatta/Tweed Heads state border, there are no signs either state will embrace greater harmony anytime soon.

Conclusions

Settlements along both Tweed River banks flourished for centuries if not millennia, seemingly without enduring conflict between Indigenous clans and language groups. But, since European settlement and, certainly more recently, rapid population growth and increasing local economic integration have made the Gold Coast/Tweed a more significant urban place, worthy of twin-city designation. While most local residents see this in terms of adjoining neighbourhoods epitomised by the Twin-Towns resort complex, there is a much broader conception of a place straddling the border. Spearritt (2009) introduced the notion of a 200 km city extending from the Gold Coast to the Sunshine Coast; an equally plausible entity is imaginable occupying the 300 km from Byron Bay in the south to Noosa in the north with the Tweed River serving as a highly significant pivot-point in this sprawling entity. Certainly, there remain significant inter-urban breaks in this imaginary city, especially south of the border where motorway driving in northern NSW mostly reveals rolling hills and occasional oceanic glimpses. But this area, especially the coastal strip, may well experience continued pressure for infill development over the coming century and produce a sprawling metropolis. Whether it becomes a fully integrated metropolitan region with its own distinct regional identity, functional governance structure and plans successfully managing sustainable growth remains to be seen.

While Australian political debate focuses occasionally on challenges involving federalism, notwithstanding periodic secessionist campaigns within states, there is no concerted pressure to move the Queensland-NSW border south of the Tweed to allow the Gold Coast/Tweed twin city to become one jurisdiction. However, in August 2020, with some states and territories wholly or partially closing borders due to COVID-19, Gold Coast's Mayor called for the border to be moved south to the Tweed itself. While clearly prompted by current practical difficulties around maintaining border-controls in a built-up area, Mayor Tate also felt it made 'natural sense. Mother Nature has got Tweed River there … once you cross … that's when most people feel … they've entered NSW' (McCutcheon 2020). NSW's premier responded that 'if anything, the border should be moved north' (McKenna 2020). While the border's current alignment seems politically and practically inconvenient, change seems unlikely, illustrating twin-city persistence 'closely linked but still separate' (Garrard and Mikhailova 2019, 17). Partly this is because practical difficulties of political and legislative difference can be worked around, and the socio-cultural challenges associated with realignment currently outweigh benefits of economic rationality. Perhaps the most that can be anticipated lies in constructing more formalised joint-planning with more significant impact in managing the twin city's whole travel-to work-area. Until then, resident Banana Benders and Cockroaches will continue living their lives on either side of the border mostly cooperatively, punctuated occasionally by intense rivalry and competition associated mostly with cricket and rugby.

Notes

1 Australians enjoy inventing nicknames for each state and territory's inhabitants, epitomising usually friendly interstate rivalry. Queensland is certainly a major banana-producer; the Cockroach moniker allegedly originated from a rugby-league commentator.

References

Aberley, D. (1993) (ed) *Boundaries of Home: Mapping for Local Empowerment*, Gabriola Island, BC: New Society Publishers.

Burton, P. (2016) City Deals: Nine Reasons This Imported Model of Urban Development DemandsDdue Diligence, *The Conversation*, accessed 19/7/2020 at: https://theconversation.com/city-deals-nine-reasons-this-imported-model-of-urban-development-demands-due-diligence-57040

CCIQ [Chamber of Commerce and Industry Queensland]. (2010) Daylight Saving: a Queensland Business Community Perspective, Brisbane: CCIQ, accessed 19/7/2020 at: https://www.cciq.com.au/assets/Documents/Advocacy/121003Daylight-Savings.pdf

Dedekorkut-Howes, A. (2014) Bioregional Planning and Growth Management. In *Australian Environmental Planning: Challenges and Future Prospects*. J. Byrne, J. Dodson & N. Sipe (eds.). London: Routledge, 234–245.

Dedekorkut-Howes, A. & Bosman, C. (2015) The Gold Coast: Australia's Playground? *Cities*. 42, Part A: 70–84.

DILGP [Department of Infrastructure, Local Government and Planning]. (2017) *Shaping SEQ: South East Queensland Regional Plan 2017*. State of Queensland.

DIP [Department of Infrastructure and Planning] (2009). *South East Queensland Regional Plan 2009–2031*. State of Queensland.

Garrard, J., & Mikhailova, E. (2019). Introduction and Overview. In *Twin Cities: Urban Communities, Borders and Relationships over Time*. Garrard, J., & Mikhailova, E. (Eds.). New York: Routledge.

GNB of NSW [Geographical Names Board of New South Wales]. 2020. Coolangatta, accessed 2/7/2020 at: https://www.gnb.nsw.gov.au/place_naming/placename_search/extract?id=JPIOjzsyMn

Green, A. (1982) *Drawn to the Coast: A Sketchbook of the Gold Coast's Heritage*. Brisbane: Boolarong Publications.

Holthouse, H. (1982) *Illustrated History of the Gold Coast*. Sydney: Reed.

Howes, M. (2016). The politics of paradise: intergovernmental relations and the Gold Coast. In *Off the Plan: The Urbanisation of the Gold Coast*. C. Bosman, A. Dedekorkut-Howes & A. Leach (eds.). Melbourne: CSIRO Publishing, 97–107.

Jones, M. (1988) *Country of Five Rivers: Albert Shire 1788–1988*. Sydney: Allen and Unwin.

Longhurst, R. (1978) *The Development of the Gold Coast as a Recreational Area to 1940*. History IVH Thesis, University of Queensland.

Longhurst, R. (1994) *The Nerang Shire – A History to 1949*. Nerang: Albert Shire Council.

MALCC [Muurrbay Aboriginal Language and Culture Cooperative]. (2020) Bundjalung-Yugambeh Dictionary, accessed 16/7/2020 at: http://bundjalung.dalang.com.au/language/dictionary?query=cooloon&type=English&numeric=Exact&dialect=All#

Allom Lovell Marquis-Kyle, Henshall Hansen Associates, Context, HJM, and Staddon Consulting. 1997. *The Gold Coast Urban Heritage and Character Study*. Gold Coast, QLD: Gold Coast City Council.

McCutcheon, P. (2020) Moving the Queensland-NSW Border South to the Tweed River Does Make Sense, *ABC News*, accessed 27/02/2021 at https://www.abc.net.au/news/2020-07-15/coronavirus-queensland-analysis-moving-border-qld-nsw/12452912

McKenna, K. (2020) Queensland Wants to Move Coronavirus Border Checkpoints into NSW – but Is It legal or Even Feasible? *ABC News*, accessed 27/02/2021 at https://www.abc.net.au/news/2020-07-21/queensland-coronavirus-what-does-moving-a-state-border-involve/12473162

McRobbie, A. (1966) *The Gold Coast Story*. Surfers Paradise: The Gold Coast Annual Company.

NSW Department of Planning, Industry and Environment (2016) *The North Coast Regional Plan 2036*, accessed 11/7/2020 at: https://www.planning.nsw.gov.au/Plans-for-your-area/Regional-Plans/North-Coast/North-Coast-Regional-Plan/A-thriving-interconnected-economy

O'Hare, D. & Burke, M. (2016) Transport: From Cream Cans and Campers to City Centres and Commuters. In *Off the Plan: The Urbanisation of the Gold Coast*. C. Bosman, A. Dedekorkut-Howes & A. Leach (eds.). Melbourne, CSIRO Publishing, 45–61.

Parker, N., Vaughan-Williams, N., Bialasiewicz, L., Bulmer, S., Carver, B., Durie, R., Heathershaw, J. (2009) Lines in the Sand? Towards an Agenda for Critical Border Studies. *Geopolitics*, 14:3, 582–587

Pigram, J. J. (1977) Beach resort morphology. *Habitat International* 2:(5–6), 525–541.

Queensland Government (2018) *History of Queensland borders before proclamation*, accessed 5/7/2020 at: https://www.qld.gov.au/recreation/arts/heritage/museum-of-lands/surveying/borders/pre-proclamation

Queensland Government (2020) Tweed Sand Bypassing, accessed 20/7/2020 at: https://www.qld.gov.au/environment/coasts-waterways/beach/restoration/tweed-river

Queensland Government/New South Wales Government (QLD/NSW) (2016) *Statement of Principles and Priorities for Cross-border Collaboration 2016–2019*, accessed 5/7/2020 at: https://www.dpie.nsw.gov.au/__data/assets/pdf_file/0004/282802/qld-nsw-statement-of-principles-and-priorities-for-cross-border-collaboration-2016-19.pdf

Russell, R. A. (1995) Tourism Development in Coolangatta: An Historical Perspective. Honours Thesis, Griffith University Gold Coast, Faculty of Business and Hotel Management.

Russell, R. & Faulkner, B. (1998) Reliving the Destination Life Cycle in Coolangatta: An Historical Perspective on the Rise, Decline and Rejuvenation of an Australian Seaside Resort. In *Embracing and Managing Change in Tourism: International Case Studies*. E. Laws, B. Faulkner & G. Moscardo (eds.). London: Routledge, 95–115.

Scott, N., Gardiner S. & Dedekorkut-Howes, A. (2016) Holidaying on the Gold Coast. In *Off the Plan: The Urbanisation and Development of the Gold Coast*. C. Bosman, A. Dedekorkut-Howes & A. Leach (eds.). Melbourne, CSIRO Publishing, 31–43.

Sharpe, M.C. (1998) *Dictionary of Yugambeh (Including Neighbouring Dialects)*. Dept. of Linguistics, Research School of Pacific Studies, Australian National University.

Small, B. 1981. *Gold Coast Sketchbook*. Adelaide: Rigby.

Spearritt, P. (2009) The 200km City: Brisbane, The Gold Coast and Sunshine Coast, *Australian Economic History Review*, 49: 1, 87–106.

Steele, W., Alizadeh, T. & Eslami-Andargoli, L. (2013) Planning across borders Why a focus on borders? Why now? *Australian Planner*, 50:2, 93–95.

Vader, J. & Lang, F. (1980) *The Gold Coast Book: An Illustrated History*. Milton, QLD: Jacaranda Press.

Young, E. & Kinsella, B. (2019) The deadly Spanish Flu and a dramatic border closure remembered 100 years on, *ABC News*, accessed 27/02/2021 at: https://www.abc.net.au/news/2019-02-05/the-deadly-spanish-flu-and-qld-nsw-border-closure-100-years-on/10781296

5 Chandigarh tri-city

Between conflict and co-operation

Ganeshwari Singh, Simrit Kahlon and Ekaterina Mikhailova

The Chandigarh-Mohali-Panchkula tri-city is spatially distributed over two states administered by provincial governments, and one Union Territory (UT) administered by India's Union Government. The planned city of Chandigarh lies at its core, serving as capital thrice over: to Punjab and Haryana states and the UT of Chandigarh. The city is surrounded by the two adjoining cities of Mohali and Panchkula, the former in Punjab state and the latter in Haryana state (Figure 5.1). Together these three spread across an area of 278 square kilometres, accommodating around 1,400,000 people. Located at the tri-junction of three distinct administrative units, the three cities are governed by three distinct administrations whose inter-relationship resembles sibling rivalry at best – a fact at least partly accounted for by the dynamics involved in the birth of these cities.

We examine the tri-city through a lens of intranational twin-city relationships characterised as durable, interdependent, unequal and prone to conflicts (Garrard and Mikhailova 2019b). We argue that strong planning regimes present in Chandigarh, Mohali and Panchkula resemble formally declared Indian twin cities (Garrard and Mikhailova 2019a), while their scantiness of joint planning distinguishes the tri-city from them.

The chapter is divided into four sections. The first section sketches the tri-city's birth and the circumstances surrounding it; the second explores how these cities grew and how the conditions or fortunes of each impacted the others; the third explores points of conflict and cooperation between these cities and how the general citizenry perceives the relationship. The fourth section examines the spatial reality of their existence as emergent during the COVID-19 pandemic.

The tri-city's birth: a contentious legacy

The tri-city's birth results as much from fortuitous circumstances as from any deliberate urban planning. As we will show, the tri-city is a typical example of instrumental twin cities whose creation served several instrumental goals of different interest groups (Garrard and Mikhailova 2019a, 107).

Chandigarh, erstwhile capital of East Punjab,[1] was independent India's first fully planned city. It was born during a tumultuous phase in world history when most former colonies, having gained independence, began constructing planned

DOI: 10.4324/9781003102526-6

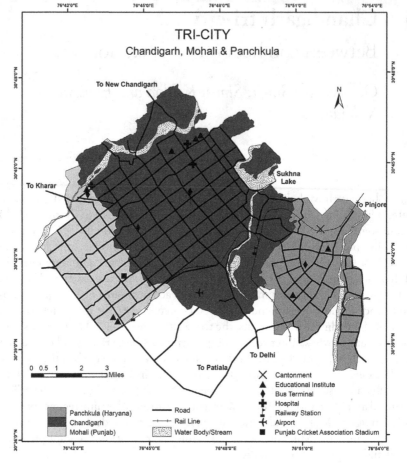

Figure 5.1 Tri-city urban layout.

Source: Map by Ganeshwari Singh.

capital cities, all interpreted as symbols of government-sponsored national-identity building (Vale 1988).

The fact that Punjab needed a new capital, having lost its erstwhile capital Lahore to newly created Pakistan, provided a perfect opportunity. Pitching the argument that no existing Punjabi city had the potential for designation as the new capital (Kalia 1987), a reality was constructed contending that only building a totally new city could really provide Punjab with something fit to become its seat of government. Named the Capital Project, Chandigarh's making was one among several development projects undertaken and constantly cited by Nehru as symbolising modernity and progress (Singh, Kahlon and Chandel 2019). To ensure best modern practices, the renowned Swiss-French architect Le Corbusier was invited to lead Chandigarh's planning and construction. The city was to be projected as symbolising progressive India and used to instil national pride and sense of belonging amongst otherwise diverse population groups.

Projected as India's 'most triumphant post-partition tale' (Mallot 2012, 31), Chandigarh was proffered as a symbol of hope and a salve helping Punjabis forget the traumatic partition and the indignity of being divested of homes and belongings. It was unsurprising, then, that Punjabis took personal pride in their ownership of Chandigarh, a matter eventually substantially dictating the course of events governing tri-city dynamics (Singh, Kahlon and Chandel 2019).

Punjab's reorganisation in November 1966 formed a crucial breakpoint in Chandigarh's history. Punjab was reorganised along linguistic contours, giving birth to the states of Punjab and Haryana and the Himachal Pradesh Union Territory. Chandigarh, until then capital of East Punjab, now became a bone of contention between the two successor states of Punjab and Haryana. Each state claimed Chandigarh as its sole capital, with neither ready to compromise. Punjab's Chief Minister even handed a blank cheque to Haryana's Chief Minister in lieu of Haryana dropping its claim. The emotional connection that East Punjabis had with Chandigarh (in common parlance, 'our new capital'), and the feeling of prestige coming through association with Chandigarh, made it impossible, indeed politically suicidal, for either state to surrender its claim. With the two states often ruled by political parties in conflict at the centre, politicians necessarily hardened their stands even further. The Central Government's involvement and interference in the Capital Project from the start enabled the Centre to declare it a Union Territory which would remain under Central Government aegis while acting as joint capital for both newly carved states. Furthermore, all assets and liabilities, including administrative posts, were divided 60–40 between Punjab and Haryana.

Changing the region's administrative status also had far-reaching spatial implications, particularly with the Capital Project's implementation still in its initial stages. Immediate tremors of change were felt quite close by on Chandigarh's spatial fringes: an area designated the 'Periphery' by the Punjab New Capital (Periphery) Control Act, 1952. In 1962 this Act was amended, extending the previously demarcated periphery area around Chandigarh from 8 to 16 km – partly to insulate the city from local spontaneous-development impulses (e.g., 'gunthewari'), thereby ensuring its character remained unsullied. A further intention behind ensuring this undeveloped belt around Chandigarh could have been to throw into relief Chandigarh's spectacular planned urban landscape against an undeveloped, largely rural backdrop. It goes without saying that land use on the periphery was intended to be strictly rural-related. Punjab's reorganisation, however, changed all this.

Following reorganisation, the periphery was divided between the three newly created administrative entities. Chandigarh's administration was now left with only 44 sq km, barely 3% of the original area. Punjab gained control of 75%, amounting to 1021 sq km, while Haryana was given 295 sq km. Worse still, the Periphery Control Act no longer gave statutory power to prevent new development within the periphery. Hoping to derive benefit from Chandigarh's proximity, both Punjab and Haryana initiated urban development within the previously 'sacrosanct' periphery.

Although violation of Chandigarh's Periphery Control Act had begun way back in 1953, when Chandimandir cantonment emerged along the north-eastern margin, Punjab was the first to violate the Act after state reorganisation. In 1967, under the plea of focusing on the industrial development of villages, Punjab identified Mohali, located near Chandigarh, for village development. In the early 1970s Punjab started developing the new town of Mohali as an industrial hub along Chandigarh's south-west margin. Hoping that Chandigarh would eventually be awarded to Punjab, the new town was developed on the same grid pattern. Mohali's Municipal Corporation website also boasts of its contiguity with Chandigarh and of Mohali being 'a mere extension of Chandigarh's road and design system without any unique planning' (Municipal Corporation SAS Nagar 2020).

Witnessing Mohali's development, Haryana followed suit, initiating development in the then village of Panchkula. Panchkula town was to be developed as a planned service-centre close by Chandigarh's eastern boundary. Spread over 19.4 sq km, Panchkula, often seen as Haryana's reaction to Punjab's Mohali (Kalia 1987), was built as a fan-shaped city and projected as a regional growth pole, serving the nearby urban centres of Pinjore, Manimajra and Chandimandir. Suffering from land scarcity closer to Chandigarh, Panchkula developed beyond the Ghaggar River.

Field observations in spring 2019 revealed that today, Chandigarh's immediate periphery often presents an incoherent belt of no-man's land. Strolling from Khalsa College (Mohali, Sector 53) to the Girls' Postgraduate College (Chandigarh, Sector 42), just 800m apart, takes one across a disorganized dump, a ditch-like stream, wild grass and unkempt roads (Figure 5.2) – scenes discordant with Chandigarh's City-Beautiful image.

Figure 5.2 Chandigarh's periphery: the unkempt road between the Khalsa Colledge (Mohali, Sector 53) and Post Graduate Government College for Girls (Chandigarh, Sector 42).

Source: Photo by Ekaterina Mikhailova, 2019.

Growth trajectories: From emulation to complementarity

Both younger cities, born of and reared by different mothers, had doubtless been conceived in the mould of the eldest sibling: Chandigarh. This was despite both Panchkula and Mohali exhibiting distinct growth trajectories marked by both lulls and spurts, spurred by different catalysts at different times. The one common growth catalyst for both, however, was their proximity to Chandigarh.

The attraction Chandigarh had for most people belonging to Punjab and Haryana states encouraged them to seek housing in Chandigarh. Initially at least, this was truer of Punjabis than Haryanvis. The prestige associated with owning a house in Chandigarh was higher among Punjabis than Haryanvis because Haryana initially was underdeveloped with little surplus money.

Chandigarh emerged in three phases. The first, lasting till 1965, comprised 30 low-density sectors (1 to 30) spread over 9,000 acres intended for 150,000 people. The second, spreading over 6,000 acres, covered high-density sectors 31 to 47, catering for 350,000 people. The third, starting in 1990, saw the construction of sectors 48–56, 61 and 63. It was initially peopled by government servants, lawyers, businessmen, landlords, retired defence personnel and celebrated artistes. Home to many institutions – Panjab University, Post Graduate Institute of Medical Education and Research (PGIMER) and College of Architecture, Punjab Engineering College, to name a few – Chandigarh faced acute accommodation need, as these institutions' employees multiplied. Punjab's merger with the Patiala and East Punjab States Union in 1956 increased pressure to accommodate more government personnel in Chandigarh. Things became so urgent that government employees either had to share government accommodation or reside in shared private accommodation. In 1959, Chandigarh's Administration stopped trying to curb land prices, scrapping the practice of selling residential plots at fixed rates. Instead, it started auctioning plots to the highest bidder, producing skyrocketing land prices in Chandigarh. The housing crisis was already worrying by 1964 (Sarin 1982). In 1971, Chandigarh's population reached 257,251 with decadal growth for 1961–1971 touching 115%, placing Chandigarh amongst India's fastest-growing cities (Table 5.1). This increased the housing shortage. Chandigarh's Administration, trying to curb overcrowding, had decreed that only one family would reside on one plot. Yet as pressure on limited

Table 5.1 Tri-city population 1951–2011

Year	Chandigarh	Panchkula	Mohali
1951	24,261	–	–
1961	119,881	–	–
1971	257,251	–	–
1981	451,610	11,239	32,351
1991	642,015	70,375	78,457
2001	900,635	140,925	123,484
2011	1,054,686	211,355	146,213

Source: Census of India.

housing stock increased, this rule could not be implemented, particularly when the Administration itself was providing shared accommodation to its employees.

Enhancing woes of households seeking Chandigarh accommodation, laws about property sale and purchase, also terms and conditions of house-ownership, were highly stringent and inimical to a buoyant housing market. In this situation, the rise of Mohali and Panchkula, providing cheaper and easier-to-access accommodation was a boon to those finding no space in Chandigarh. Mohali and Panchkula hence benefitted not just from Chandigarh's proximity but also through its distorted housing market: high land prices, stringent building bye-laws and cumbersome property-transfer procedures.

Another lacuna in Chandigarh's planning, admirably filled by Mohali and Panchkula's emergence, was the limited scope for industrial development in Chandigarh. Its master plan allowed only limited industrial space, ensuring it was exorbitantly priced. Thus, several small and medium-scale industrial units found space in Mohali and Panchkula. Mohali, as noted, developed as an industrial township with an earmarked industrial area. Many economic activities, like medium-scale and polluting industries forbidden in Chandigarh, found space in these new towns. There were hardly any industries located in the tri-city region in 1951; by 1981 there were 52 medium and large-scale units – mostly located in Mohali or Panchkula, with front offices in Chandigarh. This trend continues, with Mohali preferred for manufacturing and Panchkula for warehousing.

Mohali and Panchkula saw a further growth spurt during the 1980s when, due to Punjab's disturbed law-and-order situation, Mohali and Panchkula were perceived as safe havens for Sikhs and Hindus respectively. During this period, there was visible inter-city migration within the tri-city, with Hindus migrating towards Panchkula and Sikhs towards Mohali. Low real-estate prices across the tri-city greatly facilitated this movement. However, there were also increasing distress sales during this period. Sikhs, who were settled in Panchkula and Chandigarh, were inclined to move to Mohali, particularly following the anti-Sikh riots across the country following Prime Minister Indira Gandhi's assassination. Hindus likewise sold properties in Mohali at throwaway prices to move to Panchkula or Chandigarh. There was in fact a slowdown in the land market in the entire tri-city region due to the disturbed socio-political environment. Few people from Punjab were willing to invest in immovable property.

The tri-city's third growth spurt came in the 1990s with economic liberalisation and the entry of private players into the real-estate sector. Panchkula's development was already being adroitly managed by Haryana's Urban Development Authority. Despite possessing a much smaller area adjoining Chandigarh and weak connectivity with that city, Panchkula developed new sectors much faster than Mohali. The proximity of Chandimandir military cantonment proved a boon here, for Panchkula became a favoured destination for post-retirement settlement for defence personnel. The easy accessibility of medical services in the Command Hospital was no small attraction.

Punjab's Urban Planning and Development Authority was established only in 1995, and only in the late 1990s did Mohali see entry by private players into its real-estate sector. This was followed by a real-estate boom in the tri-city generally, particularly in Mohali during the 2000s.

During the 1990s and 2000s, many Chandigarh households were close to retirement, thus exploring options to move from rented and institutional accommodation to owner-occupied places. Chandigarh's third phase, not catered for in the original master plan, thus emerged. Nine more sectors were planned with high-density flatted development which developed in the remaining green belt dividing Mohali from Chandigarh. Mohali and Chandigarh thus moved even closer together, with people residing in Chandigarh's southern sectors, often accessing services in Mohali which was by that time better established.

Mohali, in emulating Chandigarh's character, concentrated more on developing services which people were wont to access in Chandigarh, including education and health. Even in the late 1970s, Mohali boasted two of the best tri-city schools, a Government College and a fully functional Government Hospital. In school education at least, Mohali had become a hub, accessed not just by tri-city people but also others across Punjab. Following private players' entry into both health and education, Mohali became a flourishing centre in both respects. At least three highly specialised and nationally reputed higher education and research institutions became located there, almost putting it at par with Chandigarh. Mohali also became the chosen destination for several private hospitals offering high-end services for those able to afford them. Here, Mohali benefitted from a highly skilled and vastly experienced pool of physicians and other health workers earlier employed at PGIMER, Chandigarh.

An international-standard cricket stadium emerged in 1992. Its involvement in high-class international competitions (four Cricket World Cup matches in 1996 and 2011) brought Mohali international recognition. Interestingly, the stadium itself resulted from both competition and cooperation among tri-city companies. The Punjab Cricket Association began mooting construction once the Chandigarh Administration refused to lease its own stadium to them. Designed by Ar. Khizir and Associates (Panchkula), and constructed by R.S. Construction Company (Chandigarh), it became home to Punjab's cricket team and Punjab Cricket Association's headquarters – thus housing the governing body for developing and organizing cricket in both Punjab state and Chandigarh.

Panchkula concentrated more on high-quality residential development, becoming appreciated for its commercial and recreational facilities. As a new state, Haryana created several new governmental offices and institutions, and Panchkula was an ideal location for most of these. During 2011–2015, however, Panchkula also began promoting private health-sector players.

Designating both Panchkula (1995) and Mohali (2006) as district headquarters added to the development process both in terms of spatial spread and speed. Creating the Greater Mohali Area Development Authority (GMADA)[2] for the orderly, integrated and comprehensive development of the whole district seeks to make Mohali an urban complex in its own right, comprising seven urban units plus itself, at least four sharing boundaries with Chandigarh. Panchkula district, occupying 898 sq km, includes four towns alongside Panchkula city. However, Panchkula aside, none share borders with Chandigarh. Thus Mohali's area of 1098 sq km, surrounding Chandigarh on three sides, provides considerable advantage over both Panchkula and Chandigarh. Indeed, from starting as interlopers on Chandigarh's periphery to satellite towns whose existence depended on Chandigarh to district headquarters and regional growth centres in their own right, both Mohali and Panchkula have come a long way.

Growing together: through conflict and competition

Mohali and Panchkula's desire to be like Chandigarh, Punjab and Haryana's aim to appropriate Chandigarh, and Chandigarh's urge to retain its elitist image by remaining a Union Territory and an exclusive city like no other, have all produced conflict and competition among the three cities. Not only is this visible at administrative and political levels; all three cities' residents have also become major stakeholders in the process.

Acknowledging Chandigarh's brand value, Mohali and Panchkula citizens generally introduce themselves as belonging to Chandigarh. Meanwhile, Chandigarh residents do not take kindly to any outsiders staking claim on the city. A Chandigarh respondent, emphasising his dislike for outsiders, and the city's exclusivity, says, 'Visas should be needed to visit Chandigarh'. A Chandigarh-based journalist, Aarish Chhabra (2017, 117–118), in his essay collection, jokingly suggests that 'Chandigarh natives' claiming all Chandigarh problems are due to 'outsiders' have 'the B-A-B-U (Chandigarh's Born-And-Brought-Up) syndrome', and 'proudly hide where their parents actually come from' (18). Just a generation ago, all Chandigarh's residents were outsiders.

Its exclusive administrative status is an asset highly prized by Chandigarh's residents. A 1981 survey revealed 80% of Chandigarh's residents wanted the city to remain a Union Territory. Such sentiments remain equally strong now. Chandigarh residents' self-centredness and their 'othering' towards immediate neighbours attests to intranational twin cities' persistent identities, and is crucial for understanding the tri-cities' interdependent co-existence across three jurisdictions.

Constant competition between Mohali and Panchkula to cash in on Chandigarh's proximity also fuels tri-city persistence. Mohali seems more pro-active for guessable reasons. For one, Mohali represents Punjab, Chandigarh's original home state. Letting go of Chandigarh for Punjabis is close to impossible. Besides, decision-makers of Punjab and Chandigarh UT have been more interlinked than those of Chandigarh UT and Haryana. Punjab's Governor, who is constitutional head of Punjab state, is also the constitutional head of Chandigarh UT: his formal designation reads, 'Governor of Punjab and Administrator Chandigarh (UT)'. Chandigarh's top bureaucrat, previously entitled Chief Commissioner of Chandigarh, is now known as Adviser to the Administrator. Any Union Government attempt to change this system meets with strong protest from Punjab state. As recently as 2018, Haryana's Chief Minister, Manohar Lal Khattar, argued for a statutory board to oversee and ensure homogenous tri-city development. However, Punjab's Chief Minister, Capt. Amarinder Singh, shot down the notion, stating that Chandigarh 'indisputably belonged to Punjab' (*NDTV* 2018).

In 2013, Punjab started developing another township near Chandigarh, named New Chandigarh, a name going down poorly with Haryana, which claimed Chandigarh was a brand name and copying it was 'completely unethical' (*India Today* 2013). Mohali, however, continues cashing in on Chandigarh's prestige. Having more space and better connectivity with Chandigarh, it provides adequate housing options in Nayagaon, Mullanpur, Zirakpur and Baltana to commuters from Chandigarh.

Another bone of contention between the three cities is Mohali's recently constructed international airport. Till a few years ago, the tri-city was served by an airport in Chandigarh known as 'Chandigarh airport'. In 2016, a new international airport was constructed in Mohali intended to replace Chandigarh's domestic airport. It was to be managed by Chandigarh International Airport Authority Limited, with the Indian Airports Authority owning 51% while Punjab and Haryana each owned 24.5%. However once the airport started functioning, Punjab claimed it as its own since it was built on Punjab's land, naming it Mohali Airport. Haryana took offence, arguing it should be named Chandigarh Airport (*Indian Times* 2016). In 2018, Punjab and Haryana agreed to name it after the revolutionary freedom fighter, Bhagat Singh, equally revered in both the states. However, as there is no deciding body for the tri-city capable of resolution, the issue regarding the name of the city to be suffixed with the airport's name remains unauthorized, with Punjab wanting to suffix Mohali and Haryana insisting on Chandigarh. The airport's entrance lies in Mohali; hence, for now at least, the signage reads 'International Airport Mohali', protests from Chandigarh and Haryana notwithstanding.

Recently, the three cities have tried coming together to tackle some of the issues, a move which even tri-city residents feel is much needed. Chandigarh's Master Plan 2031 argues for a symbiotic relationship with the regional urban centres, arguing for a 'Metropolitan Plan' for 2031. This should integrate the Chandigarh Master Plan 2031, GMADA Plan 2056 and the Haryana Development Plan. A report prepared by the Union Urban Development Secretary emphasised the tri-city's need to be considered a 'single urban entity' (*Times of India* 2009). In 2009, under High Court direction, a Coordination Committee, originally recommended for formation back in 1975, was finally constituted to oversee planned tri-city development and to create an integrated City Development Plan. In 2013, the three cities worked together on a Tri-City Metro Project inter-linking them by metro (*The Pioneer* 2013). The project, however, never materialised. At the time of writing, an alternative monorail project inter-connecting the tri-city is under discussion (*Metro Rail News* 2018). In 2016, a Tri-city Coordination Committee, including the three cities' Mayors, Commissioners and Chief Engineers, was formed to coordinate on various cross-border issues. However, they failed to work together (*Times of India* 2018). As of January 2020, the cities are working towards a ring road around Chandigarh for better connectivity (*The Tribune* 2020). The three states are also coordinating over law and order and tracking border-transcending criminal activity.

While there has been several tri-city planning initiatives, this section has shown that it has been hard for any of them to materialise. The disputes over these initiatives have often been more visible than their outputs. The protracted inefficiency in coordinating the overall tri-city relationship stems from continuous tri-city conflicts sustained due to their peculiar geographic layout, with Chandigarh as a centrepiece, the ingrained rivalry between the two satellite cities, a contentious politico-administrative legacy and their jurisdictional fragmentation.

The way ahead: Lessons from the pandemic

There is probably no better illustration of tri-city socio-economic and spatial interconnectedness and the consequent need for their cooperation than the way the COVID-19 pandemic emerged and spread across the tri-city. The first reported case on 18 March 2020 was a Chandigarh resident with a history of travelling to the United Kingdom. As soon as she was diagnosed, at least 70 people were quarantined. Of these, 60 were residents of Mohali, where the patient's family was running a business enterprise. Of the seven people who caught her infection, four were Chandigarh residents, two from Mohali and one in Panchkula.

As the pandemic progressed, there were regular instances of patients getting infected in any one city and carrying the infection to the other two. A health worker working in PGIMER resided in Mohali, a doctor working in Mohali lived in Chandigarh. A Non-Resident Indian couple residing in Chandigarh infected their mother in Mohali. An octogenarian housed in Chandigarh was admitted to hospital in Panchkula where she was visiting two sons residing there.

Particularly vulnerable were the rurban unplanned spaces existing on these cities' borders offering cheap rental accommodation to the less well-off, particularly those employed in Chandigarh. During the pandemic, it also emerged that Chandigarh depended for milk and vegetable supplies on surrounding villages within Mohali's jurisdiction.

Such was the social and spatial integration between Chandigarh and Mohali that the Administrations had to adopt a uniform strategy for dealing with the problem. Panchkula, meanwhile, adopted a more lenient strategy, since it was not expecting any major problem because not many citizens indulged in international travel.

Forming a continuous spatial entity, the three cities initially tackled COVID-19 by following protocols laid down by their respective state governments/administrations. Whereas Punjab announced lockdown of all commercial and industrial enterprises and other non-essential services and a curfew restricting the movement of citizens outside their homes, and Chandigarh followed suit almost immediately, Haryana initially imposed only the less stringent lockdown as a measure.

As the weeks progressed, Chandigarh officials became increasingly concerned about ensuring the tri-city followed strict pandemic-containment protocols. In a press release dated 13 April 2020, UT Administrator V.P. Singh Badnore stressed the need for a common tri-city approach. Arguing that the entire area is 'geographically contiguous and many residents move from one place to other in connection with the works and shopping', he said 'variant policies in the tri-city' would not help curb the pandemic; the only way to restrict disease-transmission was to ensure that 'Mohali and Panchkula … follow the restrictions strictly'. This is particularly demonstrative of how any issue is managed relevant to the tri-city: three cities under three different management authorities trying to work together. In May 2020, Panchkula closed its borders to Chandigarh and Mohali to prevent cross-border transmission (*The Indian Express* 2020). It even stated that all employed in Chandigarh should find residence there since daily movement to and from Chandigarh was forbidden. However, realising the decision was untenable, it was soon withdrawn. At the time of writing, the lockdown and curfew are gradually being relaxed, with intercity movement being restored, albeit with some inconvenience to commuters between Panchkula and Chandigarh, where movement is allowed only after symptom screening. Nevertheless, the pandemic's occurrence strongly drives home the tri-city's strong inter-linkages and integration, and hence their interdependence.

Newspaper articles reported news pertaining to the pandemic in tri-city terms, with headlines like 'Railways seeks data on status of migrants stuck in Tri-city' (*Times of India*, 11 April 2020), 'Tri-city in lockdown till March 31' (*The Tribune*, 22 March 2020), 'Coronavirus lockdown in Tri-city: Shortages and hardships' (*Times of India*, 26 March 2020), 'First COVID-19 death in Tri-city' (*The Tribune*, 31 March 2020). Interestingly, the captions refer to the tri-city while the main body of news reporting focus solely on Chandigarh. Some even refer to the tri-city as 'Chandigarh tri-city': 'Crackers sound jarring note as Chandigarh tri-city lights up on PM Modi's solemn plea' (*Hindustan Times*, 6 April 2020); 'No positive case in Chandigarh tri-city after 6 days' (ibid); 'Chandigarh tri-city's first coronavirus positive patient recovers' (*Hindustan Times*, 7 April 2020). This again points to Chandigarh's dominance within the tri-city.

Conclusion

Chandigarh, Panchkula and Mohali are bound together as if by an umbilical cord: but for the first city's birth, the subsequent two would not have existed. As the chronological overview of their development from the 1950s to the present has shown, the dynamics of this tri-city relationship have fluctuated over time. The early years of co-existence were more prone to dysfunctional confrontation, while recently, economic competition started dominating their relationship with several transport-cooperation projects being put forward.

Divided over three jurisdictions, with different ethno-religious and linguistic composition, the three cities are nevertheless very interdependent. Contiguous with Chandigarh, both Mohali and Panchkula became vessels to absorb Chandigarh's spill-over. Both offered ample employment opportunities as well as residences and industries. With Chandigarh failing to provide accommodation to the residents, many shifted to the new townships in search of more affordable housing.

While Chandigarh has always been the leading triplet, Mohali and Panchkula gradually gained their own identities and specialisations. Mohali developed first as an industrial centre, thereby complementing Chandigarh. Now it has become an educational hub, a sports hub and centre for medical facilities in its own right. In turn, Panchkula developed high-quality residencies, warehousing infrastructure and private healthcare centres. In other words, over time the cities have increased their distinctiveness among each other and strengthened durability of their tripartite co-existence.

Comparing the tri-city with other Indian twin cities, some resemblances emerge. First, like other planned Indian twin cities, Chandigarh, Mohali and Panchkula have each been subject to comprehensive urban planning. Second, the tri-city has been incrementally increasing its complementarity over time. However, it has developed more due to competition than cooperation, with little or no tri-city consultation or premeditation. No wonder the major contrast between the tri-city and other Indian twins lies in the tri-city's shortage of planning beyond individual twins – at the supra- and inter-city levels. The limited number of tri-city planning activities, their modest results along with numerous disputes over joint planning initiatives testify that, so far, inter-city conflicts continue hampering collaboration efforts. The other feature distinguishing the tri-city from other intranational Indian twin cities (Garrard and Mikhailova 2019a) is an ambivalent relationship with the rural hinterland. While Chandigarh remains largely self-centred, Mohali has exploited the locational advantage available to the district (which surrounds the tri-city urban complex from three sides) and initiated large-scale urban development in this space.

As the COVID-19 pandemic has shown, tri-city coordination remains limited, in fact creating inter-city tensions and complicating their residents' lives. Our scrutiny suggests that the tri-city's only way forward lies in cooperating together to ensure sustained growth and development for the entire region.

Notes

1 After India's independence and Partition, the erstwhile state of Punjab was partitioned into East and West Punjab. East Punjab was that part of pre-partition Punjab which was given to India. Chandigarh thus was the capital of East Punjab. After the re-organisation of East Punjab, Punjab and Haryana, were the main successor states and the hill territories of Punjab were transferred to Himachal Pradesh.
2 GMADA was created by Punjab to prepare an Integrated Plan for an 1190 sq km peripheral region around Chandigarh.

References

Chhabra, A. (2017) *The Big Small Town: How Life Looks from Chandigarh*. S.A.S. Nagar: Unistar Books.

Garrard, J., and E. Mikhailova (2019a). Indian twin cities, in: Garrard J., Mikhailova E. (eds) *Twin Cities: Urban Communities, Borders and Relationships Over Time*, Routledge, London and New York. P. 104–118.

Garrard, J., and E. Mikhailova (eds) (2019b). *Twin Cities: Urban Communities, Borders and Relationships Over Time*, Routledge, London and New York.

India Today (2013) 'Haryana will get Chandigarh, Punjab can claim Lahore or Shimla, says a peeved Hooda'. [Accessed 21/08/2020]. Available: https://www.indiatoday.in/

india/story/haryana-will-get-chandigarh-punjab-can-claim-lahore-or-shimla-says-peeved-hooda-171617-2013-07-25

Indian Times (2016). 'Now Punjab and Haryana fight over ownership of Chandigarh Airport, after Punjab calls it Mohali Airport'. [Accessed 21/09/2020]. Available: https://www.indiatimes.com/news/india/now-punjab-and-haryana-fight-over-owner-ship-of-chandigarh-airport-after-punjab-calls-it-mohali-airport-261831.html

Kalia, R., (1987). *Chandigarh: The Making of an Indian City.* Oxford: Oxford University Press.

Mallot, J. E., (2012). *Memory, Nationalism, and Narrative in Contemporary South Asia.* New York: Palgrave Macmillan.

Metro Rail News (2018) Chandigarh now plans a Monorail, after dumping Metro rail project. [Accessed 20/12/2020]. Available: https://www.metrorailnews.in/chandigarh-now-plans-a-monorail-after-dumping-metro-rail-project/

Municipal Corporation SAS Nagar, (2020). About Us. [Accessed 13/05/2020]. Available: http://mcmohali.org/Aboutus.aspx

NDTV (2018). 'Chandigarh belongs to Punjab': Amarinder Singh rejects Haryana's plan. [Accessed 15/08/2020]. Available: https://www.ndtv.com/chandigarh-news/chandigarh-belongs-to-punjab-amarinder-singh-rejects-haryanas-plan-1881180

Sarin, M., (1982). *Urban Planning in the Third World: The Chandigarh Experience.* New York, NY: Mansell Publishing.

Singh, G., Kahlon, S. and Chandel, V.B.S., (2019). Political discourse and the planned city: Nehru's projection and appropriation of Chandigarh, the capital of Punjab. *Annals of the American Association of Geographers* 109(4), 1226–1239.

The Indian Express (2020) 'Haryana's Panchkula to seal borders with Chandigarh and Punjab'. [Accessed 28/08/2020]. Available: https://indianexpress.com/article/cities/chandigarh/haryana-panchkula-chandigarh-punjab-border-sealed-exemptions-corona-virus-lockdown-6389097/

The Pioneer (2013). Officials of UT, Haryana, Punjab to talk on details. [Accessed 08/09/2020]. Available: https://www.dailypioneer.com/2013/state-editions/officials-of-ut-haryana-punjab-to-talk-on-details.html

The Tribune (2020). 'Punjab, Haryana agree to complete ring road around Chandigarh'. [Accessed 25/08/2020]. Available: https://www.tribuneindia.com/news/chandigarh/punjab-haryana-agree-to-complete-ring-road-around-chandigarh-30295

Times of India (2009). Panel seeks united Tricity development. [Accessed 10/09/2020]. Available: https://timesofindia.indiatimes.com/city/chandigarh/Panel-seeks-united-Tricity-development/articleshow/4511485.cms?from=mdr

Times of India (2018). 'Cattle menace: Tricity coordination committee fails to see light of day' [Accessed 20/09/2020]. Available: https://m.timesofindia.com/city/chandigarh/cattle-menace-tricity-coordination-committee-fails-to-see-light-of-day/amp_article-show/65437553.cms

Vale, L. J., (1988). *Designing national identity: Recent capitols in the post-colonial world.* M.Sc. Thesis, Massachusetts Institute of Technology.

6 Embryonic twin cities

Reggio Calabria and Messina in Italy

Dario Musolino and Luigi Pellegrino

Reggio Calabria and Messina are two important South Italian cities on the Strait of Messina divided by just a few kilometres of sea. Their proximity might bring Italian people to see them as twin cities. Historical commonality in a unique region may add to the impression.

However, the current situation is not that of twin cities: socio-economic interactions are limited, and institutional coordination is lacking. Many characteristics typical of twins as outlined in volume 1 (Garrard and Mikhailova 2019a) are either weakly present or missing. The Strait is crossed daily by flows of passengers and freight, but mostly it is transit traffic, with supra-local origin and destination. Nowadays, even local people seem uninvolved in the integration idea.

Twin cities can integrate on three dimensions (Meijers et al. 2014): functionally, institutionally and culturally. The functional dimension relates to relationships, i.e. mutual interactions in terms of flows of people (e.g. commuters), goods, resources, etc., thanks particularly to interconnected transport networks and services. Transport is fundamental: improvement here can make independent twins 'highly integrated as a daily urban system' (Metrex 2010, 38). However, functional integration cannot be achieved unless the two city structures are initially complementary, in fact a 'double-town' model, rather than a 'town couple', thereby efficiently completing each other (Buursink 1994). The institutional dimension instead concerns cooperation and coordination between institutional actors, ranging from occasional ad hoc agreements (e.g. cultural events) to full inter-municipal cooperation, even joint public services/utilities and common bodies. The third cultural dimension concerns generating a common identity for the twinned area.

Different dimensions can be given different intensities. Cardoso and Meijers (2020) emphasise that the tighter and deeper the integration, the wider and bigger are the resulting socio-economic advantages in terms of agglomeration economies and economic productivity, possibly overcoming potential negative returns deriving from excessive concentration of activity in a single, large area. Full integration in functional, institutional and cultural terms is the ideal. However, this does not necessarily mean that twin cities should amalgamate: this might produce 'loss of local urban identity and the dissolution or erosion of borders (e.g. not only physical) between twin cities', as observable, e.g. with

DOI: 10.4324/9781003102526-7

Figure 6.1 Italian twin cities.

Source: Map by Marion Planque and Ekaterina Mikhailova.

Buda and Pest in Hungary, and Chatham and Rochester in the UK (Garrard and Mikhailova 2019a, 15).

Reggio Calabria and Messina are still far from integration along any of these three axes; thus, they are far from drawing advantage from it. Nonetheless, their integrative potential is strong, based particularly on their complementarity: the key reason we call them embryonic twin cities. Italy has no internal twins of its own, though its borders partly accommodate an external twin with Switzerland (Como/Chiasso) and Slovenia (Gorizia/Nova Gorica) which has been studied (Gabrovec 2013; Lipott 2013). Only recently have Reggio Calabria and Messina been seen as twinned possibilities (Figure 6.1) (Musolino 2018; Pellegrino 2019), with characteristics potentially rendering them 'engineered twin cities' observable on some European borders (e.g. Copenhagen-Malmo), or the many coordinated planned Indian twin cities (Garrard and Mikhailova 2019a). In fact, as evident later, their twin-city potential depends fundamentally on improving transport networks and services in the Strait of Messina, required not just to shrink distance, but also to make it easy, safe and efficient to move between the two urban areas.

Reggio and Messina: two important cities with fluctuating relationships

Reggio Calabria and Messina sit on the Strait of Messina, in Southern Italy. Reggio Calabria on the eastern side, in Calabria; Messina on the western side, in Sicily. The Strait's narrowest part is about 3.2 km, while the two city centres are about 12 km apart. Passenger-transit time is 10–35 minutes, depending on the port used; Reggio Calabria-Messina is most used by commuters.

Table 6.1 Demographic and economic size of Reggio Calabria and Messina

	Population		Firms		Employees	
	Total (th.)	*% (regional)*	*Total (th.)*	*% (regional)*	*Total (th.)*	*% (regional)*
City core (municipality)						
Reggio Calabria	178.7	9.3	9.9	9.0	24.8	9.0
Messina	229.6	4.6	13.6	5.0	34.4	4.8
Together	408.3	5.9*	23.5	6.1*	59.2	5.9
Metropolitan area						
Reggio Calabria	541.3	28.1	30.0	26.3	70.3	25.6
Messina	620.7	12.5	38.8	14.3	96.3	13.3
Together	1,162.0	16.9*	67.8	17.8*	166.6	16.7

* On the total of Sicily and Calabria. *Source:* Elaborations based on Istat data (dati.istat.it); Population: 2019; Firms: 2011.

Each is important demographically and economically; they are similarly sized (see Table 6.1). Together, they have about 408,000 inhabitants: as a single city, they would become Southern Italy's third largest urban area, after Naples and Palermo. Including the wider metropolitan provinces of Reggio Calabria and Messina brings the total to 1,162,000, potentially Italy's 16th largest administrative region.

The two municipalities, and their metropolitan areas, host numerous firms, almost 68,000 at the metropolitan scale, accounting for about 18% of all Calabrian and Sicilian firms. In employment terms, the municipalities collectively total about 59,000, while both provinces host about 166,600 (17% of Calabria's and Sicily's total).

Both municipalities are provincial, though not regional, capitals. Reggio Calabria has Metropolitan City status, one of Italy's ten according to the national law. Messina is also a Metropolitan City, though based on regional rather than national law.

Considering their economic weight and 'institutional role', we could say that these cities are at the same level in the national economic and institutional hierarchy, and not characterised by 'apparent subordination or dominance' (Garrard and Mikhailova, 2019a, 16), with one city dependent on or out-punching the other.

Nature created a natural barrier between these cities, which is rare for twin cities (even embryonic ones): the sea. Many are river-divided, but few oceanically, other than those like Copenhagen-Malmo, Helsinki-Tallinn and (until 'Brexit') Dover-Calais, separated by considerable distance and whose togetherness is produced by recent fast-transit developments.

During their long history, the sea did not prevent the two sides from flourishing in interactive ways, as historians have shown (Gambi 1965). In the Greek age, for example, under the tyranny of Anaxila, there was created what might now be called a joint urban area composed of Rhegion (currently, Reggio) and Zancle (currently, Messina). During the Byzantine age, the two cities became very important for silk production and trade. However, since the 16th century, negative events like earthquakes (most recently in 1693, 1783 and 1908) and the plague (1743) have all helped limit the frequency and intensity of relationships. In addition, the South Italian economic decline, starting from Italian unification in 1871 (the average Southern GDP per capita declined so much that currently it is about half of the average GDP per capita in Central and Northern Italy), probably helped to reduce the economic activity and the business interactions between the two sides of the Strait of Messina.

Nevertheless, Messina and Reggio Calabria have had dynamic and fluctuating relationships throughout their history as observed with other twin cities (Wilbraham, 2019). From the late-19th century ferries started operating between them. After the Second World War, mutual interactions further intensified, into the 1960s (for example, around 40 hydrofoil journeys connected the two cities, while today there are about 20!). This caused Gambi (1965) to define the cities as a region (*Regione dello Stretto*) independent of Sicily and Calabria. For him, the sea became almost a river, on whose banks two urban fabrics developed.

However, from the 1970s, integration slowed again and began reversing. There are no analyses of why. Two possibilities tentatively suggest themselves: first, we can suppose that, since the end of the 1980s, macroeconomic factors, like the decreasing current and capital (investments) government spending, particularly in Southern Italy, probably affected the two urban economies, which are strongly dependent, directly or indirectly, on the public sector (Cersosimo 1993). Second, the well-known presence of organised crime is a likely factor. It was particularly evident in Reggio Calabria, and extremely oppressive and violent in the late 1980s, with the mafia war in Reggio Calabria killing about 700 people. This probably frustrated economic and entrepreneurial initiative, and economic and social dynamism, within the two cities and regions. Another explanation was the increasing duplication of some urban functions/institutions, which we will shortly explore.

Nowadays, few see the two cities as a unique region, let alone twin cities. In the new millennium, integrating Reggio Calabria and Messina, while raised occasionally by local actors, has been completely absent from regional and national debate. Almost no one inside or outside these towns sees them as twins thereby precluding one of key identifiers of twin-citydom as defined in volume 1 – self-identification and identification by others (Garrard and Mikhailova 2019a, 4).

Two cities scarcely integrated, in institutional and functional terms

Reggio Calabria and Messina are independent. They have no common governance at supra-urban and regional levels; municipal policies are rarely coordinated, and their interactions are low. They are embryonic twins, meaning they are at an initial, potential, stage of integration.

Furthermore, they and their respective provinces belong to different regional governments, whose status is differently defined in the Italian constitution: Sicily is an autonomous region with a special statute, while Calabria manages with an ordinary statute. The establishment of the two Metropolitan Cities, part of a long-lasting, and not always coherent, process of fragmentation and decentralisation, started in the 1970s with the reform of the ordinary regions. It is also affected by different institutional and regulatory frameworks. The central government's most recent local government reform (L. 56/2014) established Metropolitan Cities in ordinary regions; in Special-Statute regions, this institution is a prerogative of regional government.

Reggio Calabria and Messina municipalities in recent decades have designed and followed consistently independent urban-development strategies and policies. They made no attempt, nor apparently wanted, to coordinate. Nor was there any broader institution empowered to enforce coordination.

Moreover, until very recently, there has been no integrative push from central or regional government. Attention in recent decades has mostly focused on constructing the Messina Bridge, a project of supra-local relevance aiming to stably connect Sicily and the mainland. The long national political debate on its construction actually started in the 1980s, and probably overshadowed the integration issue in national public opinion, as some have argued (Ziparo 2016). Furthermore, some studies suggested that the two cities would benefit little because, given its location far from the two urban cores, there would not be a significant reduction in terms of generalised transport costs for travelling between them compared to ferries (Advisor 'Collegamenti Sicilia-Continente' 2001). Overall, the lack of political will affected all governmental levels. Some scholars identify an 'institutional wall' between the two sides.

Moreover, Reggio Calabria and Messina are distant not only in institutional but also functional terms. Overall, the two urban areas and their citizens work, live, behave independently of each other, hardly qualifying them for the twin-city label (Garrard and Mikhailova 2019a, 14): 'alongside governmental action or inaction, much of the reality or otherwise of "twinning" is created by the responses and behaviour of ordinary citizens in their day-to-day mutualities'.

Figures about commuting flows are highly relevant, showing the limited interactions between the cities (Musolino and Pellegrino 2020). The percentage of commuters crossing the Strait compared to total commuters in these municipalities and provinces is extremely low: cross-Strait commuters' number under 5,000. Moreover, in recent decades, they have declined in absolute and percentage terms. Inter-province commuting during 1991–2011 shrank by about 45%. Most cross-Strait passengers are people journeying between Sicily and the mainland: total liner passengers in transit at Messina in 2019 were around 10,700,000 (Autorità di Sistema Portuale dello Stretto 2019).

The two socio-economic urban systems have developed independently in recent decades, without cooperation or synergies in either public or private service sectors. As the 'institutional distance' analysis mentioned earlier shows, public actors did not coordinate at all. This is evident in the universities. The Mediterranean University of Reggio Calabria was founded in 1968, much later than the 16th-century University of Messina. In recent decades, it greatly expanded its student numbers and has achieved excellent results in research. However, instead of focusing on disciplines untaught in Messina, and following a synergic and complementary development strategy, it added new departments and courses duplicating courses taught in Messina. Before then, Messina was the university favoured by Reggio Calabria law students, many commuting daily across the Strait.

Health services provide another public-sector example. In 2016, Reggio Calabria opened a high-level cardiac-surgery centre. This occurred despite Messina possessing better-quality health services, and being historically specialised in the health sector with an important University hospital, already offering cardiac surgery.

Private-sector integration looks equally poor. Cross-Strait investment flows (i.e. business ownership and control) are extremely low. Messina-owned subsidiaries located in Reggio Calabria and Villa San Giovanni municipalities represent 1.4% of all subsidiaries, while in the opposite direction, the percentage is 0.8% (Musolino 2018). This appears surprising considering the two urban economic systems' proximity and size.

Overall, decades-long low coordination and synergy among public and private actors have caused low and declining interaction between the two sides. Transport services, particularly across the Strait but also more generally within the two provinces, have hardly proved more coordinated. Figures and empirical evidence vividly illustrate their inadequacy. We are far from the optimistic and flourishing situation described over fifty years ago by Gambi (1965). Delfino et al. (2011) drew attention to three aspects:

- Low daily frequency. Maritime passenger trips connecting the two core urban areas account only for 13% of total daily trips. Meanwhile, maritime freight trips, connecting other cross-Strait ports but far from the core urban areas, are very frequent (254 of 291 daily trips);
- No night-passenger trips. Again, maritime freight transport connections between the two ports outside the core urban areas (Villa San Giovanni, in Calabria, and Tremestieri, in Sicily), do include night trips;
- Poor physical, modal and fare integration, both regarding their own urban and metropolitan transport networks and between the two metropolitan transport systems. Only 5 of 463 road public-transport lines cross the Strait, as almost all road public-transport lines connect places within the two provinces.

This points to how the two cities developed duplicating, rather than complementary, systems, thereby reducing inter-city interaction and coordination. Meanwhile, both central government policies and maritime market structure probably also hampered development. Note, for example, the steadily declining financial resources flowing from the central government to Southern Italy since the 1990s (Svimez 2019), especially to the Mezzogiorno municipalities, following territorially unfair allocation criteria. This reduces Reggio and Messina municipalities' chances of strengthening their public-transport systems (ibid.). Note also the monopolistic structure of maritime-transport services crossing the Strait (Fera and Ziparo 2016; Nuzzolo 2001), thereby hampering possible rival players, especially regarding passenger transport. These Strait transport-system shortcomings help to explain the limited interactions using a supply-side approach.

Potential for integration: the complementary nature of the two areas

The analysis thus far shows little developmental coordination between the two cities. Nonetheless, as we now argue, closer examination reveals their potential for partnership (Musolino 2018): their economic bases have complementary characteristics; similarly their natural historical and cultural resources, and their physical geography. All this could enable the sort of formal twinning observable among many Indian twin cities enabling exploitation of complementary distinctiveness (Garrard and Mikhailova 2019b)

Table 6.2 focuses on quantitative indicators about sectoral specialisation – location quotients.[1] In employment terms, each city has some specialisations. Reggio Calabria has sectoral specialisation in wholesale and retail trade, as well as transport, especially road-transport services. By contrast, Messina specialises in manufacturing, tourism and health.

Exploring each sector in depth, the impression of complementary distinctiveness becomes clearer. One example is tourist attractiveness (e.g. landscape, cultural and natural resources). Messina seems more specialised here, but both

Table 6.2 Location quotients of the Reggio Calabria and Messina economies (sectoral data at the provincial scale; 2017)

	Reggio C	*Messina*
Mining	0.82	**1.13**
Manufacturing	0.86	**1.10**
Power supply, gas, air conditioning	0.82	**1.13**
Water supply, sewage, waste management	0.92	1.06
Construction	0.92	1.06
Retail and wholesale trade	**1.10**	0.93
Transport and logistics	**1.30**	0.79
Accommodation services, restaurants	0.85	**1.11**
Information and communication services	0.81	**1.14**
Finance, banking, insurance	0.98	1.01
Real estate	0.77	**1.17**
Professional, scientific, technical activities	1.00	1.00
Rental, travel agencies, business support services	1.03	0.98
Education	1.07	0.95
Health and social assistance	0.84	**1.11**
Arts, sport, entertainment	1.00	1.00

Source: Elaborations on Istat data (database ASIA).

sides have distinctive attractions. Reggio Calabria possesses the only mountain area (Aspromonte) with the national park around it. Messina has the Eolian Islands, with highly developed tourism, a volcanic area (Etna) and the famous and spectacular coastal town of Taormina. Both sides are rich in 3S tourism (Sun, Sand, Seaside), but Sicily's side is more developed and highly attractive to foreign tourists. Meanwhile, less seasonal green and mountain-tourist assets are mostly on Calabria's side. Aspromonte has the only ski resort, one of Southern Italy's biggest.

Cultural resources show similar complementarity. The National Museum of Reggio Calabria is one of Italy's most important, hosting an archaeological collection from sites in Magna Graecia, including two very important Ancient-Greek 'wound statues', *Bronzi di Riace*. Messina's *Museo Regionale* instead houses art works from other ages, e.g. by artists like Antonello da Messina.

Overall, the Strait area's cultural and natural heritage is rich and diverse, capable of reaching widely varied tourist demand. Yet tourist-sector governance and cultural/natural heritage management are not integrated, excepting some on-off cooperative inter-municipal projects organised in recent years (*Settimana della Cultura*). The relevance of 'hit-and-run' tourism, especially on the Calabrian side (Musolino et al. 1999), shows the negative effects stemming from the lack of joint strategy.

Transport shows similar shortcomings. Table 6.2 suggests there are more transport-related establishments and employees in Reggio than in Messina, but specialisations, functions and infrastructure are sufficiently diverse that they potentially possess a completely integrated system (Delfino et al. 2011). Calabria has the Strait's only airport, but this is under-exploited and badly connected with the Sicilian side, thereby losing passengers to Catania airport (Calabrò, Della Spina, & Viglianisi, 2016). Local politicians sometimes discuss improving the airport's Sicilian connections with innovative proposals (e.g. using a hovercraft), but it remains unresolved. Thus, the airport has low traffic, few flights connecting to main Italian cities, and no international flights.

The Strait has many ports, with rather diverse functions. On the Sicilian side, there is Messina, important for both freight and passengers crossing the Strait, which recently increased even cruise traffic (Autorità Portuale dello Stretto 2019); a small port dedicated only to crossing freight traffic (Tremestri-eri, officially part of the Port); and the port of Milazzo (oil and raw materials) on the Tyrrhenian coast, which is an industrial port serving the refinery and other heavy-industrial plants, and a port for passenger traffic with the Eolian Islands. Calabria has the Port of Villa San Giovanni, used for crossing passenger and freight traffic, and the small commercial port located in Reggio Calabria. It also has the transhipment port of Gioia Tauro: Italy's biggest transhipment container port, and one of the Mediterranean's biggest (it is a hub connecting regional and global networks, particularly serving traffic coming from Far East countries like China). The central government's 2016 reform of Italian Port Authorities changed the Strait's governance of its complex system of transport infrastructure. However, instead of creating a unique Port Authority for all the Strait's ports, the reform created two authorities: one competent for the Sicilian ports and those of Reggio Calabria and Villa San Giovanni; another for the port of Gioia Tauro. This decision actually emerged from institutional conflict among regional and central government levels, again revealing the difficulty in producing common interests between the two sides.

Another example is evident in manufacturing. Overall, and judging from Table 6.2, Messina looks somewhat more specialised in manufacturing. However, deeper inspection shows Reggio Calabria is more specialised in the railway industry (one of its few important manufacturing plants is owned by the multinational Hitachi, specialising in rolling stock). It also specialises in the food and beverage sector, and in the wood and furniture sector (related to Aspromonte's copious timber supplies). By contrast, Messina specialises in heavy industries (oil, chemical, plastics, etc.) and in the textile/clothing sector.

Therefore, on the one hand they have the potential to integrate, if seen in terms of sectoral complementarity and in both quantitative and qualitative terms. On the other, 'institutional distance' prevents exploitation of the advantages of such structural characteristics, while also enhancing competition and duplication, with mutual cannibalisation: essentially, a battle of the have-nots.

Nevertheless, some political decisions recently emerged, which deserve mention to complete our picture of integrative potentialities. In 2015 the two municipalities agreed to apply jointly to include the Strait among the UNESCO World Heritage Sites; so far, this has gone no further. Also in 2015, Calabria's regional government decided to constitute a Permanent Interregional Conference for policy coordination in the Area of the Strait of Messina, a representative body from Sicilian and Calabrian regional institutions, with proposing and advisory capacities. In 2017, Reggio Calabria and Messina's mayors formally agreed to cooperate over integration in several fields (like accessibility, mobility, tourism). In 2019, alongside both regional government heads (Sicily and Calabria), they signed another agreement to identify the integrated area's borders and jointly manage the entire integration process, though the body has yet to appear. July 2019 saw the first step towards fare integration in the Strait area: now, one public-transport ticket will cover both cities. All these decisions, except the last, appear intentional rather than enactive, yet to be followed by actual steps.

Integration's future: scenarios and strategies

In 2019, a Delphi investigation explored future integrative scenarios for Reggio Calabria and Messina, and the strategies and policies for pursuing this ambitious project in the light of the recent new initiatives. This utilised experts on the Strait (Musolino and Pellegrino 2020; Pellegrino 2019). The Delphi technique is a long-used qualitative research method based on repeated and systematic consultations, over several rounds, with an expert panel on a specific topic, aiming to bring out one or more dominant viewpoints (Turoff 1970). However, it has been little used in Italy. Italian geographical and social science literature is poor in studies using the technique, particularly concerning an issue like this.

Nevertheless, Delphi findings about future integrative possibilities seem encouraging. Experts in the first investigative round identified five possible areas for integration: (1) transport services; (2) utilities; (3) constituting a Metropolitan City of the Strait of Messina (e.g. joint municipalities for Reggio Calabria and Messina); (4) constituting a Region of the Strait of Messina (a joint province for Reggio Calabria and Messina); (5) constructing a stable physical connection (e.g. the Messina Bridge, or perhaps a tunnel).

However, when asked in the second round to rank these scenarios for likelihood, a predominant viewpoint clearly emerged. Transport-services integration (especially, maritime-transport services and urban public transport) was deemed most likely, followed by constructing a stable connection. The three other scenarios, although interesting and desirable, were deemed less likely. Moreover, transport-services integration is the scenario thought likely to emerge soonest: about 30% of experts think it might occur in under five years, and 80% in no more than ten. This is doubly interesting because, in Italy, the average time required for public works is quite long. Operationalising transport works, for example, on average takes almost five years (DPS 2014).

Delphi experts also deemed transport integration as the sector most worth pursuing to facilitate more general integration. However, they raised several relevant issues: the importance of increasing local people's awareness and consent; the need for political will and strategies at the central government level (e.g., they suggested central government should introduce territorial continuity for the entire Strait, not just for Sicily); the need to integrate other public services/utilities, like health, education, water and waste management, etc.

The Delphi experts particularly highlighted needs to increase daily frequency of maritime services; better coordination of local public-transport companies in the two urban areas, and between Calabrian and Sicilian railway services; also to improve fast maritime-passenger connection between Messina and Villa San Giovanni. Experts also wanted tariff integration between different modes of transport.

Overall, Delphi investigators saw transport services as the main bottleneck preventing integration. Based on a supply-side approach, experts saw this as crucial to giving integration new impetus, enhancing interactions and exploiting complementarities. Integration here was most easily achievable in the fairly near future. Enhanced-transport investments would be cheaper than investing in large-scale infrastructure and can be pursued without first building any joint institutional framework. Experts were sceptical about institutional integration, although they welcomed any relevant moves at municipal or provincial levels. Lastly, procuring local public support was seen as worth attempting, even if not most important.

Conclusions

Overall, Reggio Calabria and Messina are clearly not fully interdependent, interconnected twin cities. Rather they represent an initial stage of possible integration.

Interactions are poor, given the size of the two metropolitan areas, and have been decreasing in the long term. Moreover, Calabrian and Sicilian local institutions are 'very distant', at any geographical level; they seemingly do not have the willingness to truly proceed towards a joint institutional framework. Leastways, until recently. Recent initiatives, even if appreciable and necessary to prepare the legal framework for integrating the two cities, are hardly yet decisive.

However, there is much potential for inter-city integration. Analysing their complementarities highlights this. Their economic structure, their characteristics and assets are different enough for them to complete each other if integrated, taking advantage of widely ranging external and internal economies, thereby accessing all the economic advantages of being one big metropolis with almost half a million inhabitants, rather than of two smaller cities. This has not just enhanced economic advantages but also increased political clout derived from being an urban area as big and weighty as Palermo and Catania.

Furthermore, as the experts suggest, integrating Reggio Calabria and Messina is realisable in a fairly brief time frame, particularly if local governments focus their efforts primarily on strengthening transport services, including those crossing the Strait. Integration might be what is needed to reverse the seemingly unstoppable downward path these cities have followed in recent decades. Reggio Calabria and Messina have been declining steadily: their economy has weakened, and they have been losing relative importance, both in their regions (Calabria and Sicily) and in the Mezzogiorno and national contexts (Musolino 2018). Their competitiveness and attractiveness, aspects extremely important for local development, are very low, and they suffer from anomalous factors, negatively affecting economic development, like the Mafia's presence, as several studies have shown (Albanese and Marinelli 2013; Daniele and Marani 2011).

Therefore, integration could be a strategy, an historical opportunity, to bring these two cities out of the vicious circle they apparently inhabit. As other border cities show (Garrard and Mikhailova 2019a, chapters 12–14), integrating previously divided places can restore centrality to areas that had become peripheral in their national or sub-national contexts.

Notes

1 Location quotient (LQ) is an indicator generally used in economic geography to measure sectoral specialisation, usually based on the number of employees by sector. It is a ratio calculated for each sector, whose value can be or higher than 1 (meaning specialisation) or lower than 1 (de-specialisation). For example, in our case, as concerns Messina, the formula for measuring specialisation in sector *j* is as follows:

[(Employees sector j Messina) / (Total employees Messina)]
[(Employees sector j "Reggio C. + Messina") /
(Total employees "Reggio C. + Messina")]

We have highlighted in Table 6.2 the sectors where the LQ is at least higher than 1.10, meaning a significant specialisation. As far as the denominator is concerned, we have taken the sum of the two metropolitan areas ("Reggio C. + Messina") as relevant geographical scale, with which meaningfully comparing the sectoral specialisations of each of the two cities.

References

Advisor "Collegamenti Sicilia-Continente" (2001). Rapporto finale, The Executive Summary. Roma, 28.02.2001.

Albanese G, Marinelli G (2013) Organized crime and productivity: Evidence from firm-level data. *Riv Ital degli Econ* 18:367–394.

Autorità di Sistema Portuale dello Stretto (2019) Relazione annuale del Presidente.

Buursink J. (1994), Dubbelsteden, in: *Acta Geographica Lovaniensia, Leuwen*, Vol. 34, 175–180.

Calabrò, F., Della Spina, L., and Viglianisi, A. (2016) Il miglioramento dell'accessibilità per l'incremento della competitività dell'aeroporto dello Stretto: il contributo della cultura della valutazione. In: G. Fera, & A. Ziparo (eds), *Lo stretto in lungo e in largo, prime esplorazioni sulle ragioni di un'area metropolitana integrata dello stretto di Messina*. Centro Stampa d'Ateneo: Gennaio, 173–180.

Cardoso R., and Meijers E. (2020) Metropolisation: The winding road toward the citification of the region. *Urban Geography*

Cersosimo D. (1993), Un modello di economia dell'emergenza. In Mazza F. (eds), *Reggio Calabria. Storia, cultura, economia*. Soveria Mannelli: Rubbettino, 347–401.

Daniele, V. and Marani, U., (2011). Organized crime, the quality of local institutions and FDI in Italy: A panel data analysis. *European Journal of Political Economy*, 27 (1), 132–142.

Delfino, G., Iannò, D., Rindone, C., and Vitetta, A. (2011). *Stretto di Messina: Uno studio della mobilità intermodale per i passeggeri*. Villa San Giovanni: ALFAGI Edizioni.

Dipartmento per lo Sviluppo e la Coesione Economica (2014), *I tempi di attuazione e di spesa delle opere pubbliche. Rapporto 2014*.

Fera G., Ziparo A. (Eds.) (2016), *Lo stretto in lungo e in largo, prime esplorazioni sulle ragioni di un'area metropolitana integrata dello Stretto di Messina*. Centro Stampa d'Ateneo: Gennaio.

Gabrovec M. (2013), Open borders with uncoordinated public transport: The case of the slovenian-italian border *European Journal of Geography – Journal of the European Association of Geographers*, Vol. 4(3).

Gambi L. (1965) *Calabria*. In E. Migliorini (Ed.), *Le Regioni d'Italia*. Torino: UTET, XVI–568.

Garrard J. and Mikhailova E. (Eds.) (2019a), *Twin Cities. Urban Communities, Borders and Relationships over Time*. Routledge: London, New York.

Garrard J. and Mikhailova E. (2019b), Indian twin cities. In: Garrard J. and Mikhailova E. (Eds.), *Twin Cities. Urban Communities, Borders and Relationships over Time*. Routledge: London, New York, 104–118.

Lipott S. (2013), Twin cities: Cooperation beyond walls. The case of cross- border cooperation between Italy and Slovenia, *Paper presented at: Relocating Borders: A comparative approach, Second EastBordNet Conference*, 11–13 January, Humboldt University, Berlin, Germany.

Meijers E., Hoogerbrugge M. & Hollander K. (2014), Twin cities in the process of metropolisation, *Urban Research & Practice*, 7:1, 35–55.

METREX Expert Group on Intra-Metropolitan Polycentricity (2010), *Intra-metropolitan Polycentricity in Practice – Reflections, Challenges and Conclusions from 12 European Metropolitan Areas*. Glasgow: METREX.

Musolino D. (2018), Characteristics and effects of twin cities integration: The case of Reggio Calabria and Messina, 'walled cities' in Southern Italy. *Regional Science Policy & Practice*, 1–20.

Musolino D. and Pellegrino L. (2020), Le twin-cities dello Stretto e la prospettiva dell'area integrata: un approccio quali-quantitativo. *Rivista Economica del Mezzogiorno*, a. XXXIV, No.3, Bologna: Il Mulino.

Musolino D., Timpano F., Chilà M., Castrizio D., & Gattuso D. (1999). Ricerca sul turismo nell'area di Reggio Calabria. Rapporto finale. Consorzio InnovaReggio e Azienda di Promozione Turistica di Reggio Calabria, unpublished manuscript.

Nuzzolo A. (2001), Il sistema dei trasporti. *Meridiana*, No.41, 101–113.

Pellegrino L. (2019), Verso una bi-regional city? Un'applicazione del Policy Delphi alla realizzazione dell'area integrata dello Stretto. Tesi di Laurea (M.Sc. Thesis), Università della Valle d'Aosta, Aosta, Italy.

Schultz H. (2002), Twin Towns on the Border as Laboratories of European Integration, Frankfurter Institut für Transformationsstudien, *Discussion Paper 4/02*.

Svimez (2019), *Rapporto 2019 sull'Economia e la Società del Mezzogiorno*. Il Mulino. Bologna.

Turoff M. (1970), The Design of a Policy Delphi. *Technological Forecasting and Social Change*, vol. 2.

Wilbraham R. (2019), NewcastleGateshead: A Dynamic Partnership. In J. Garrard and E. Mikhailova (Eds.), *Twin Cities. Urban Communities, Borders and Relationships Over Time*. Routledge: London, New York, 51–64.

Ziparo A. (2016), Il nodo dei trasporti nell'assetto dell'area. In G. Fera, & A. Ziparo (Eds.), *Lo stretto in lungo e in largo, prime esplorazioni sulle ragioni di un'area metropolitana integrata dello stretto di Messina* (169–172). Centro Stampa d'Ateneo: Gennaio.

Part II
International Twin Cities
A. In Europe

7 Prussian border twin towns

The urban geopolitics of an amorphous territorial state

Thomas Lundén

Introduction: towns, territories and their borders

European towns were historically defined as enclosed territories with chartered rights, with dense populations encompassing different occupations; they served trading purposes and provided supportive services for surrounding areas – all as highlighted in central-place theory (Christaller 1933; Lösch 1940–1944). Town and countryside were distinguished by customs posts, where agricultural products and transit goods were taxed, or by walled-city gates, closed during plagues or sieges.

Urban location often resulted from both 'natural' and societal factors, the balance depending on available technology and type of dominant governance. Along seashores, trading posts becoming towns were located on bays, river mouths and other places conducive to shipping technology. Inland towns almost automatically developed at fords or bridgeable places along rivers, usually with centres on one side and suburban developments on the other. Urbanization could also depend on the location of bounded resources: minerals, energy etc, producing industrialization. But few towns appeared without political support, most spectacularly as new towns or strategically located urban fortifications.

Territories are areas enclosed by boundaries marking the end of a certain rule, or for independent states, sets of rules usually forming constitutions. Boundaries are definitions, lines separating regions in terrestrial space.

We might analytically distinguish three different urban-border locations:

- *'Nature'-originated borderization*. While borders defined by water divides or mountain tops usually exclude urbanization, rivers encourage it by delineating areas of local culture and cooperation (Ratzel 1903, 566). Kjellén (1916, 55f) added that trafficked rivers were badly suited to separating states, but, judging from the ongoing World War, he noticed them becoming 'natural trenches', predicting this experience could influence the eventual peace treaties by rehabilitating rivers as borders. He was wrong about 1918 (excepting the Rhine), but prescient about the Oder-Neisse line.
- *Plebiscite/referendum-originated borders*. The 1919 Versailles Treaty allowed some self-determination in territorial delimitation (but biased against Germany). Popular voting behaviour rarely directly reflected simple 'ethnicity',

DOI: 10.4324/9781003102526-9

'nationality' or 'religion' since plebiscites often occurred amidst uncertainty, pressurised by one or both contending states, even states-to-be.

- *Strategically determined borders.* The 1945 Potsdam Agreement defined the Oder-Neisse river border with its strategic exceptions and the resulting German-German divide as evident later in the chapter.

For state governments, territorially peripheral urban places have definite disadvantages: susceptible to external economic, social, even military influences from proximal territories beyond their legal and social reach. But this factor can become advantageous for neighbouring states. Fortified towns often emerged near enemy territory. Towns truncated by effectively impervious borders existed mainly for geopolitical reasons, varying with the timing of urbanization, and relative time-sequence: preceding, or resulting from, bordering (Lundén 2019). Urban-border location is thus a mixture of geopolitical influences. Border towns are explicable theoretically by different but interdependent social-science approaches:

- Towns located by authoritative decisions from (contending) states, either by one state compensating territorial loss or manifesting territorial advance (via fortifications).
- Existing towns bordered by imposed border-based ethnic or nationalist principles.
- Riverine towns (usually with cross-river counterparts) where the river defines the border.
- Border towns as markets, especially due to price-differentials resulting from domestic regulation.
- Industrial towns located in politically divided areas, especially related to mineral resources.

The geopolitics of territorial change in Prussia and Germany and its urban effects[1]

With rapidly changing Prussian/German territorial configuration, border towns were either the victims or the effect of new delimitations. Prussia had two contrasting origins. The eastern part, at the Baltic Sea's south-eastern corner, once home to Baltic-speaking Prussians but Germanized under the German Order, came under Polish dominance. However, in 1618 through inheritance, it was attached to the Brandenburg Electorate with its Berlin-located Hohenzollern ruler. After the 1648 Peace of Westphalia, Brandenburg's territory almost reached Stettin, which became the capital of the Swedish Duchy of Pomerania. Several complicated territorial and organizational changes resulted in Frederick I in 1701 becoming king of Prussia, now encompassing all the kingdom's territories. The Prussian territorial state, existing formally from 1701 to 1871, formed a territorial conglomerate, mostly increasing over time by agglutinating smaller states, often only a capital and its hinterland, before Prussia's inclusion (as the main component) into Imperial Germany.

Almost by chance, some Prussian towns became situated on or near the boundary with another, often hostile, state. Some had, or acquired, trans-border twins. Following unification in 1871, Germany reached well-established neighbours in a powerful position. However, after 1919, its new borders became contested, often with irredentist demands. World War II's outcome almost totally changed Germany's border geometry. Furthermore, while the events of 1989–1991 only marginally changed the territorial configuration, they transformed the cross-border relations of existing border towns.

Border towns in Brandenburg and Prussia

The earliest examples of one-sided border towns facing Prussia are Demmin and Wolgast, on the Peene River, part of Swedish Pomerania from 1648. Like Stettin, Demmin was lost to Prussia in 1720, and Prussia annexed Wolgast in Swedish Pomerania (Danish, 1715–1720) in 1815. Anklam in 1720 became divided between Prussia and Sweden, as we shall see.

When Prussia annexed Silesia in 1742, Kattowitz was a small village near the Polish border, but it grew with industrialization, acquiring chartered urban rights in 1865. Poland's first partition in 1772 left Toruń (German: Thorn) in Poland, but separated from its hinterland in Prussia, and by Prussian blockade from its contacts with the Polish Gdańsk exclave and the rest of Poland. With this partition, Prussia covered East and West Prussia, South-Eastern Pomerania, Silesia and Brandenburg. It had remarkably few border towns, excepting Swedish Pomerania, plus the emerging Upper Silesian industrial towns bordering Poland, and on the Polish side, the small town of Wischowa (German Fraustadt), which Prussia annexed in 1795. The Polish border made an indent along the Weichsel (Wisła) River, leaving Toruń in Poland and Bromberg (Bydgoszcz) in Prussia 40 kilometres apart. In the north-east, Prussia covered the entire Danzig bay excepting the town of Gdańsk. In the second partition in 1793, Prussia acquired Toruń/Thorn.

After Prussia finally incorporated Silesia in 1795, new urban agglomerations emerged in Silesia near the border with Habsburgian Bohemia and Russian-Congress Poland. Gdańsk/Danzig was lost to Prussia but, in 1807, was declared a Free City under French supremacy, but then was included in Prussia in 1815. The Warsaw Grand Duchy, also established by France in 1807, had a few small towns, e.g. Rawicz near the Prussian border, while Chelmno (German: Kulm) on the Polish and Graudenz (Polish: Grudziądz) on Prussia's side of the Vistula formed a neighbourhood within 15 kilometres. After the Napoleonic Wars in 1815, Thorn became situated on the border near Russian-Congress Poland. The border-crossing was at Ottloschin (Polish Otloczyn) with Alexandrowka on the Russian side, about 10 kilometres away (Universität Oldenburg 2019). In Russian-Congress Poland only Kalisz (German: Kalisch) by the Prona River was close to the Prussian border.

With Rheinprovincz's formal inclusion in 1824 (already enacted by the Vienna Congress in 1815), Prussia gained a large territory around the lower-German part of the Rhine, bordering the Netherlands and Luxembourg, with Koblenz bordering the Duchy of Nassau and Aachen bordering the Netherlands. While

Aachen had no Dutch urban equivalent, its neighbouring coal-mining townlet of Herzogenrath had long faced Kerkrade in the Netherlands. Already in 1816, the Prussian and Netherlands kings signed a treaty defining the border along New Street (a still-existing delimitation). Five years later, 50 customs officers were operating notwithstanding, or due to, a vivid legal and illegal cross-boundary interaction, remaining until World War I (Ehlers 2000). The town of Eupen, 25 kilometres south of Aachen, with a long history of changing affiliation, was also included within Prussia, bordering Belgium on two sides.

With the Danish Duchy of Holstein's late inclusion in 1866, Prussia got a new border twin, Altona, suburb of the Hanseatic city of Hamburg. Altona historically had special rights for religious minorities and was socially very different from Hamburg (Lundén 2015, 243). This became the last annexation before the German Empire's establishment following France's defeat in 1871.

Imperial German, Weimar Republican and Nazi German border towns until 1939

The German Empire's foundation made former state borders inside the Empire obsolete. However, victory over France created a new border in Alsace and Lorraine, partly within industrial agglomerations; and the south German states' inclusion also incorporated several towns bordering Switzerland (Lörrach at Basle and Konstanz facing Kreutzlingen) and Austria (Freilassing facing Salzburg, and Passau without a twin). Apart from the very exceptional twin of Herzogenrath-Kerkrade, pre-World War I Germany had no real twins with the exception of Konstantz and Kreutzlingen, while Lörrach and Freilassing mainly served as cross-border suburbs of Basle and Salzburg.

Germany's defeat and the ensuing Versailles Treaty and plebiscites brought enormous territorial changes and new border-town situations. In the south-west, France regained Alsace and Lorraine. The Rhine became a Franco-German boundary, dividing Strasbourg from Kehl and depriving the city of half of its hinterland (Christaller 1933/1966 III:D). The German industrial Saar came under League of Nations mandate but was economically integrated into France, with the town of Saarbrücken bordering France, but the 1935 referendum quickly restored German control. Before that, the German town of Zweibrücken bordered the Mandate area. In the west, Germany lost Eupen to Belgium in a contested plebiscite (O'Connell 2013).

In the north, a plebiscite gave Denmark Northern Schleswig, leaving Flensburg in Germany (Klatt 2007). In eastern-mainland Germany, the Silesian industrial agglomeration became divided, leaving ethnic minorities on both sides. The partition made Katowice Polish while neighbouring Beuthen and Zabrze (in 1915 renamed Hindenburg) remained German. In the east a totally new geopolitical situation emerged. Following one of US President Woodrow Wilson's theses on national self-determination, Poland in the 1919 Versailles Treaty was granted a slice of land ('the Polish corridor') splitting Germany into two while Danzig was put under League of Nations protection as a Free City (Morrow and Sieveking 1936).

Lösch (1940/1944) notes that the 1920 plebiscites in Ostpreußen (East Prussia) separated ethnic German towns from ethnic Polish hinterlands, suffocating them. Schneidemühl (c20,000) and Marienwerder (12,000) ended up bordering the Polish Corridor from the west and east respectively. The Ostpreußen enclave bordered Poland and Lithuania plus the autonomous Free Town of Danzig, formally Polish, but with a strong German majority. The new delimitation left several towns on or near the international boundaries, but only two emerging border twins: Tilsit-Panemunė and Danzig-Gdynia. The East Prussian village of Eydhtkuhnen bordering Verzhbolovo in the Russian Empire was an important railroad junction, but insignificant in urban terms: not until the 1920s did increasing traffic render Eydhtkuhnen a full-fledged town, with Kybartai on the Lithuanian side. The insignificant town of Schirwindt, with Lithuanian Kudirkos-Naumestis (Vladyslavov in the Russian Empire) on the other side of the small river, had about 2000 inhabitants in 1944 but was totally destroyed and never rebuilt.

German territorial advances during 1938–1945 created a rapidly changing European political map, with new borders and border towns. The 1938 Munich Agreement allowed Nazi Germany to annex the Sudetenland after invading in October 1938, making the garrison town of Theresienstadt (Terezín) a border town with Czechoslovakia's remains before Nazi occupation in March 1939. After the joint German-Soviet assault on Poland in September 1939, Germany re-established the pre-1919 eastern territories but added a slice of Russian-Congress Poland, including Łódź, renamed Litzmannstadt, bordering the remains of western Poland which was put under special military administration.

Post-World War II: German urban borders and lost territories

World War II's ending in Europe in May 1945 created a new geopolitical situation, articulated in Potsdam 1945 but formally signed in the 1947 Paris Peace. Allied-occupied Germany was severely truncated in the east, dividing its Ostpreußen exclave between Poland and Russia, making the Soviet part a domestic exclave of the Russian SFSR. Poland's territory was moved westwards, including all of Silesia, Danzig/Gdansk, its preliminary western border facing the Soviet Occupation Zone and mainly following the Oder and Neisse rivers (Kruszewski 1972). Several river towns were thus divided between Poland and (Eastern) Germany, with important river-mouth exceptions.

Germany's division into four Allied-occupied areas, plus similarly dividing Berlin, developed into the split into two Germanies and two Berlins, the eastern part(s) becoming the German Democratic Republic, the western the Federal German Republic, with West Berlin formally under western Allied control. The German-German border had few, if any, border towns, Lübeck being closest, with divided Berlin (1961–1989) a very special geopolitical issue (Ute Gabanyi 2017).

The USSR's 1991 disintegration recreated an old border-town situation: the Kaliningrad region became an exclave of the Russian Federation, bordering Poland and Lithuania.

Some examples over time

Anklam: a divided border-river town in 18th-century Germany

Anklam is a small town in northern Germany, near the Peene River mouth into the Baltic Sea. Originally part of the Hanseatic League and located within the Duchy of Pomerania, it became part of Sweden following the 1648 Peace of Westphalia. In 1715 Sweden lost part of the area to Denmark and in 1720 to Prussia; the border being drawn in the Peene River (Figure 7.1). But the northern riverside, the Peenedamm, remained in Danish Pomerania (it was returned to Sweden in 1720). Both states had river rights. While the bridge was formally divided midway, Prussia claimed the whole as part of its fortifications (Meier 2008, 270–275).

While all merchants lived in the Prussian town centre, skippers, carpenters and shipyards were located on the Swedish side. Children from the Swedish side crossed the bridge to school. The commandant had an agreement with Prussia about consultations. Most problems were solved practically. Sweden had other worries with contention about fishing and fish-vending rights; these were legally decided in a room in the Dutch Mill housing Sweden's border guard. Swedish and Prussian guards stood 'only six steps apart'; even during wartime, civilians could pass unhindered (Gross 1998, 277). In 1815, Prussia acquired Swedish Pomerania, but the internal administrative division and some legal rules remained for several decades (ibid., 280).

Figure 7.1 Anklam, depicted by the Swedish officer G.H. Barfot in 1758 showing the town on the Prussian side and Peenedamm on the Swedish side of the river.

Source: Archive of Timmermansorden, Stockholm, by permission.

Tilsit/Sovietsk–Panemunė: changing allegiances and border relations

Until 1919, Tilsit was located in German East Prussia on the Memel with the suburb of Pogegen across the river. Tilsit became a border town in 1919 (Lachauer 1995, 29) because the Versailles Treaty (X: 99 art. 28) put the Memel area with Pogegen under international control and French administration, Tilsit remaining in German East Prussia. One urban geographical study in 1935, strongly influenced by Nazi vocabulary, described the division along the river as 'a cut by brutal force', destroying commerce and transport (Kirrinnis 1935, 103). In 1923 Lithuania annexed the Memel area, producing substantial German migration into Tilsit (ibid., 197f.). Germany and Lithuania concluded a treaty on small border traffic after several years of conflict, effective from December 1928.

In March 1939, after Nazi German pressure on Lithuania, Germany annexed Memel and acquired both riversides. Following the 1945 Soviet victory, the Nemunas River became a domestic Soviet boundary between the Lithuanian SSR and the Kaliningrad exclave of the RSFSR. With Lithuanian independence in 1991, Sovietsk, with c43,000 inhabitants, became a border town again, but already in March 1991, after the unilateral Lithuanian independence declaration in March 1990, the *Panemunė* side was equipped with customs officers (Lachauer 1995, 39). With Lithuanian independence recognized in September 1991, Sovietsk and *Panemunė* became international border towns. German Eydhtkuhnen, renamed Chernyshevskoye, now serves as a small border town opposite Lithuanian Kybartai. A new bridge was planned linking *Panemunė* with Sovietsk, beside the Queen Louisa Bridge, but was delayed until December 2020, when it was ready for opening (Vylegzhanina 2020).

Danzig and its unequal twins: a border solution to a geopolitical dilemma

Following Poland's first partition in 1772, Prussia annexed all of Danzig Bay, excepting Gdańsk, a Polish exclave divided into two areas separated by Prussian territory. While important for Poland as a Baltic outlet, it was under heavy Prussian pressure via customs inspections and intimidations. To erode Danzig's position as a market and handicraft centre, Prussia rapidly founded the 'Immediatstadt Stolzenberg' just south of the border, based on four small suburbs, totalling around 8000 inhabitants. Königsberg-to-Berlin postal connections were redirected to avoid Polish Gdańsk territory (Loew 2011, 141). When Prussia annexed Danzig in 1793, the towns became domestic twins. After Napoleon invaded Prussia, the 1807 Peace of Tilsit designated Danzig a Free City, including Oliva, Stoltzenberg and the Hel Peninsula. Prussia annexed the area in 1814–1815.

Following the 1919 Versailles Treaty, Danzig with its hinterland (including the seaside resort of Zoppot) came under League-of-Nations protection as a Free City, locally ruled but with Poland controlling functions like foreign affairs, military and customs. Free City status left three urban settlements in border situations:

- *Marienburg* (Polish: Malbork) in Germany (Westpreußen, later Ostpreußen), c35,000 inhabitants in 1935 with a small suburb on Danzig's side of the Nogat River.
- *Tczew* (German: Dirschau) in Poland's Corridor bordering Danzig territory: c16,000 inhabitants in 1921, with a 30% German minority.
- *Zoppot* in Danzig (Polish: Sopot) with approximately 30,000 inhabitants in 1930, bordering the landed estate of Kolibki (German: Kolibken) in the Polish Corridor which in 1935 was annexed to the town of Gdynia.

Overwhelmingly ethnic German, with a small but largely Germanized Kashub-speaking minority, Danzig was hardly a trustworthy place for Poland's maritime relations. So the insignificant hamlet of Gdynia was expanded to become a maritime outlet, acquiring town status in 1926. Its population grew from approximately 2000 (1921) to 115,000 (1939), Gdynia (or rather, Kolibki, its small suburb) directly bordered Danzig's town of Zoppot, relations were hostile and legally complicated. Even before the Nazi takeover, the Berlin government and media alleged Polish policies were threatening and expansionist, involving claims to Pomerania; Danzig's government complained to its League of Nations protector that Gdynia's expansion menaced Germany (Post 2015, 31–32). As World War II started, Germany annexed Gdynia (as well as Danzig) renaming it Gotenhafen. Following the Potsdam Agreement 1945, the cities of Gdynia, Sopot and Gdańsk formed one Polish urban agglomeration.

Szczecin, Świnoujście, Police, Nowe Warpno and Stalinstadt: nation-state containment at the Oder

The Potsdam Conference on 1 August 1945 decreed that,

> pending final determination of Poland's western frontier, the former German territories east of a line running from the Baltic Sea immediately west of Swinamunde [*sic*] … thence along the Oder River to the confluence of the western Neisse River and along the Western Neisse to the Czechoslovak frontier … shall be under the administration of the Polish State and for such purposes should not be considered … part of the Soviet zone … in Germany.
>
> (VIII: B, section 2)

The agreement defined a preliminary demarcation (21 September 1945 in Schwerin), separating areas under Polish and Allied (i.e. Soviet) administration, left four German towns near the Baltic under Polish administration as exceptions from the Oder River–based definition: Stettin, Swinemünde, Neuwarp, and Pölitz. It was considered inexpedient to have the two sides share control of the ports and river access to the Baltic Sea, thereby leaving Russia an important strategic advantage.

Partially excepting Pölitz, the towns were rapidly made Polish and called Nowe Warpno, Police, Świnoujście and Szczecin. On the side remaining in Germany, only one small town, Löcknitz, plus the resorts of Heringsdorf/Ahlbeck, were within c10 kilometres of the new border. Nowe Warpno's twin, Altwarp village, was just 2 kilometres away across Stettiner Haff. During ensuing years, the border remained relatively closed, even after the June 1950 Poland-GDR agreement (Aischmann 2008, 192–193). With Świnoujście having a Soviet military garrison, even Polish citizens' access to Wolin and Uznam islands (including Świnoujście) was restricted until 1957 (Jędrusik 2013, 144).

The Pölitz/Police case is peculiar. For almost a year (4 October 1945 to 28 September 1946), Pölitz and surrounding parishes were actually an exclave of the Soviet Zone of Germany, bordering Poland on all sides. The exclave had German mayors belonging to the two recognized parties, KPD and SPD. Telephone, telegraph and postal connections were attached to Mecklenburg-Vorpommern. But most land connections with Germany were impossible, as the Polish authorities required visas that were rarely granted (Aischmann 2008, 135–149).

Upstream the Oder, the GDR East German government decided in 1950 to build a large ironworks at Fürstenberg, a small Oder town, selected apparently for its safe distance from the Western enemy. From 1953 to 1961 the new town was called Stalinstadt but then merged with Fürstenberg to form Eisenhüttenstadt (Nicolaus and Schmidt 2000, 47). Notwithstanding its river-border location alongside Poland, there was and is no local urban cross-river connection.

Divided towns on the Oder and Neisse rivers

The Potsdam agreement left several towns on the river border dividing Poland from Soviet-Zone Germany (GDR from 1949): Frankfurt an der Oder/Słubice, Guben/Gubin and Görlitz/Zgorzelec. As in other Polish-incorporated areas, the German population on the river's eastern side was forcibly ejected, alongside most symbols of German history. The border was effectively sealed until c1950, when the Zgorzelec Agreement facilitated formal contacts, under strict government control (Jajeśniak-Quast and Stokłosa 2000). In 1972 the border was opened for local inhabitants, producing strong local traffic bent on purchasing products unavailable at home. With the East European economic crisis and Poland's rising Solidarity movement, the border was closed again (Stokłosa 2014). The 1989 events in Poland and Germany, and German unification in 1991, gradually improved matters. Border controls eased, direct inspections ceasing in 2007 when Poland entered the EU. However, German restrictions on Polish workers remained until 2011 (Jańczak 2011, 43; Jańczak 2018; Szytniewski 2013).

At Frankfurt (Oder), the eastern-riverbank settlement, Dammvorstadt became Słubice once it came under Polish administration, then repopulated with people from across Poland (Jańczak 2011, 43). The towns have an organized cooperation, a successful cross-river bus line and Collegium Polonicum in Słubice resulting from cooperation between Frankfurt's Viadrina and Poznań University (Jańczak 2007, 88), and, since 2015, a common distant-heating network. The municipal diets regularly meet jointly with simultaneous interpretation facilities (Budde 2019). But social integration is low, notwithstanding some NGO collaboration, probably due to different public-involvement levels, and contrasting governmental systems (Poland being unitary, Germany federal) producing hierarchical asymmetry (Jańczak 2011, 44).

Further upstream, Guben/Gubin is the only town-pair where the old centre, with its town hall, was on the eastern, now Polish side. A 1999 newspaper article (Rothe 1999) reported negatively on social and economic conditions, with high unemployment, ageing populations, bad reputations for crime and racism, illegal immigration into Germany, smuggling and trafficking. But the towns have cooperated since 1991; a joint water-purification plant has been built, and a river-island has been developed into a meeting-place. Since 2015, municipal representatives meet regularly, and local events are announced on both sides. With EU Interreg funding support, the towns have developed the concept Eurotown (see Eurostadt Guben/Gubin).

On the Neisse/Nysa river, Görlitz and its ex-suburb, Zgorzelec on the Polish side had one well-guarded bridge under the GDR. As elsewhere along the border, local cross-border traffic was virtually non-existent, excepting 1972–1980. Following German unification, the pedestrian *Altstadtbrücke* bridge opened in 2004. An old railway bridge, the Viaduct bridge, originally for Dresden-Breslau (now Wrocław) traffic, reopened in 1957, run by Koleje Dolnoslaskie with three daily trains in 2019 from Dresden to Wegliniec c15 km east of Zgorzelec with further connections to Wrocław. Like the other two river-twins, the German twin suffered depopulation; but thanks to a generous donation, the town's old parts have been renovated and seemingly attract tourists and pensioners. Görlitz has one of Germany's highest levels of extremist-party voting and the lowest for the SPD (see Jańczak 2018's review of political asymmetry in the borderlands). Görlitz hosts one campus of the Hochschule Zittau/Görlitz, an applied-sciences university.

Conclusions

This chapter has explored the geopolitical reasons for border towns, particularly urban cross-border twins, using examples from Prussian and German history. Their territories have changed more than most European states, with boundaries shifting in location and in geopolitical, economic and ethno-cultural importance. The eastern borderland looks the most vulnerable and changeable area, but until 1945 the French and to some extent Belgian and Danish borders were profoundly contentious. In political geographical and geopolitical terms, there was at different times, particularly around 1919, vivid discussion about the 'best boundaries' from different perspectives (e.g. Brigham 1919). However, border location in relation to major towns was clearly not on the agenda. Kjellén (1916, 56) points out that the border mountains of the Erzgebirge (facing Austrian Bohemia) and the Vosges (facing France post-1871) are less favourable as German defences because of the difference in the slope between the neighbouring states, but adds (62) that a strong state like Germany can easily endure a bad border.

With the extremely rapid changes to Prussian territorial configuration, border towns were either victims or effects of new delimitations. However, the most striking conclusion from studying the area's maps is the lack of *urban* border situations. Brandenburg-Prussian expansion generally occurred via agglutinating smaller entities, each state usually consisting of a capital and surrounding hinterland. Border towns thus constituted exceptions, created by specific geopolitical conditions, and provide interesting insights into rules and traditions at different times. Cross-border relations varied according to levels of border openness, and how much difference there was between relevant states. Under the Brandenburg electorate and Kingdom of Prussia, many borders were with ethnic German duchies, some of which Prussia eventually swallowed.

Industrialization, especially increasing coal power, induced urbanization in certain borderlands. After German unification in 1871, border relations increasingly occurred with well-organized neighbours, and especially post-1919 involving irredentist demands, even overlapping ones, based on ethnicity. Two World Wars and their aftermaths totally changed the geopolitical situation, producing new border relations: sometimes interrupting contacts, only to be further changed by the Cold War's termination, German reunification and Soviet dissolution.

With German reunification, Denmark and Poland's EU accession and the Schengen Agreement, formal possibilities for cross-border urban contacts from Germany have increased. But as Jańczak (2007, 2011, 2018), Stokłosa (2014), Szytniewski (2013) and Balogh (2014) have noted, local interaction depends heavily on overlapping interests, e.g. the existence of kin-state minorities or needs for common political action to solve joint problems (e.g. in the natural environment). In the German-Polish case, there is no ethnic overlap (except for Polish people 'sleeping abroad but working at home', i.e. moving into neighbourhoods on the German side with empty or low-cost apartments due to Germans migrating westwards. Urban geopolitics is not only about legislation, but also about societal engagement.

Notes

1 Among other sources, my study of Prussian border twin towns relies on three historical maps: Karte der Freien Stadt Danzig (n.d.), Karte von Ostpreußen (1939) and The Century Atlas and Gazetteer of the World (1900).

References

Aischmann, Bernd (2008), *Mecklenburg-Vorpommern, die Stadt Stettin ausgenommen. Eine zeitgeschichtliche Betrachtung.* Schwerin: Thomas Helms.

Balogh, Péter (2014): *Perpetual borders, German-Polish cross-border contacts in the Szczecin area,* Stockholm: Meddelanden från Kulturgeografiska institutionen Nr 145, Stockholm University.

Brigham, Albert Perry (1919). Principles in the Determination of Boundaries. *Geographical Review* 7, no. 4, 201–219.

Budde, Vanja (2019). Frankfurt (Oder) und Słubice wachsen weiter zusammen'. *Deutschlandfunk* 22.05.2019. https://www.deutschlandfunk.de/europaserie-frankfurt-oder-und-slubice-wachsen-weiter.1769.de.html?dram:article_id=449438 [Accessed 2019-11-08]

Christaller, Walter (1933), *Die zentralen Orte in Süddeutschland*, Jena: Gustav Fischer [*Central Places in Southern Germany*. Englewood Cliffs: Prentice Hall, 1966].

Ehlers, Nicole (2000), De Muur van Kerkrade – Geschiedenis van een straat. In: *Geografie* 9/6/2000, 5–7.

Eurostadt Guben/Gubin. https://www.guben-gubin.eu/de [Accessed 2019-11-08]

Gross, Ralf (1998), 'Die Grenzlage Anklams 1715–1815', in *Geographische und historische Beiträge zur Landeskunde Pommerns*. (Eds) Ivo Asmus, Haik Thomas Porada und Dirk Schleinert. Schwerin: Thomas Helms, 276–280.

Jaješniak-Quast, Dagmara, Katarzyna Stokłosa (2000), *Geteilte Städte an Oder und Neiße. Frankfurt (Oder) – Słubice, Guben – Gubin und Görlitz – Zgorzelec, 1945–1995*, Berlin: Arno Spitz.

Jańczak, Jarosław (2007), Europeanization of trans-border Communities. The Polish-German case, [in:] Jarosław Jańczak (ed.), *Rediscovering Europe: Political Challenges in the 21st Century EU*, Poznań: Wydawnictwo Naukowe INPiD UAM, 77–88.

Jańczak, Jarosław (2011), 'Cross-border governance in Central European border towns', in Jarosław Jańczak (ed.) *De-Bordering, Re-Bordering and Symbols on the European Boundaries*. Berlin: Logos Verlag, 37–52.

Jańczak, Jarosław (2018), 'Symmetries, asymmetries and cross-border cooperation on the German–Polish border. Towards a new model of (de)bordering, *Documents d'Anàlisi Geogràfica*, Vol. 64, No. 3, pp. 509–527.

Jędrusik, Maciej, (2013), 'Usedom/Uznam. The political economy of divided islands', in Godfrey Baldacchino (ed.), *Unified Geographies, Multiple Politics*. Basingstoke and New York: Palgrave Macmillan, 137–156.

Karte der Freien Stadt Danzig (n.d.), 4.e verbesserte Auflage. Danziger Verlags- Gesellschaft Maßstab 1:100 000.

Karte von Ostpreußen (1939), *Maßstab 1: 300 000*.

Kirrinnis, Herbert (1935), *Tilsit, die Grenzstadt im deutschen Osten*. Tilsit: Sturmverlag.

Kjellén, Rudolf (1916), *Staten som lifsform*. Stockholm: Geber.

Klatt, Martin (2007), Fliessende Grenzen in einer Grenzstadt. Sprache, Kultur, gesellschaftlicher Status und nationale Identität in Flensburg des langen 19. Jahrhunderts', in B. Sruck, C. Duhamelle & A. Kossert (eds), *Grenzregionen. Ein europäischer Vergleich vom 18. bis zum 20. Jahrhundert*. Frankfurt/M/New York: Campus Verlag, 315–332.

Kruszewski, Z. Anthony (1972): *The Oder-Neisse Boundary and Poland's Modernization. The Socioeconomic and Political Impact*, New York, etc.: Praeger.

Lachauer, Ulla (1995), *Die Brücke von Tilsit, Begegnungen mit Preußens Osten und Russlands Westen*. Reinbek bei Hamburg: Rowohlt.

Loew, Peter Oliver (2011), *Danzig. Biographie einer Stadt*. München: C. H. Beck.

Lösch, August (1940–1944). *Die räumliche Ordnung der Wirtschaft*. Jena: Gustav Fischer [*The Economics of Location*. New Haven: Yale University Press, 1954].

Lundén, Thomas (2015), 'Geopolitics and religion – a mutual and conflictual relationship. Spatial regulation of creed in the Baltic Sea Region', *International Review of Sociology: Revue Internationale de Sociologie*, vol. 25:2, July 2015, 235–251.

Lundén, Thomas (2019), 'Border twin cities in the Baltic Area – anomalies or nexuses of mutual benefit?', in John Garrard and Ekaterina Mikhailova (eds), *Twin Cities Urban Communities, Borders and Relationships over Time Series: Global Urban Studies*, New York & London: Routledge, 232–245.

Meier, Martin (2008), *Vorpommern nördlich der Peene unter dänischer Verwaltung 1715 bis 1721: Aufbau einer Verwaltung und Herrschaftssicherung in einem eroberten Gebiet*. München: R. Oldenbourg.

Morrow, Ian F.D, assisted by L.M. Sieveking (1936), *The Peace Settlement in the German Polish Borderlands. A Study of the Situation To-day in the Pre-war Prussian Provinces of East and West Prussia.* London: Oxford University Press.

Nicolaus, Herbert & Lutz Schmidt (2000), *Einblicke. 50 Jahre EKO Stahl.* Eisenhüttenstadt: EKO Stahl GmbH.

O'Connell, Vincent, 2013. '"Left to their own devices" Belgium's Ambiguous Assimilation of Eupen-Malmedy (1919–1940)' *Journal of Belgian History.* 43 (4): 10–45.

Post, Gaines (2015), *The Civil-Military Fabric of Weimar Foreign Policy.* Princeton: Princeton University Press.

Ratzel, Friedrich (1903), *Politische Geographie oder die Geographpie der Staaten, des Verkehrs und des Krieges.* Second revised edition, München and Berlin; R. Oldenbourg.

Rothe, Frank (1999), 'Guben und Gubin - auf den ersten Blick trennt die beiden Städte nur ein Buchstabe', *Der Tagesspiegel* 10.12.1999. https://www.tagesspiegel.de/politik/guben-und-gubin-auf-den-ersten-blick-trennt-die-beiden-staedte-nur-ein-buchstabe/110052.html [Accessed 2019-10-27]

Stokłosa, Katarzyna (2014), 'The Border in the Narratives of the Inhabitants of the German-Polish Border Region', in Katarzyna Stokłosa and Gerhard Besier (eds), *European Border Regions in Comparison: Overcoming Nationalistic Aspects or Re-Nationalization?* London and New York: Routledge, 257–274.

Szytniewski, Bianca (2013) The dynamics of unfamiliarity in the German-Polish border region in the 1970s, 1980s, and 1990s, in eds. Arnaud Lechevalier, Jan Wielgohs, *Borders and Border Regions in Europe Changes, Challenges and Chances.* Bielefeld: Transcript 183–200 [Accessed 2020-01-11].

The Century Atlas and Gazetteer of the World, ed. (1900) J.G. Bartholomew. London: John Walker.

Universität Oldenburg (2019) https://ome-lexikon.uni-oldenburg.de/orte/danzig-gdansk [Accessed 2019-09-16]

Ute Gabanyi, Anneli (2017). 'The Berlin Wall' in A. Gasparini (ed.). *The Walls between Conflict and Peace.* Leiden: Brill, 161–182.

Vylegzhanina, Ulyana. (2020) The largest border crossing in Russia was built in the Kaliningrad region. *Rossyiskaya gazeta* 15.12.2020. https://rg.ru/2020/12/15/reg-szfo/v-kaliningradskoj-oblasti-postroili-krupnejshij-v-rf-pogranperehod.html [Accessed 2021-01-04]

8 Two generations of Eurocities along the northern section of the Spanish-Portuguese border

Juan-Manuel Trillo-Santamaría, Valerià Paül and Roberto Vila-Lage[1]

The Spanish-Portuguese border currently hosts various 'twin cities', especially along its northern section, embracing Galicia (Spain) and Northern Portugal (officially, Região Norte), commonly labelled 'eurocities'. These cross-border inter-urban projects have become possible in the middle term because of EU integration processes, but importantly also due to the long-standing shared history, culture, environment, society, language, etc., between both involved regions.

The term 'eurocity' intentionally mimics cross-border projects at other levels, e.g. euro-regions and euro-districts, the 'eu-' prefix indicating that Europe (and/ or the EU) is the relevant arena (Sohn 2018). A possible precursor is 'Eurode', a cross-border urban project between the Dutch municipality of Kerkrade and the German municipality of Herzogenrath, coined in 1997. However, there, the prefix 'eu-' is followed by a place-specific toponym '-rode' meaning 'reclamation of land' (Ehlers 2007).

The term 'eurocity' first appeared in the 1997 agreement between the Provincial Council of Gipuzkoa (Spain) and the inter-municipal community Bayonne-Anglet-Biarritz (France) to describe the so-called Basque eurocity which spans about 60 km. This is a cooperation framework for 25 municipalities, constituting an urban continuum between San Sebastián and Bayonne along the Biscay Bay coast. Later, the term migrated from the Spanish-French border to the Spanish-Portuguese border, where it gained momentum from 2007 (Figure 8.1). To the best of our knowledge, the Iberian Peninsula is the only place where the term 'eurocity' is now employed.[2]

As the term 'region' is applicable on both macro- and micro-scales, we assume a eurocity is a euroregion at the local scale where primary actors are towns, and ones not needing to be physically adjacent. What matters is the idea of building a joint project wherein local inhabitants are actively engaged (Trillo-Santamaría et al. 2015). In contrast to euroregions that often are transfrontier cooperation bodies with a legal personality, eurocities do not require a particular legal setting or any specific type of cross-border governance.

This chapter explores two eurocity waves in the Galicia and Região Norte area, which we see as two twin-city/eurocity generations. The first is based on classical bilateral agreements (the eurocity Chaves-Verín), while the second operates as a network of twin cities under the umbrella of one governing structure – European Grouping of Territorial Cooperation (EGTC) Rio Minho – led

DOI: 10.4324/9781003102526-10

Figure 8.1 Location of the Iberian eurocities.

Source: Map by Juan-M. Trillo-Santamaría, Valerià Paül and Roberto Vila-Lage.

by supra-municipal political bodies. Our study rests upon literature analysis and intensive fieldwork over recent years, especially the direct involvement of two of us advising and leading cross-border projects, meetings and research on the *raia húmida* (see later in this chapter).

The chapter divides into six parts. First, we introduce the Galician-Northern Portuguese border. The second describes the general background of cross-border cooperation across the whole region. Thirdly, the eurocity Chaves-Verín is analysed. Fourth, we examine eurocities in the *raia húmida* and EGTC Rio Minho developments. We conclude by comparing the two eurocity generations, with an epilogue framing the eurocities situation in the context of COVID-19.

A border with strong commonalities: an open path for cooperation

The Spanish-Portuguese border, commonly known as the *raya* (in Spanish) or *raia* (in Portuguese-Galician), dates from the 12th century, with Portugal's *de facto* independence following the battle of São Mamede (1128) and a self-proclaimed kingdom (1139). The Treaty of Alcañices in 1297 between Castile and Portugal is this border's first legal recognition, thereby underpinning the assumption that this is one of the world's oldest boundaries. However, as elsewhere in Europe, the boundary was not fully delimited and demarcated until the later 19th century, with the Treaty of Limits (1864), ratified in the General Act of Demarcation (1906) (Trillo-Santamaría and Paül 2014).

Before the 12th century, the Iberian Peninsula's north-western region – broadly, where Galicia and Região Norte are located today – was assumed to be a common land, first amalgamated under the Roman Gallaecia province and persisting into the early Middle Ages. Portugal's emergence did not end human interactions across that broad region. Generally speaking, analogous natural environments have framed similar cultural interventions, relating to parallel landscapes all along the current border area. In fact, the landscape differences observable today come from contemporary dissimilar trajectories on each side of the border, as regards spatial planning and environmental management.

The border has never prevented social interactions. Even during the 20th-century Iberian dictatorships (Salazar in Portugal 1933–1974; Franco in Spain 1939–1975), smuggling remained very important, and anti-regime activists maintained intense contacts. Historically, Galicians and Portuguese have commonly crossed for social purposes, like mixed marriages, joint celebrations, fairs or pilgrimages. Despite representing barely 20% of the Spanish-Portuguese boundary, the area hosts 6 out of 16 main Spanish-Portuguese border-crossing points, accounting for almost 44% of total average daily traffic (ADT) in 2016. Mostly these flows cross the international bridge Tui (Galicia)-Valença (Portugal), with an ADT of 15,015 vehicles per day (2016), representing 43% and 19% of all Galician-Portuguese and Spanish-Portuguese flows, respectively (OTEP 2018).

Social exchanges have been facilitated by a common language, somewhat unusual in border areas. Although socio-linguistically considered different languages, Galician and Portuguese are philologically and linguistically conceived as only one: the *raia* is home to transition dialects between Galician and Portuguese.

Despite this area's commonalities, it is usually divided between the *raia húmida* ('wet border') and the *raia seca* ('dry border'). The former is marked by the Minho River's final stretch (about 70 km), while the latter, much longer (about 225 km), is a land border. The *raia húmida*, located in an urban corridor connecting the westernmost Portuguese and Galician cities, hosts higher density population and more intense economic activities (Paül et al. 2020). Meanwhile, the *raia seca*, like most Spanish-Portuguese border areas, is a marginal rural area with declining population; there are only two urban exceptions: Verín (Galicia) and Chaves (Portugal).

Galicia-Região Norte cross-border background

Galicia and Região Norte have pioneered cross-border cooperation on the Spanish-Portuguese border and constitute a benchmark for other experiences. Although no bilateral political agreements explicitly mention the term

Table 8.1 Data of the case-study area (2018).

	Population (inhabs.)	Land area (km²)	Density (inhabs./km²)
Euroregion			
Galicia	2,701,743	29,577	91.3
Região Norte	3,572,583	21,286	167.8
Total	6,274,326	50,863	123.4
Chaves-Verín			
Chaves	39,345	591.2	66.5
Verín	13,817	94.1	146.8
CHAVES-VERÍN	53,162	685.3	77.6
Tâmega Eurodistrict*	110,551	3,814.9	29.0
Rio Minho EGTC area			
Tui	16,902	68.3	247.4
Valença	13,823	117.1	118.0
TUI-VALENÇA	30,725	185.5	165.7
Monção	17,902	211.3	84.7
Salvaterra de Miño	9,691	62.5	155.0
MONÇÃO-SALVATERRA	27,593	273.9	100.8
Cerveira	8,877	108.5	81.8
Tomiño	13,464	106.6	126.3
CERVEIRA-TOMIÑO	22,341	215.1	109.9
Wider areas			
Rio Minho EGTC	363,836	3,317.2	109.7
Lower Minho Valley area**	189,648	1,911.7	99.2
Vigo metropolitan area***	478,508	744.9	642.3

* Under discussion.
** Unofficial.
*** As officially defined in Galicia.*Sources*: www.ine.pt and www.ine.es (accessed 28/5/2020).

'euroregion', this cross-border area is commonly known as Galicia-North of Portugal euroregion (Table 8.1).

Galicia is an autonomous region, holding some devolved executive and legislative powers, while Região Norte is a deconcentrated administrative entity depending directly upon the Lisbon central government. Despite strong disparities in competences, both regions have found ways to cooperate since the early 1980s, acceleratingly so after both countries joined the European Communities in 1986. In 1991, the first Spanish-Portuguese cross-border body was created: the Galicia-North of Portugal Working Community.

At the euroregional level, an EGTC emerged in 2010. This new Galicia-North of Portugal EGTC (EGTC-GNP) does not replace the Working Community; the latter is assumed to be the political forum for discussion, while the former has become the technical branch to implement agreements. In 2000, several Portuguese and Galician municipalities formed an association entitled the Eixo

Atlântico (Atlantic Axis) that has become a Working Community member and generated multiple studies and publications, mainly on urban issues.

Local-level cooperation in Galicia-Região Norte also confronts other dissimilarities in the capacity of territorial authorities. Contrary to regional-level developments, Portuguese municipalities are stronger and much wider (in population and land area) than their Galician counterparts. The Portuguese Act 75/2013 allowed creation of four Inter-Municipal Communities in Portugal's northernmost region, all strongly willing to cooperate with their Galician counterparts. However, in Galicia (and Spain generally) such inter-municipal governance has been effectively forbidden since the Spanish Act 27/2013. In this sense, Galician provincial councils (two in Southern Galicia) have attained new roles in cross-border cooperation, somewhat contesting Galician government leadership in this agenda.

Alongside institutionalised cooperation, other cross-border agreements (e.g. between business associations and trade unions) are worth mentioning. Common social, sport and cultural initiatives and events are regularly held, like the radio station *Ponte nas Ondas*. Meanwhile, since 2002, a public foundation exists between Galicia and Região Norte's six public universities: the Euroregional Studies Centre (CEER). This participates in the Iacobus programme, managed by EGTC-GNP, which aims to facilitate and fund euroregional academic mobility rather as the 'Erasmus' exchange programme does for EU university students.

Beyond the EGTC-GNP, two other EGTCs have emerged, both directly related to eurocities: EGTC Chaves-Verín (2013) and EGTC Rio Minho (2018).

Eurocity Chaves-Verín: the forerunner of local border cooperation

Chaves and Verín are not located on the boundary, they are 22 km apart: Chaves is 8 km from the borderline, and Verín 14 km. They started institutionalised cross-border cooperation in 2007 when Verín joined the Eixo Atlântico Association and, as a part of its bid, foresaw a specific agreement with Chaves. A year later, a Strategic Agenda was presented (Domínguez 2008), with three main axes: simplified residents' access to services and facilities (called 'euroc-itizenship'), sustainable development and economic development. In 2014 the eurocity Chaves-Verín was legally created with ratification of the EGTC by both national governments. Its headquarters are located in the old border post. This re-purposed building is intended as a multi-purpose meeting place for border people, including rehearsals for musicians. Tourist information is also provided.

EU structural funds, namely INTERREG, have been obtained in 2007–2020 to sustain the eurocity projects and activities with various project leaders. Two were led by Verín, one by the EGTC Chaves-Verín. Meanwhile, two more projects on hot springs had the eurocity as a partner. During 2007–2020, the eurocity has organised multiple events, mostly around culture, sport, training, education and tourism, all intensively disseminated online (via websites, Facebook and Twitter). There has been a monthly common cultural agenda, alongside common tourist guides and tour packages, and participation in Fair Trades (e.g. FITUR in Madrid).

Among the projects since 2007, two are noteworthy: one on the eurociti-zen card, aiming to forge a sense of belonging to the eurocity, and the other on cross-border public transportation. The eurocitizen card is freely available for all Chaves-Verín residents, granting equal access to services (libraries, swimming pools, museums) on both sides of the border; it also entails discounts at some events and specific businesses. By 2018, almost 18% of the eurocity population had the card, with around 190 people using it each day. To facilitate mobility, the eurocity has taken the lead through a B-solutions[3] pilot project to launch regular passenger transport between Chaves and Verín. The service is free, running only on special days of cultural events and weekly fairs. It started in mid-2019 but was suspended in March 2020 because COVID-19 led to border closure. Before its launch, buses connected both cities to the boundary line, but without crossing it, and schedules were uncoordinated, due to the mismatch between EU, national, regional and local transport regulations.

A survey on eurocity mobility undertaken by eurocity technicians in 2019 shows the frequency of boundary-crossing: one to three times monthly, 65%; one to two times weekly, 10%; three or more times weekly, 25%. The survey also reveals that people cross for fuel, shopping and personal matters (25% each); alongside work and leisure/nightlife (12.5% each).

During these years, the eurocity has forged strong international connections, for example, joining the Iberian Network of Transfrontier Entities, the Associ-ation of European Border Regions or participating in the Open-Days/European Week of Cities and Regions in Brussels. Importantly, its work has been recog-nised at EU level: in 2015 the eurocity was awarded the RegioStar, and some policies have been considered benchmarks by Zillmer et al. (2018, 2020).

Since inception, the eurocity has attempted to enlarge its coverage area, implying that neighbouring municipalities are perceived as potential partners. This has been repeatedly discussed but without clear outcome. Under an ongoing INTERREG project, there is currently a study connected to a Strategic Agenda for the 2030 Eurodistrict Tâmega – the Tâmega being the river straddling the area. This eurodistrict's area might include, beyond the eurocity, five additional Portuguese municipalities and six in Galicia (Figure 8.1). Arguably, an EGTC at this level might replace the current eurocity (Ladeiras 2018).

The three eurocities of the *raia húmida*

The Minho River marks the international boundary between Spain (Galicia) and Portugal in its lower stretch until it opens into the Atlantic. Thus, a fortress system along the riverbank was constructed across the centuries by nation-states to control their respective territories. However, a transcendental symbolic recon-figuration has occurred in recent years. The fortresses have been reconceived as a cultural landscape attracting visitors, and have been reinterpreted through the discourse of cross-border cooperation and European integration.

Until the 1990s, there was only one river bridge: the 1886 railway and pedes-trian bridge connecting Tui and Valença. These two cities were the first to estab-lish a eurocity on the *raia húmida* in 2012. As new EU-funded bridges emerged

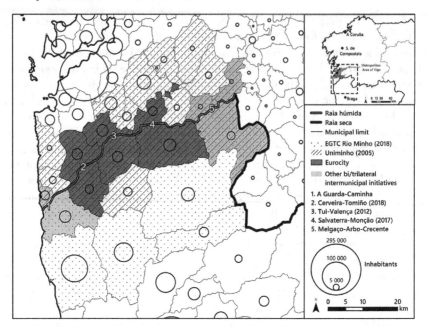

Figure 8.2 EGTC Rio Minho area.

Source: Map by Juan-M. Trillo-Santamaría, Valerià Paül and Roberto Vila-Lage.

in the 1990s, new cooperative forces have apparently flourished, with two other eurocities appearing: Cerveira-Tomiño and Monção-Salvaterra (Figure 8.2). A recent survey shows mobility is basically weekly or fortnightly, for shopping (39%) or leisure (36%); people not feeling this is an international experience, but simply moving nearby (Carballo 2020). The main crossing points for border people are, in order of importance, for Cerveira-Tomiño, Monção-Salvaterra and Valença-Tui: 60% of respondents stated they had friends on the other side; while 32% had relatives (Carballo 2020). Another survey, specifically for Tui-Valença, showed around 60% of Tui inhabitants crossed at least weekly, alongside 80% of Valença inhabitants (Pousa 2017).

The cooperation agreement creating the eurocity Tui-Valença was signed in early 2012 by the two mayors. This perceives the border not as a barrier, but an opportunity for socio-economic development. The objectives comprise the local economy, trade, institutions, technologies, tourism, society and education. By way of strong symbolism, the agreement's eighth anniversary (10 February 2020) was celebrated by both local government boards staging a common session in Valença, to be subsequently followed by similar annual sessions alternatively in Tui and Valença.

During these years, this eurocity has focused on two areas: sports (28 activities in 2018; 25 in 2019) and tourism. Notable are a eurocity travel guide (*Two Cities, Three Languages and Two People, United by a River, an Emotion and a Willingness*) and a tourist train linking both cities. Public institutions like Valença's swimming pool and Tui's theatre and music school are shared. More interestingly, Portuguese voluntary firefighters from Valença cross to fight fires in Tui when needed, with Tui's local government offering modest payment in exchange (Pousa 2017). This is justified because in Spain, firefighters are civil servants, while in Portugal, they are often voluntary. Hence, local people are organised to combat fires. In Tui, Galician fire stations are more distant than the Portuguese ones, so calling up voluntary Portuguese firefighters is faster.

The eurocity is well known to local inhabitants: around 91% know it exists and strongly approve of cross-border cooperation (4.4 out of 5 of the Likert scale) (Pousa 2017). This eurocity has run for years with funding from both local governments. Since 2017, an INTERREG project has been awarded to build the Strategic Agenda 2019–2021 and a planning document, *Future Eurocity 2030*; a eurocity website is planned. Tui and Valença also participate in another INTERREG project aiming to improve the river's tourism capacity. Moreover, the municipalities of the eurocity Monção-Salvaterra de Miño take part.

As early as 1985 Tomiño and Cerveira signed a symbolic 'Friendship Agreement'. Until 2004, the only way across the river was by ferry, but then the 'Friendship Bridge' linking both towns was inaugurated – there is currently a project aiming to build a pedestrian bridge around the so-called Friendship Park. In 2014, the mayors revived the agreement and launched the Strategic Agenda process. This was the final result of a participatory process taking account of stakeholder and civil-society opinions. There are four main cooperation areas: (a) reinforcing the Minho River as a resource (mainly for sustainable development and tourism); (b) enhancing cross-border mobility and territorial development; (c) developing common management of public services; and (d) economic development. An INTERREG project under the 2014–2020 framework has funded this Agenda's implementation in recent years. In 2018, Tomiño and Cerveira, following their counterparts, decided to become an 'officially' designated eurocity, Cerveira-Tomiño, the agreement being signed on the Friendship Bridge.

To facilitate the Agenda's development and promote cross-border public participation, cross-border ombudsmen (one from each municipality) were created: an initiative awarded a best practice award in citizen participation by the International Observatory on Participatory Democracy in 2019. Shared facilities include Vila Nova de Cerveira's swimming pool, and the Municipal Music School of Goián (Tomiño), equally available to people from both sides. The eurocity also promoted free sport activities in 2018 and 2019 with over 400 participants (e.g. hiking and a Friendship Triathlon). However, there are problems when school-student groups want to cross to participate in events and/or use services: being minors leaving the country, parents must sign specific legal permission. The ombudsmen have directly tackled this problem, addressing specific recommendations to local, regional, national and European institutions.[4]

Another notable initiative is participatory budgeting. Since 2016, the eurocity Tomiño-Cerveira can spend approximately €20,000 per year to fund cross-border projects proposed by partners from both sides of the border and selected by local votes. A website gathers all the information for this initiative: https://participac-erveiratomino.eu.

In 2018, the eurocity ran a citizens' survey on border issues.[5] Several points emerged: (a) almost 75% were aware of cross-border cooperation between Tomiño and Cerveira, with sports activities best known; (b) cross-border cooperation is very popular, with 95% assessing it as positive or very positive; (c) 95% see EU funds as an opportunity to improve life quality for border people; (d) 96.6% of respondents see the border as an opportunity; (e) 93% are positive or very positive about shared services.

The final eurocity created on the Portuguese-Galician border is Monção-Salvaterra de Miño. Its inauguration in March 2015 celebrated the 20th anniversary of the bridge linking both towns. Like the others, a cooperation agreement was signed. Although the agreement was supposedly about 'town-twinning', the term 'eurocity' has commonly been employed since the start in 2015, even in branding. Since early May 2017, the eurocity headquarters have been located in the former Portuguese border post.

So far, eurocity activities have been modest, mainly around tourism, culture and sport. Shared-service usage is envisaged and, as with Tui-Valença, Salvaterra's municipality will offer financial contributions to Monção's voluntary firefighters for any actions in its area. In any case, the lack of an INTERREG project supporting the management and funding of eurocity initiatives is perceived as an issue needing attention: the partners are seeking European funding.

EGTC Rio Minho

In 2006, a cross-border association consisting of an Inter-Municipal Community on the Portuguese side (with 6 municipalities) and the Provincial Council of Pontevedra (representing 16 of the Province's 61 municipalities) was created, and labelled Uniminho (Association of the Transboundary Minho River Valley). The three eurocity municipalities participated in this association, which was created basically to channel INTERREG projects. Uniminho was legally passed in the framework of the 2002 Spanish-Portuguese Treaty on Transboundary Cooperation.

Since the early 2010s, it was widely assumed that one or more EGTCs would be required for the *raia húmida* area to go beyond merely managing EU funds. Some eurocities internally discussed their institutionalisation as EGTCs, echoing Chaves-Verín's experience. However, the 2015 elections produced a new Provincial Council majority that impelled reinforcing cross-border cooperation via an EGTC beyond bi-municipal agreements, replacing the previous Uniminho. In that period, the Inter-Municipal Community had doubled its geographical scope (with the creation of the new Alto Minho Inter-Municipal Community); thus, the new EGTC includes municipalities not just within the Minho River basin but also within the Lima River basin. Accordingly, from the Galician side, the EGTC area covers several municipalities not strictly on the riverbanks but more functionally related to Vigo, clearly within its metropolitan area (Figure 8.2).

In 2017 an INTERREG project was granted to develop a Strategy for the future EGTC territory and in 2018 the EGTC was created and named Rio Minho. The Strategy was developed by the Euroregional Studies Centre CEER and its six founding universities between 2017 and 2019. Beyond developing different types of public participation and involving several academics in the process, we should mention the discussion held between the different scales within the EGTC to elaborate the Strategy.

Firstly, the inter-mayoral strategic meetings involved the three eurocities plus two other 'twin cities' not declared eurocities but understood to be strong cross-border cooperating neighbouring municipalities: Arbo-Crecente-Melgaço and Caminha-A Guarda, latterly with O Rosal (Figure 8.2). Interestingly, these 'twin cities' are tri-municipal rather than bi-municipal. Thus, a tension emerged between the scale of the individual 'twin cities' (including the eurocities), and the scale of the EGTC as a whole. In any case, most cross-border shared services considered essential for the future of the *raia húmida* in the Strategy passed in 2019 are at this bi-/tri-municipal level, not for the whole EGTC area.

Secondly, in the Strategy's first draft, available in early 2018, before most of the participation procedures, there was constant distinction between these municipalities of the Minho River valley, seen as experiencing close cross-border dynamics, and the other municipalities covered by the EGTC. After considerable discussion, the definitive Strategy does not distinguish between these 'two cross-border speeds'. Although this is a political agreement, from an academic perspective, its implications are uncertain given that the most intense cross-border dynamics occur in the strict Minho valley area. Thanks to intense participation procedures and political negotiations, priorities in the Strategy action plan changed significantly from the document's initial draft to its final version. This is the case, for example, for sustainable tourism, which changed from fourth priority in early 2018 to second in mid-2019.

Conclusions

Galician-Portuguese eurocities have developed and managed specific common projects and initiatives, mostly by means of EU funding, to bring people from both sides of the *raia* closer. As different surveys show, border inhabitants are both knowledgeable and empathic towards eurocities and local cross-border activities, and also towards cross-border neighbours. This contrasts with lower enthusiasm at the euroregional level (Trillo-Santamaría 2014). As usual in Europe, social, sport and cultural events are easier to manage and fund, leaving specific border problems, like cross-border spatial planning and common health services, still to be tackled. In any case, the way found in Chaves-Verín of offering a public cross-border transport service opens up future innovative solutions. This accords with the ongoing process of drafting new EU legislation to resolve legal and administrative obstacles in cross-border contexts.

In a context of well-established cross-border cooperation at the euroregional level for over three decades, Chaves-Verín represents a first generation of eurocities. Although originally supported by the Eixo Atlântico – Chaves-Verín recently abandoned the Association – this eurocity responds to the two municipalities' need to develop projects. At the very beginning, it lacked higher-level government support. In contrast, the three *raia húmida* eurocities represent specific local bilateral initiatives conforming to a wider territorial project embracing the whole Minho River valley, the EGTC Rio Minho. In this case, the EGTC has been strongly supported by supra-municipal governance structures on both sides, mainly the Pontevedra Provincial Council and the Alto Minho Inter-Municipal Community. In fact, the area's eurocities had originally sought to establish their own EGTCs, but this idea was finally abandoned once the EGTC Rio Minho was created. This specific layer of cross-border NUTS-3[6] scale governance offers a broader scope of projects and, importantly, funding for developing common projects. This is why we identify a second generation of eurocities.

In this regard, the EGTC eurocity Chaves-Verín has apparently accomplished its main objectives and anticipates expanding not only its territorial scope but also its partners. The bid for an EGTC Eurodistrict Tâmega might imply enlarging the area covered by the EGTC. We interpret this project as mirroring the EGTC Rio Minho, also because the new EGTC might imply superseding the previous EGTC eurocity. Hence, we would have a second example of this second generation of eurocities or, more accurately, of an eurodistrict. Importantly, this might imply including wide rural areas: highly depopulated and aged, and with particularly stagnant economies, causing a necessary reconfiguration of a currently urban-based agenda.

Looking at the two generations of eurocities through the lens of twin-city characteristics (Garrard and Mikhailova 2019), we admit that all Spanish-Portuguese eurocities, particularly the Minho River eurocities, are interdependent. All of them have strong territorial identities and in the foreseeable future will persist in their separate existences, as they belong to different nation-states. At the same time, their cross-border territorial branding initiatives have been highly successful and admired. This is thanks to the common language, culture and traditions of Galician and Portuguese people further strengthened with mixed marriages and cross-border friendships, cross-border work and regular cross-border visits for leisure or shopping. Finally, we find ongoing formal and informal negotiations to be the most conspicuous common feature of twin cities and eurocities. In this cross-border region, informal negotiations precede the formal ones, based on historical social practices all along the border. The formal negotiations have several co-existing layers – from the Galicia-North of Portugal euroregion to single eurocities – complementing each other.

Epilogue

Any kind of eurocity evolution on both *raias*, at least at the short- to mid-term, would eventually depend on post-COVID-19 crisis management. In fact, the crisis has been highly problematic in the border area, producing intense local-national tensions, because central-government decisions are usually at odds with border people's lives. In fact, Spain's declaration of the State of Alarm (14 March 2020) and the Portuguese State of Emergency (18 March 2020) accompanied the closure of the Spanish-Portuguese frontier (16 March 2020): the freedom of movement assured by the Schengen Agreement was suspended until 1 July 2020. Only nine border controls were maintained in the whole Spanish-Portuguese border, two in the Galician-Northern Portugal euroregion: Chaves-Verín and Tui-Valença.

The daily contacts of eurocity dwellers were suddenly suspended. Although cross-border workers could legally cross the boundary, a specific problem arose in eurocities whose bridge had been locked. This was the case with Cerveira-Tomiño and Salvaterra-Monção. Normal travel times of five to ten minutes could become one hour; 2 km driving could become 100 km. This naturally impacted mobility costs. In response, the EGTC Rio Minho mounted a specific study of COVID-19's socio-economic impact in the *raia húmida*.[7] It proposed measures to palliate the crisis, like opening more border crossings (eventually happening on 15 June 2020), at least for cross-border workers, and to coordinate the progressive end of lockdowns by an equivalent system of gradual re-openings: cross-border economic activities are strongly interdependent. In a more general perspective, the EGTC Chaves-Verín organised a virtual meeting of all Spanish-Portuguese eurocities to discuss pandemic-related problems and solutions on 5 May 2020. In a final document, they requested special status for these cross-border areas, including free movement and EU/INTERREG-specific funds to help recovery. Indeed, and partly due to lobbying from these cross-border entities during these months, Spain and Portugal signed a common trans-frontier agreement to deal with COVID-19 recovery measures in late August. And on 10 October 2020, at the 21st Spanish-Portuguese Summit, Spain and Portugal's governments approved a Common Strategy for Cross-Border Development. This Strategy aims to cooperate in five strategic areas: mobility; infrastructure and digital connectivity; education, health and social services; economic development and innovation; and sustainability and biodiversity conservation. For these purposes, they seek to obtain and manage European recovery funds. During these months, national and regional newspapers have publicised the effects of the crisis on borderlanders' labour and personal lives, thereby following eurocities' and EGTCs claims.

In the middle of the pandemic's third wave, with increasing infections and fatalities, Portugal decided to close the boundary again on 30 January 2021. Closed bridges, strict controls for border-crossing, and severe social and economic impacts, especially for cross-border workers, have re-appeared. Re-closing the international boundary, ironically without any EU-coordinated decision or a Madrid-Lisbon common strategy in March 2020 and again in February 2021, shows how fragile the eurocities are. Until the 'territorial trap' (Agnew 1994) is overcome, it seems impossible to de-centre decision-making from state capitals to their periphery. Analysing eurocities' resilience in face of central government decisions is an exercise of hope: at the edges of the nation-states, eurocities result from bottom-up politics and policies, gathering opinions, wills and energies from border stakeholders – both public and private actors – seeking to develop their own future for their own border regions.

Acknowledgements

Public servants from the EGTCs Chaves-Verín and Rio Minho are acknowledged.

Notes

1 R. Vila-Lage holds a predoctoral research grant from the Spanish Ministry of Science, Innovation and Universities with reference FPU18/04624; he was also granted a research stay in Northern Portugal funded by the IACOBUS programme.
2 In the Iberian Peninsula, the term 'eurocity' appears in four different languages: *eurohiriaren* (Basque), *eurocité* (French), *eurocidade* (Portuguese-Galician) and *eurociudad* (Spanish).
3 B-solutions is 'a 4-year initiative to tackle legal and administrative border obstacles/difficulties along EU internal borders' (b-solutionsproject.com). It is promoted by EU's Directorate-General for Regional and Urban Policy and managed by the Association of European Border Regions.
4 Available in English: https://eurocidadecerveiratomino.eu/wp-content/uploads/2019/04/20190207_ombudsman_report_recommendations_for_the_removal_of_the_barriers_to_child_and_juvenile_border_mobility_en.pdf [accessed 28.05.2020]
5 The survey is available at https://eurocidadecerveiratomino.eu/wp-content/uploads/2018/08/enquisa-servizos.pdf [accessed 28.05.2020].
6 NUTS is the French acronym for Nomenclature of Territorial Units for Statistics. As defined by Eurostat, this 'is a hierarchical system for dividing up the economy territory of the EU'. Regional policy is based on this classification. More information is at: https://ec.europa.eu/eurostat/web/nuts/background [Accessed 28.05.2020]
7 Unpublished document kindly sent to the authors by the EGTC Rio Minho. On 28 April 2020 the EGTC presented a declaration in this regard, sent to regional, national and European institutions. Available: http://smartminho.eu/declaracion_conjunta_aect/ [Accessed 28.05.2020]. Currently, the Galician Government is drafting a plan to promote local cross-border cooperation after COVID-19 impacts.

References

Agnew, J. (1994). The Territorial Trap: The Geographical Assumptions of International Relations Theory. *Review of International Political Economy*, 1, 53–80. DOI:10.1080/09692299408434268

Carballo, A. (2020). *A construcción dunha rexión transfronteriza: a raia húmida do Miño*. Unpublished Phd Thesis. Santiago de Compostela: USC.

Domínguez, L. (coord.) (2008). *Chaves-Verín: a Eurocidade da Agua*. Vigo: Eixo Atlântico do Noroeste Peninsular.

Ehlers, N. (2007). *The Binational City Eurode: the Social Legitimacy of a Border-Crossing Town*. Aachen: Shaker.

Garrard, J. & Mikhailova, E. (2019). Introduction and Overview. In Garrard, J. & Mikhailova, E. (eds.). *Twin Cities. Urban Communities, Borders and Relationships over Time*. Routledge: London, 1–20.

Ladeiras, A. (coord.) (2018). *Eurociudad Chaves-Verín: situación y perspectivas*. Fundación Galicia Europa. Available: https://www.poctep.eu/sites/default/files/0212_estudio.pdf [Accessed 28.05.2020]

OTEP (2018). OTEP Homepage: https://www.mitma.gob.es/informacion-para-el-ciudadano/observatorios/observatorios-de-transporte-internacional/observatorio-transfronterizo-espan%CC%83a-portugal [Accessed 28.05.2020]

Paül, V. et al. (2020). An Omitted Cross-Border Urban Corridor on the North-Western Iberian Peninsula?. In Santos H. et al. (eds.) *Science and Technologies for Smart Cities*, 27–35. Cham: Springer.

Pousa, A. (2017). *Eurocidade Tui-Valença: un espazo de cooperación emerxente no Baixo Miño*. (Unpublished End of Degree Thesis). Santiago de Compostela: USC.

Sohn, C. (2018). Cross-border Regions. In Paasi, A. et al. (eds.). *Handbook of Regions and Territories* (298–310). Cheltenham: Edward Elgar.

Trillo-Santamaría, J. M. (2014). Cross-Border Regions: the Gap between the Elite's Projects and People's Awareness. Reflections from the Galicia-North Portugal Euroregion. *Journal of Borderlands Studies*, 29(2), 257–273. DOI: 10.1080/08865655.2014.915704

Trillo-Santamaría, J. M. & Paül, V. (2014). The Oldest Boundary in Europe? A Critical Approach to the Spanish-Portuguese Border: the *Raia* between Galicia and Portugal. *Geopolitics*, 19(1), 161–181. DOI: 10.1080/14650045.2013.803191

Trillo-Santamaría, J. M. et al. (2015). Ciudades que cruzan la frontera: un análisis crítico del proyecto Eurocidade Chaves-Verín. *Cuadernos Geográficos*, 54(1), 160–185.

Zillmer, S. et al. (2018). *EGTC Good Practice Booklet*. Brussels: Committee of the Regions. Available: https://cor.europa.eu/en/engage/studies/Documents/EGTC-good-practices.pdf [Accessed 28.05.2020]

Zillmer, S. et al. (2020). *EGTC Monitoring Report 2018–2019*. Brussels: Committee of the Regions. Available: https://portal.cor.europa.eu/egtc/news/Pages/egtc-monitoring-report-final-study.aspx [Accessed 28.05.2020]

9 The past upon which the future dwells

Lines and divisions in former Yugoslavia

Dorte Jagetic Andersen

To understand cross-border twin towns in former Yugoslavia, we should consider recent history. Partition in the 1990s, mainly along ethno-national lines, turned the region's administrative borders into state borders, dividing areas previously economically, socially and culturally well-integrated. Bosnia-Herzegovina is most influenced by partition, divided both externally and internally along ethnic lines; the Danube region is another area significantly influenced by partition. Cross-border twin towns could therefore be expected to emerge along both Bosnia and Herzegovina's outer borders and in the Danube area. Yet there are few: only Slavonski Brod/Brod and Kostajnica/Hrvatska Kostajnica, both on the Croatian-Bosnian border, and Zvornik/Mali Zvornik, on the Serbian-Bosnian border, qualify.

To help understand the relationship between recent partition processes and the area's twin towns, we must include a third player: the divided city. This term describes cities influenced by urban partition along ethnic lines, Jerusalem and Belfast being the best examples (Calame and Charlesworth 2009). Former Yugoslavia's border regions provide fruitful soil for such divisions (Castan Pinos and Andersen 2015), more so currently than for twin towns. Mostar, physically divided between Croats and Bosnians by the river Neretva, is a commonly recognised divided town in the region; yet much that characterises divided cities holds true also for Sarajevo, equally for Mitrovica in North Kosovo.

This chapter compares twin cities with divided cities as urban phenomena, drawing explicitly on two examples. The first exemplifies twin towns: Brod in Republica Srbska and the Croatian town, Slavonski Brod, towns divided by the river Sava. The second exemplifies a divided city: Vukovar, a Croatian border town located by the Danube. The three towns are only 80 km apart. Whether towns divide ethnically or inter-relate as twins is no historical accident; it depends on how relationships form, divisions are activated and thus on the potentialities in the local landscape for interaction and collaboration across divisions. Whether twin cities overcome inward orientation, becoming more open and empathic towards each other, or maintain cross-border divisions, including ethnic ones, obviously depends on multiple factors.

Central in the post-Yugoslav context is how divisions are conceived, constructed and problematised locally, including questions about whether and how far divisions are perceived necessary to peace and stability (McCall 2014).

DOI: 10.4324/9781003102526-11

Because of recent history, inter-group relations in much of former Yugoslavia are influenced by divisions. As Arieli suggests in Chapter 10 of this volume, 'While entrenched shared narratives contribute to societal resilience during conflict, in post-conflict periods their filtering function minimizes prospects for transforming perceptions of former cross-border enemies to neighbors and regional partners.' This enables discussion of the difficulties of overcoming divisions alongside the successes involved when ameliorative efforts work. Accordingly, the chapter investigates divisions by exploring symbols and memories in the urban landscapes of Slavonski Brod and Brod, and Vukovar. It shows the two locations experiencing very similar historical processes, including placement in the Austrian-Hungarian empire, experiences of early industrialisation, their respective positions in both the Kingdom of Yugoslavia and the Socialist Federal Republic of Yugoslavia and their intense involvement in recent conflicts. Nevertheless, divisions today articulate very differently in internally divided Vukovar from how they do in the twin towns Brod/Slavonski Brod. In Vukovar, asymmetry between ethnic groups influences everyday life because 'the right to the city' has become determined by 'a right to remember'; whereas in (Slavonski) Brod, asymmetries mainly concern economic and physical environments with less wartime-induced need to explicitly emphasise spatial ownership. This creates conditions for divisions to be, if not erased, then at least lessened as obstacles to cooperation and integration.

I will first outline the importance of ethnic divisions in former Yugoslavia showing how they articulate as 'lines', not only physically but also symbolically, materially and spatially. I then explore the two cases in turn, showing how symbolic, material and practical lines express in the urban landscapes. The chapter concludes by comparing the twin towns with the divided town, thus exploring how regional differences impact divisions in former Yugoslavia.

Asking questions about twinned, doubled and divided towns in the context of the former Yugoslavia immediately raises questions about how national identity is articulated and handled within urban settings. Aggressive nationalisms and partition of national groups reflect both the nature and consequences of the area's 1990s conflicts (Lampe 2000). Based in an ethnonationalist ideal of a world territorially divided into ethnically homogenous states, national communities became defined in terms of their monoethnic origins with in-born rights to authority over a territory's political, economic and social affairs (Connor 1993)[1]. Hence, the conflicts spurred general distrust towards 'the ethnic other',[2] distrust not just utilised by states to legitimise ethnic cleansing but underpinning civil society and everyday existence (Castan Pinos and Andersen 2015). Making matters worse, challenges posed by ethno-nationalism were, paradoxically, combatted with the self-same logic, first producing the problems because the international community most commonly used partition to resolve the area's conflicts. Differences would be sorted by confining them, not only geographically but also in terms of loyalty and belonging. State-mapping and consequent separation of differences were carried out along ethnic lines, Bosnia and Herzegovina being the most prominent example of such 'peace amelioration' (McCall 2014, 52).

Even if lines drawn in the mid-1990s were geopolitical constructs, they are today lived locally, influencing local populations' everyday lives. When investigating divisions in the region's urban landscapes, our real question is thus how the lines articulate in people's everyday existence. In, 'Lines, Traces and Tidemarks' (Green 2018), Sarah Green suggests we think about borders as 'lines' appearing as 'traces'. Green defines a trace as the absence of something formerly present, a material remnant indicating something which, though not visible, still provides tangible, often material, evidence of what was once there (2018, 77). Additionally, 'trace' indicates time passing in ways geometric lines cannot; cutting through both space and time, referring to a past as well as affecting current everyday lives.

Therefore, the traces metaphor helps understand how lines on maps can appear in material form, thus helping us 'think about the entangled relation between symbolic, material, and legal forms' (ibid, 70). We know this 'power of absence' from processes of identity formation. Forming identity depends not on the presence of 'the other', but on the presence of an image of the other who is absent, 'there' not 'here', an absence marking 'who we are', without necessarily ever having been so (Hall 1996). Connecting the logic of identity-markers to the border as a line exposes their importance also for national and ethnic divisions of the logic of absent-presence. The line understood as trace is used to put things in their rightful place, to carve out distinctions used to place things, including people, into categories of here and there, in and out, us and them, thus providing means to sort out national and ethnic belonging.

Because of recent conflicts influencing the area, urban inhabitants' everyday practices in both cases (i.e. who people interact with and how they move in urban space) connect with 'a social memory' centring on the sorting of ethnic belonging. Lines of division are given special meaning through the notion of *narod*, which is simultaneously a political (national) and a cultural (ethnic) identity. Bosnia and Herzegovina excepted, each independent state has its dominant *narod* with an exclusive right to power in the nation-state. Even when the notion distances identity somewhat from traditional notions about ethnicity (Kolstø 2016, 6), *narod* has helped legitimise the existence and importance of lines of divisions locally, making them almost impossible to erase.

Hence, everyday practices stabilise, and a geographical place is made in the urban landscape via ever-present traces reminding people of the lines of division, sorting and confining them into spaces where they can move, thus making maps they may inhabit. Lines dividing ethnic groups, whether state borders dividing two cities or divisions internal to the city, thus refer to traces of what was once 'made in walking'; yet the line itself settles the activity, 'the walking', in a map, spatially sorting differences. Lines of old are thereby relived in people's movements and discussions, delimiting space for themselves and others. Memories of conflicts and divisions can resonate among later generations living with and learning to understand each other, each other's motives and intentions by recalling traces of the past, which they then perform into being in the here-and-now.

Understanding twin and divided towns in former Yugoslavia, we must therefore also understand how inhabitants' urban daily lives are significantly confined by spatial practices: drawing lines, making places, mapping out ethnic groups. Because conflict and mistrust are still not entirely forgotten, trust in the map remains important alongside physical movements, and in urban social memory. Living in ethnically divided environments is in other words like living out a 'map', performing lines of division telling us where things 'should be located'. However, there is no recipe for which spaces and maps are made; rather, the distance between lines on the map and how they are interpreted allow for multiplicity. As de Certeau (1985) states:

> History begins at ground level, with footsteps ... Of course, the walking process can be marked out on urban maps in such a way as to translate its traces (here heavy, there very light) and its trajectories (this way, not that). However, these curves, ample or meagre refer like words, only to the lack of what has gone by. Traces of a journey lose what existed: the act of *going by* itself.
>
> (129; emphasis in original)

As will be evident shortly, social memories mapping space along ethnic divisions and how such mapping 'occurs' is crucial for towns' ability to twin, double or divide.

In *The Bridge over the Drina*, Ivo Andrić tells us about the Mehmed Paša Sokolović Bridge in Višegrad in Bosnia. It bears silent witness to the enormous geopolitical changes this region has experienced while also facilitating passage for the great mix of populations crossing it daily. Hence, the bridge, with its crossings, comes to symbolise continuity and stability in an otherwise discontinuous and unstable everyday existence, beginning when the Ottomans constructed it in the mid-1600s through to its partial destruction during World War I.

The twin towns of Brod and Slavonski Brod are also connected by a bridge, the Savski Bridge (Figure 9.1) linking Bosnia and Herzegovina with Croatia across the river Sava. This bridge is younger than the Mehmed Paša Sokolović Bridge and does not constitute the same architectural wonder. However, for many years, the history of Slavonski Brod, the Croatian town, was told with an eye to the symbolism of bridges. As Romic et al. state, 'because of its strategic role in connecting the newly conquered territory of Bosnia and Herzegovina to Austrian-Hungary [sic] where both towns were located from 1878–1929, it was known as "Bosnia and Herzegovina's door"' (Romic et al. 2017, 36). Like Višegrad, the two Brods experienced a turbulent history, symbolised in multiple name changes. The first Brod, established on what is today the Sava's Croatian side, was simply 'Brod', but as the Turks built a fortress on the other riverbank, this became Turkish Brod, making the other Brod into 'Brod on the river Sava'. Later, Turkish Brod became Bosanski Brod, and in 1994, during the conflicts, it became Srpski Brod. However, after the conflicts, Bosnia's constitutional court found the name unconstitutional and in 2004 the town reverted to being Brod, as both towns had been in socialist Yugoslavia.

Given heavy influence from historical change and power struggles, especially between Austria-Hungary and the Ottomans, one might expect the narrative

Figure 9.1 The Savski Bridge connecting Slavonski Brod and Brod. Photo: Wiki Commons.

of conflictual divisions to be more dominant in the twin towns. However, the townscape has been and remains influenced by attempts to articulate connection, among others by promoting the image of the bridge as a door, an opening for interaction.[3] When this story of connectedness is told, trade connections are fore fronted, not only between the two towns but within the Balkans as a whole.[4] Another such symbol, now to the wider world, is the river Sava, connecting the towns to the entire South-Eastern European region. The river is arguably the main reason Slavonski Brod became important industrially in both the first and second Yugoslavia; also under-pinning both towns' considerable urban development under Austria-Hungary: 'Due to their favourable geographical-political position, the buoyancy of the Sava river and the road-rail bridge ... built in 1879, Slavonski Brod ... became a large centre of trade, crafts and industry' (Romic et al. 2017, 36).

Despite clearly existing, lines of division are imagined by the majority in the two towns as something which can be overcome, or at least crossed daily. Before the more recent conflicts, Slavonski Brod and Brod also complemented each other quite well. Already, from the late 19th century and during industrialisation, they developed in parallel. As Romic et al. (2017) conclude, 'double town' would best describe their relationship before the Bosnian conflict: 'Apparently this term not only denotes pairs of binational border towns which turned into one town, (Buursink 2001, 16) which Brod and Slavonski Brod were during Yugoslav socialism, but it also has no connotation that implies any similarity between the towns' (34). Despite the administrative and ethnic borders coinciding, the river shared no memory of ethnic conflicts or trauma; even with modest cooperation and exchange, the urban settlement's two parts co-existed peacefully, each providing for the other.

During the Bosnian conflict, 1992–1995, both connectedness and differences between Slavonski Brod and Brod became important because of their border location between Serbian paramilitaries supported by the Yugoslav People's Army (JNA) and Croatian equivalents supported by the international community. The area suffered considerable casualties (similar to Vukovar's); yet, when the Bosnian conflict escalated in 1992, the international community had already recognised Croatian independence. Despite JNA occupation of areas within today's Croatia (including Vukovar, see later in this chapter), it became a refuge from the Bosnian conflict. Slavonski Brod became a hub for Bosnian refugees fleeing the war via the Sava. Many refugees stayed, some also having a Croatian-Catholic background and thus being eligible for Croatian citizenship.

After the conflicts and the break-up, the relationship entered a different phase. Despite potential complementarity, we now have an area with hard borders, articulated through direct promotion of old and new historical trauma, and thus lines of division emphasising a conflictual history. Slavonski Brod has maintained its population, partly because of the wartime refugee influx from Bosnia, and is today Croatia's sixth biggest urban area; together with Brod, it has around 80,000 inhabitants. Moreover, the town managed to integrate somewhat successfully into the new post-socialist socio-economic structures. It has profited from Croatia's European Union accession, both during and after accession, drawing on EU structural-development funds. Brod, on the river's Bosnian side, suffered considerably during the conflicts, not least because of the Sijekovac massacre where the HOS, a Croatian paramilitary organisation, attacked a Serbian Orthodox church. Brod experienced depopulation until the bridge's reconstruction in 2000, which re-enlivened Brod but far less than its cross-river twin.

Hence, the conflicts emphasised an asymmetry already detectable during early industrialisation. Slavonski Brod/Brod may thus be considered part of what Schultz calls 'the third wave of twin towns':

> After war and partition, bridges were detonated, cross-border rails demolished, common public services disrupted. The ethnically-segregated population of these border twin cities became the outrider for prejudices and hostile stereotypes against the neighbours, living with their backs to each other.
>
> (quoted in Romic et al. 2017, 36)

Romic et al. (ibid.) use the term 'ex-double town', considering how the towns complemented each other much better before the conflicts than subsequently:

> The twin towns located on the border between Bosnia and Herzegovina and ... Croatia, which used to be an almost unified urban area and local community situated in the inland part of Yugoslavia, suddenly became

border towns/divided towns. The mutual cooperation of their institutions, organisations and citizens were hindered or stopped not only because of the transformation of border lines – from open towards closed model – but also because of political and armed conflict between Serbs and Croats, the two dominant ethnic groups in that area.

(32)

One could expect that inter-town asymmetry, alongside their new physical separation, would emphasise lines of division also in local narratives. However, this seems not to be the case (cf. Romic et al. 2017).[5] Rather, the bridge still connects actually and symbolically, reminding inhabitants both of their belonging and the fragility of peaceful coexistence. The consequences of recent conflicts are now also reflected in the bridge's story: destruction during the Bosnian conflict and rebuilding in 2000, thereby becoming an international border. Interaction has become harder across the closed border and cross-border interaction at official level is limited, yet the bridge remains symbolically open; asymmetry results from geopolitical changes rather than locally-articulated divisions. Locals quite commonly cross the bridge, particularly youngsters from Brod who go to Slavonski Brod, the bigger town with more possibilities for cinema visits, shopping, or other consumer activities (Romic et al. 2017). However, as yet, cross-border interaction seems to have gone little further.

Like Slavonski Brod and Brod, Vukovar was an important industrial town in socialist Yugoslavia. Around the size of Brod before the war, the town and surrounding areas were part of socialist industrial-development plans during these years, foremost because of its location by the Danube and the Borovo Company's presence, located on an industrial site outside Vukovar, producing shoes for the entire Yugoslav state. Vukovar's 19th-century and early-20th-century history also resembles Slavonski Brod's. Even when somewhat smaller in scale, Vukovar and surroundings were economically striving areas because of their river location, with a very mixed population because of the area's attractiveness to settlers from the entire Austrian-Hungarian Empire and later the Yugoslav Kingdom. Judging from archaeological findings, the whole area around the Danube had attracted settlers for centuries.

Yet, there is no bridge, let alone talk of building bridges, connecting Vukovar with Serbia across the Danube. The area around the river is rural, and the local connection to Serbia is a small country road-crossing located approximately 30 km outside town, hidden away and rarely used other than by local farmers. The main connection to Serbia in these parts of Eastern Slavonia is the motorway connecting the two countries' capital cities. Despite (and maybe because of) many years of state unity in Yugoslavia, and even more years as an integrated living space for a mixed population, there is heated discussion between Croatia and Serbia about the border's proper location: following the course of the Danube as it does now as supported by Serbia; or following a line tracing cadastral municipal borders in the 19th century, as claimed by Croatia. The island of Vukovar in the middle of the Danube opposite Vukovar town underlines these disputes, being claimed by both sides.

As both narrative and symbol, Vukovar explains why these disputes are hard to settle. Compared to Slavonski Brod and Brod, its recent history and thus prevailing narrative relating to wartime events, are very different, rendering it perhaps Croatia's most ethnically divided town. In 1991, it unwillingly hosted a battle between the JNA and the so-called Croatian 204th Vukovar brigade, consisting of approximately 1800 members recruited from the National Guard Corps, police, plus volunteers. Officially, the conflicts in East Slavonia began in August 1990 when 12 policemen were killed in Borovo. Unlike other places at the border with Serbia, people in Vukovar chose not to surrender to the JNA and were besieged by Serbian paramilitary forces backed by the JNA for 87 days until 18 November, after which the town fell under Serbian occupation. Serbs occupied most of the Croatian border region, deporting large numbers of its citizens, until retaken by Croatia in Operation Storm in 1995. During the siege, Vukovar was destroyed almost completely.

Related to these events, Vukovar's story today is one of struggle for independence, glorifying heroes who fought for it, ultimately glorifying war. Unlike Slavonski Brod and Brod, where lines of divisions live alongside connections, Vukovar has become 'a memory container', one-sidedly gathering a nation to mourn its victims and celebrate its heroes (Figure 9.2). Vukovar's story has come to symbolise Croatian homeland war and suffering, a central episode in the Homeland War narrative. Ljubolevic (2016, 38) has noted that it 'is a milestone of the master narrative of the Croatian identity and of the entire nation-building process ... Vukovar holds a central place in Croatian victimhood ... a special symbolic

Figure 9.2 White Cross, where the Vuka meets the Danube. On the cross, it says: 'Vukovar a miraculous town, Vukovar is pride, Vukovar is defiance, It is a tear in one's eye, Sorrow at one's heart ... and a smile on one's lips, Vukovar is both past and future'.

Source: Photo by Dorte Jagetic Andersen.

place of Croatian suffering in the war and its striving for independence'. Few days pass without it being mentioned on national media referencing its role in the national narrative. Vukovar even hosts a march every 18 November; 50,000 Croats travel there to parade the streets and celebrate national independence. Nothing comparable occurs in other Croatian towns, including Slavonski Brod.

Controlled by nationalist politicians in the 1990s, heritage-selection processes were tainted by power and dominance, both aimed at gaining power and portraying who was in power. War memory has been communicated as owned by a particularly well-defined ethnic group, the Croats, irrespective of how the different ethnic groups were affected by the war. Vukovar thereby became national property in Croatia, drawing lines in the urban landscape and emphasising ethnic division with both practical and emotional consequences. Hence, Vukovar's status in the national narrative enhances ethnic divisions in the town and its surroundings.

The town is full of material markers of its violent past, constantly reminding of the consequences of ethnic fragmentation and thus affecting human bonds. The most dramatic impacts of partition are loss of life and physical injury, and in Vukovar, the memory of these are many. Almost 30 years after the event, Vukovar's townscape strikingly confronts you head-on with images of war and devastation: Literally every fourth building is left empty and full of bullet holes. Inhabitants fleeing Vukovar during the siege and the years under Serbian authority received support from the US and EU to rebuild their houses in the years following the conflicts. Nevertheless, many families left the town, thereby bequeathing the townscape a ghostlike appearance. Another striking aspect of Vukovar's streets are the many tourist buses on the parking lots and the many school classes wondering around in lines, from monument to monument. Unlike the generational interaction in the two Brods where consumption is at the centre, in Vukovar youngsters are confronted with wartime events, ensuring this memory lives on.

Visitors from the entire nation demand of Vukovar that the town population relives the atrocities; somewhat paradoxically, their ability to do so keeps the town afloat financially (cf. Clarke et al. 2017). War heritage and the selection of what is remembered and forgotten form lines of division, referring to a specific canon used by some and for specific reasons, a canon lived and practiced in the here-and-now (Smith 2006). Hence, collective memories of place representing a nation's innermost spirit influence people's sense of space and place. Given Vukovar's status in Croatian nation-building narratives, its spatial identity is linked to narratives of war:

> Post-war Vukovar is a town captured in a post-conflict trauma, doomed to either repeat the violence of ethnic division or to practice resistance through forgetting. Here cultural identities are cast in stone and for those living in and appreciating the town, hope seem scarce; Vukovar mourns a future that will never come … Leaving Vukovar, one is left wondering how many years it will take until the Serbs remaining in Vukovar will also take to the streets on November 18th [the Croatian bank holiday remembering the Homeland War].

(Andersen 2019: 65)

Ethnic divisions influence Vukovar's physical landscape, especially when adjoining areas are included; Serbs and Croats live separately, having each their own institutions and cafés (Matejcic 2012).

In Vukovar and the surrounding region, several heritage institutions such as the Museum of Vuedol Culture, founded in 2015, work hard to change the town's status in social memory by focusing on a more distant past and the Danube as a habitual space uniting people living on either riverbank (Kardov 2004). The attempt is to tell a story like that of Slavonski Brod and Brod about connections to the broader Balkan area even when there is not much interest in this among tourists to the area. And like Brod, a massacre haunts Vukovar. The JNA bombed the hospital despite international law forbidding such locations from being military targets. Patients were hidden in the basement. When the JNA took control of the town, the remaining patients were killed outside town in 'the Ovčara massacre'. Among Vukovar's population, there is a wish for this to become a story of military atrocities towards humanity, rather than Serb aggressions against Croats (Andersen and Prokkola 2018). However, despite lacking national identifications for the massacre's victims – and, indeed, most survivors of the siege – it is almost impossible to tell the story as one unifying humanity rather than dividing ethnic groups.

Heated discussion in Croatia about whether the area, with its still-considerable Serbian minority, should have double-language road signs (Latin and Cyrillic) illustrates how conditions for common perceptions of the past are extremely difficult to achieve. Social segregation and intolerance continue as ever-present realities despite the many reconciliating and healing projects being developed. As a Croatian reporter noted, 'If looks could kill, there would be many dead people every day in Vukovar. This is still a divided place' (Matejcic 2012), something still sensed by the author during field visits in 2015 and 2016. It will be years before any bridges cross the Danube here.

Conclusions

Peaceful coexistence is still far from everyday reality for people in former Yugoslav border regions. Considering what happened in this area in the 1990s, it is hard to imagine lines and division becoming less important anytime soon. When scholars emphasise that the twin-cities model 'turned out to be rather successful as a form of cross-border cooperation' and 'seems to be not only a means of getting through difficult social and economic conditions and solving local problems (... rather important in itself), but also a testing-ground for new forms of international cooperation' (Anischenko and Sergunin 2011), we should recognise how some former-Yugoslavia borderlands pose certain obstacles for the model.

Everyday walking and talking in former Yugoslavia depend heavily on recent interpretations of past events, which are obviously influenced by overarching narratives about a 'need for partition' and thus emphasise divisions. Yet, towns and regions can articulate divisions and form identities and belongings in different ways; even those border areas most marked by violent conflict do not repeat narratives of lines and divisions in identical ways. Comparing Slavonski

Brod and Brod with Vukovar along the five axes recognised as typical of twin cities around the world (Garrard and Mikhailova 2019), both show very asymmetrical interdependence: in one case between two urban areas on either side of a state border; in the other between two ethnic groups living in a town and its surroundings. In both cases, the other ethnic group (Serbians and Bosnians respectively) is highly dependent on the Croatian ethnic group, economically, politically and socially, generating very unequal relationships. Moreover, both cases are certainly influenced by tensions between inwardness and openness. However, Vukovar is by far the more inward-looking community. In Slavonski Brod and Brod, greater willingness appears likely to achieve openness in the future, and the 1992–1995 Bosnian war is read as an interruption, with the local authorities and business communities attempting to cater for interaction, despite huge difficulties involved. Vukovar on its side has emerged from the Balkan wars, as a place unwilling to open and to negotiate with 'the other', an interaction, which is presumed to threaten the entire Croatian nation.

Hence, even in a confined area like Eastern Slavonia, memories are multiple, complex, reliant on locally situated processes wherein divisions materialise over time and on different bases of selection about which material to make visible and forefront. Asking youngsters to relate to war heritage during visits makes for a very different spatial imaginary than the utilisation of shopping centres and cinemas. Lines are, in other words. walked, appropriated and told differently, even when comparing very adjacent places; locally, there seems to be space open for walking against recent history, playing the story in different tunes.

Notes

1 The internal dynamics certainly explain what caused the conflicts; however, to explain their escalation and the destabilisation of the entire region, external influences must also be considered (cf. Woodward 1995).
2 Remembering how ethnicity remains a concept whose practical application is problematic (cf. later in this chapter and in Jenkins 1997).
3 In conversation with interlocutor, July 2020.
4 In conversation with interlocutor, January 2020.
5 This was also confirmed in conversation with interlocutor, January 2020.

References

Andersen, D. J. (2019) 'Making Space for the Future: Reconstructing Spatial Identities after Armed Conflict'. In Stoklosa, K. (ed.) *Borders and Memories. Conflict and Cooperation in European Border Regions*, Berlin: LIT-verlag.

Andersen, D.J. and Prokkola, E.-K. (2018) 'Heritage as Bordering: Heritage Making, Ontological Struggles and the Politics of Memory in the Croatian and Finnish Borderlands.' *Journal of Borderlands Studies*, online edition, https://www.tandfonline.com/doi/full/10.1080/08865655.2018.1555052

Anischenko, A.G. and Sergunin, A.A. (2011) 'Twin Cities: A New Form of Cross-Border Cooperation in the Baltic Sea Region.' *Baltic Region* 11: 19–27.

Calame, J. and Charlesworth, E. (eds.) (2009) *Divided Cities, Belfast, Beirut, Jerusalem, Mostar, and Nicosia*. Philadelphia: University of Pennsylvania Press.

Castan Pinos, J. and Andersen, D.J. (2015) 'Challenging the Post-Yugoslavian Borders: The Enclaves of Sastavci and Dubrovnik' in Janczak/Osiewicz (ed.) *European Enclaves in the Process of De-bordering and Re-bordering*. Berlin: Kaldor.

Clarke, D, Cento Bull, A. and M. Deganutti. (2017) 'Soft Power and Dark Heritage: Multiple Potentialities', *International Journal of Cultural Policy* 23(6): 660–674.

Connor, W. (1993) *Ethnonationalism: The Quest for Understanding*. New Jersey.

de Certeau, M. (1985) *The Practices of Everyday Life*. Berkeley: University of California Press.

Garrard, J. and Mikhailova, E. (eds) (2019) *Twin Cities: Urban Communities, Borders and Relationships over Time*. London/New York, Routledge.

Green, S. (2018) 'Lines, Traces and Tidemarks'. In Demetriou, O. and Dimova, R. (eds.) *The Political Materialities of Borders: New Theoretical Directions*. Manchester: Manchester University Press.

Hall, S. (1996). 'Introduction. Who Needs 'Identity'? In Hall, S. and du Gay, P. (eds.) *Questions of Cultural Identity* (1–17). London: Sage Publications.

Jenkins, R. (1997) *Rethinking Ethnicity: Arguments and Explorations*. London: Sage Publications.

Kardov, K. (2004). Silencing the Past: Vukovar between the Place and Space of Memory. In Miroslav Hadžić (ed.) *The Violent Dissolution of Yugoslavia – Causes, Dynamics and Effects*: 227–238, http://www.bezbednost.org/upload/document/the_violent_dissolu-tion_of_yug.pdf (accessed 26.08.2019)

Kolstø, P. (ed.) (2016) *Strategies of Symbolic Nation-Building in South-Eastern Europe*. London: Routledge.

Lampe, J.R. (2000) *Yugoslavia as History—Twice There Was a Country*. Cambridge: Cambridge University Press.

Ljubolevic, A. (2016) 'Speak Up, Write Out: Language and Populism in Croatia' in Hancock (ed.) *Narratives of Identity in Social Movements, Conflicts and Change*. Bingley.

Matejcic, B. (2012) Vukovar Still Imprisoned by its Bloody Past, The Balkan Fellowship for Journalistic Excellence website, http://fellowship.birn.eu.com/en/alumni-initia-tive/alumni-initiative-articles-vukovar-still-imprisoned-by-its-bloody-past (accessed 26.08.2019).

McCall, C. (2014) *The European Union and Peacebuilding*. Basingstoke: Palgrave Macmillan.

Romic, R.P., Trninic, D., Savic, D. (2017) 'Divided Brothers: Ex-Double Towns of Brod and Slavonski Brod,' *Socialni studia* 1/2017, 31–53.

Smith, L. (2006) *The Uses of Heritage*. London: Routledge.

Woodward, S.L. (1995) *Balkan Tragedy: Chaos and Dissolution after the Cold War*. Washington, DC: Brookings Institution Press.

Part II

International Twin Cities

B. In the Middle East and Africa

10 Aqaba and Eilat

Twenty-five years of 'good neighborly relations' in a post-conflict environment

Tamar Arieli

This chapter evaluates patterns of municipal cross-border cooperation (hereinafter CBC) between the cities of Aqaba (Jordan) and Eilat (Israel), situated side by side at the head of the Gulf of Aqaba. This cross-border municipal environment had a special place in the vision for 'good neighborly relations' outlined in the 1994 Israel-Jordan Peace Treaty. Notwithstanding this vision, the ongoing regional conflict and the context of cold Israeli-Jordan relations is reflected in the dearth of civilian initiatives and the dominance of political and security authorities.

Aqaba-Eilat municipal cross-border ties are reviewed through documentary analysis and field research including participant observation at municipal and cross-border settings – seaports, border-crossing points, tourist operations and environmental activities. Conversations with stakeholders (residents, entrepreneurs, national and local government and security officials) allow a sector-based analysis of perceptions and interests focused on the border. These reveal the delicate, almost underground, cross-border interactions which somewhat rescale this cross-border municipal region.

National borders and CBC

Borders are interfaces between countries and societies, reflecting degrees of conflict and accommodation, cooperation and integration (Newman 2003). Classic perceptions of borders view them as barriers with border regions as geographic and political peripheries (House 1980). Yet cross-border regions worldwide have become strategic actors affecting policy towards facilitating CBC. CBC is indeed basic to European and North American civilian and economic contexts, yet its rationale and mechanisms are identifiable notions and realities of open borders which starkly contrast to policies and practices of borders in conflict and post-conflict environments.

Conflict significantly deters cross-border interaction (Minghi 1991), but, when managed or resolved, previously alienated borders can progress to co-existence and interdependence (Martínez 2005). As 'trans-boundary diplomacy' filtering up from border regions, CBC may even impact national and cross-national discourse and policy (Henrikson 2000). Yet, border transformation potential can be exaggerated in the context of intractable conflict which profoundly impacts

DOI: 10.4324/9781003102526-13

individuals and society (Crocker et al. 2020; Kriesberg 1993). Post-conflict environments are the distinct legacy of intractable conflict, and their social and political characteristics are especially significant for realistic analysis of CBC prospects.

Post-conflict environments retain shared societal narratives which evolved over extended periods of conflict as collective adaptations to hardship and trauma, supporting necessary societal mobilizations. These collective beliefs serve as prisms through which new conflict-related information and experiences are interpreted and incorporated (Bar-Tal 2007). While entrenched shared narratives contribute to societal resilience during conflict, in post-conflict periods their filtering function minimizes prospects for transforming perceptions of former cross-border enemies to neighbours and regional partners. Andersen in Chapter 9 noted how divisions and conflict are retained over generations in urban social memory, continuously reflected in patterns of physical movement and communication.

National authority and security agencies retain dominance over civil life during post-conflict periods. This dominance severely inhibits civilian initiatives and limits cross-border mobility. This is another remnant of conflict, reflecting historic patterns and memories of closed borders, non-communication, and mistrust. These features of post-conflict periods continue to inhibit cross-border initiatives even after conflicts are formally resolved.

Yet post-conflict environments are not unique in displaying challenges to developing CBC, which does not inevitably advance even within European contexts, despite extensive support (Jańczak 2013). Regional institutions and cross-border interdependence may be perceived as threatening by national authorities (Scott 1999; Van Houtum 2000). Hence, Gilles et al. (2013) suggest 'bringing the state back' to border discussions and recognizing state power to design and ultimately control cross-border interaction, rather than inflating the significance of regional and local-level agencies. As this case study will demonstrate, post-conflict border regions can be surprisingly supportive of such insights regarding the significance of national supervision and security dominance over cross-border cooperation.

Municipal CBC

Border-adjacent cities are central to globalization discourses. Local communities in binational urban regions have unique opportunities to promote local interests through CBC, affecting wider political environments and policy. Yet municipal-level CBC relies on institutional linkages, as Ganster and Collins (2016) observe in the US-Mexican context. Informal cooperation between local individuals may precede formal political ties, yet are often opportunistic and incidental, and thus of limited wider significance. Institutionalized CBC develops gradually, involving formality and declarations of will and policy (Sohn 2014). Therefore, minimal political cross-border relations in post-conflict environments affect prospects for informal initiatives, driving them into underground channels.

Realities of twinned cities vary, reflecting national and local circumstances and resources, entrepreneurship and interpersonal relations which frame prospects for a range of cross-border inter-municipal relations and their regulation (Arieli and Cohen 2013; Keating 2014; Nugent 2012; Peck and Theodore 2012). Advanced city twinning is expressed through extensive bonding with cross-border neighbours, involving some merging of identities (Joenniemi and Jańczak 2017). A balance of local and national interests and circumstances essentially determines the outcome, depth and sustainability of binational city relations.

This multiplicity of geographical, historical, political, social and cultural variables explains the dominance of case-study research of binational city phenomena. Unsurprisingly, these case studies lead to conceptualization of twin-city relations as inherently plural (Joenniemi 2014, 3) and have not produced a universal analytic framework (Garrard and Mikhailova 2019; Joenniemi and Jańczak 2017). It is therefore interesting how many of these studies somewhat apologetically present the disappointing realities of continued rivalry and only sporadic CBC of minimal political clout, as if diverging from some notion of binational or twin-city ideal. This is evident in studies of municipal settings worldwide, in national as well as international contexts of neighbouring cities, regarding both ethnically homogeneous and heterogeneous cross-border populations. These realities frame expectations for developing relations between Aqaba and Eilat, in their distinctly peripheral location on the Israel-Jordan post-conflict border.

Israel and Jordan

Israel and Jordan maintain cold peace, defined as stabilized but not quite normalized relations, lacking developed transnational ties (Press-Barnathan 2006). Both remain committed to their 1994 Peace Treaty but demonstrate only minimal political dialogue overshadowed by diplomatic crises and the unresolved Israeli-Palestinian and Israeli-Arab regional conflicts. These undermine full implementation of the Treaty's political and societal vision of 'good neighborly relations' (Israel-Jordan Treaty of Peace, 1994). Notwithstanding significant mutual benefits from ongoing state-level gas, water and security cooperation and continued facilitation of economic, environmental and social cooperation, relations are minimal at both political and societal levels.

Jordanian public opinion consistently opposes normalized relations with Israel and has often voiced demands to abrogate all cooperation, even the Treaty itself. This underpins the recent Jordanian government's decision not to extend the special regime governing the Zofar and Naharayim border areas, thereby uprooting Israeli farms and returning land, previously leased to Israel, to Jordanian control (Kardoosh 2019). Israel's recent consideration of annexing parts of the Jordan Valley hardly considered Jordanian opposition, and further destabilized the Treaty. The establishment of Israel-United Emirate relations sidelines Jordan's long-standing position that Israel should end its West Bank occupation as a condition for normalized relations, even within an existing peace treaty. Furthermore, Jordan fears that expanding Israel-Gulf-state relations endangers Jordan's unique position as custodian of the Al-Aqsa Mosque compound and Muslim and Christian holy sites of Jerusalem (Ersan 2020).

The shared border is central to the Treaty's vision of peaceful neighbourly relations. The late King Hussein saw it as a 'Valley of Peace', facilitating shared interests of infrastructure and economic projects, thereby anchoring cross-border relations, and enhancing mutual security through environmental cooperation, water-sharing, and energy development (Israel-Jordan Treaty of Peace, 1994). Yet parallel Israeli and Jordanian infrastructure still function side by side, despite their potential to maximize efficiency and reduce redundancy through coopera-tive planning. Reasons for limited CBC include the reluctance of Jordan's mid-dle class, mainly of Palestinian origin, to normalize relations in the shadow of the Israeli-Palestinian conflict, Israel's superior economic capacities and Jordan's fear of dependence, Israel's western economic and cultural orientation and bureau-cratic obstacles in both countries (Shamir 2004).

Notwithstanding, cooperative initiatives exhibit both state and civilian CBC. Security, water and energy are the main areas for national-level cooperation, which has significantly expanded with Israel's recent development of water desalination and natural-gas discoveries. This cooperation is executed with lit-tle public exposure and is supported by both countries' highest state echelons despite multiple political crises. Civilian local cooperative initiatives and ties develop only slowly and far from the public eye due to their perceived illegiti-macy (Arieli 2012; Cohen and Ben-Porat 2008). Hence, despite distinct shared interests, small businesses in Aqaba and Eilat maintain minimal and almost undercover CBC.

Contrasting trends of resilient yet secretive national cooperation and rela-tively minor fragile local initiatives of CBC paradoxically reflect both Jorda-nian and Israeli interests. Jordanian national agencies are extremely sensitive to widespread public campaigns which oppose normalization with Israel, and therefore rarely initiate CBC and actively avoid its exposure. Meanwhile, Israeli government agencies overwhelmingly defer to security-apparatus expertise (Michael 2007), heavily regulating cross-border activity to avoid security risks, almost regardless of damaging civilian and even national interests in terms of their potential contribution to regional stability. Exposure of CBC is limited in deference to Jordanian sensitivities (Arieli 2012).

The extreme inhibition of CBC and its exposure further minimizes the poten-tial for CBC in both countries. Bitter disappointment about the meagre 'fruits of peace' continuously feeds Jordanian public opposition to normalizing relations with Israel, deterring engagement in CBC, despite its potential. This is the con-text for evaluating cross-border municipal relations in the Gulf of Aqaba.

Eilat and Aqaba

Aqaba and Eilat are situated in the desert region of the Gulf of Aqaba, with only 7 km between their city centres. Despite geographic proximity, the cities have no shared history, and their populations are distinct, ethnically, religiously, and socio-economically. Their infrastructure of parallel seaports, airports, hos-pitals, emergency forces, environmental agencies, and highways mirror-image each other.

Aqaba enjoys rapid growth and development, due to government policy and investment and the economic activity around Jordan's only seaport. Its population of 148,000 (2015) is almost three times that of Eilat (52,000 in 2018). Jordan's Aqaba Special Economic Zone Authority (ASEZA) attracts local and foreign investment in developing infrastructure and constructing high-end residential and vacation projects, dramatically changing Aqaba's waterfront since 2001. Aqaba has become an Arab regional transport and logistics hub.

Eilat's development has stagnated due to government policy, limited investment and coastal space, and relatively marginal seaport activity. Its population has not grown significantly during the past two decades, reflecting the undiversified municipal economy highly dependent on the tourism sector, as well as the city's relative isolation from the social and economic centres of Israel. The port of Eilat is only marginally active due to its distance from Israel's centre and competition with the more developed ports of Haifa and Ashdod.

The potential for cross-border municipal cooperation between Aqaba and Eilat was distinctly outlined in Article 23 of the 1994 Peace Treaty, alongside a vision of joint development for the two cities:

> The Parties agree to enter into negotiations…not later than one month … of ratification of this Treaty, on arrangements that would enable the joint development of…Aqaba and Eilat with regard to such matters, inter alia, joint tourism development, joint customs, free trade zone, cooperation in aviation, prevention of pollution, maritime matters, police, customs and health.

Inter-municipal relations since 1994 have been recently analysed by Arieli (2016) and Soberon and Rofé (2016). They emphasize the dominance of political circumstances over local initiatives, despite obvious ecological interdependence and potential for CBC. Indeed, 25 years of Israel-Jordan diplomatic relations offer perspective on patterns of communication and municipal cooperation in the complex post-conflict circumstances and call for distinguishing vision and potential from ground realities.

Expectations for Aqaba-Eilat municipal cooperation in the shadow of regional conflict are based on evolving local perceptions of interdependence in both cities, the desert climate, limited alternative economic opportunities and their distinctly peripheral location relative to their national political-economic centres. Both cities have developed port service sectors (Aqaba more so than Eilat) but neither is an industrial centre. The Red Sea Gulf is the focal point for tourism for both cities, catering to both local and foreign clientele and hence the significance of environmental cooperation in their relations.

Evolving cooperation

Notwithstanding the municipal emphasis of Article 23, only sporadic intercity contact developed during the decade after its signing. The first initiative was a 1996 social-educational project, involving computer communication and occasional reciprocal visits of Jordanian and Israeli high-school students (Sherf and Katzir, interviews, 2007). Challenges of project-funding and obtaining visas for Jordanian students caused lengthy intervals between activities which, alongside participant turnover, were detrimental to project success.[1]

The Red Sea Marine Peace Park (RSMPP) of the U.S. Agency for International Development, Middle East Regional Cooperation Program (USAID/ MERC), was an early initiative to facilitate peace through Israel-Jordan joint research, monitoring, managing, and preserving of the Gulf of Aqaba. This U.S.-funded three-year project, launched in 1999, involved deeper institutionalization of CBC in its ties to academic institutions such as the Aqaba Marine Biology Station and the Eilat Inter-University Institute for Marine Sciences, teaming Israeli and Jordanian scientists to create a high-resolution map of the Red Sea Gulf reef ecosystem and foster sustainable use of the coral reef ecosystem resources.

The Red Sea Marine Peace Park project was concluded in 2003, with an international symposium on the integration of marine science and resource management in Aqaba. There, in the presence of Jordanian, Israeli and American officials as well as local and international scientists and environmentalists, a Memorandum of Understanding (MoU) was signed between Israel's Ministry of Environmental Protection and the Aqaba Special Economic Zone Authority. The agreement was a commitment to continued scientific cooperation in the Gulf of Aqaba. The MoU did not include joint management plans, and after its signing, Israeli and Jordanian stakeholders resumed patterns of parallel coastal management (Portman and Teff-Seker 2017). Nevertheless, Israel-Jordan marine science cooperation has outlived the original program and continues to be modestly realized through data-sharing and joint academic proposals, particularly in coral mari-culture (Shashar, Inter-University Institute for Marine Sciences, interview, May 2020). This ongoing cooperation is especially significant in the context of the widespread Jordanian academic community's commitment to anti-normalization campaigns.

The first municipal-level cooperative initiative addressed the shared chronic problem of mosquitos. Coordination of mosquito-control efforts took only days to install and encountered no opposition. It has since expanded into regular cross-border monitoring of pests, stray dogs and animals as a public health and safety measure. Eilat's mayor visited Aqaba in December 2004, ten years after the Peace Treaty signing. Aqaba's mayor reciprocated soon after (January 2005). These full-day visits included tours, meetings between local officials and exchanges of gifts and contact details (Samorai, Department of Regional Cooperation, Eilat, multiple communications, 2007–2020).

Seven joint inter-municipal sub-committees and a steering committee were founded in 2005 to facilitate regular communication in monitoring regional threats – earthquakes, flood and fire – and to address shared interests in infrastructure and economic development. These committees were the culmination of Article 23 in expressing local recognition of interdependence. They were impressive both in their institutionalized structure within both municipalities and scope of activities (Al-Moghrabi, former Head of Planning and EIA Section, ASEZA [Aqaba], and Samorai, multiple communications 2007–2010).

The structural imbalance between the two cities' local-government regimes significantly challenged the committees' work. ASEZA is an autonomous Jordanian governmental authority encompassing Greater Aqaba, while Eilat is a municipality, subordinate to Israel's Ministry of Interior. Nevertheless, the municipal Department of Regional Cooperation established in Eilat is unparalleled amongst other Israeli municipalities. Its coordination of the inter-municipal committees was supported by the Israeli Foreign Ministry and yet burdened by the bureaucracy of multiple government agencies (Israeli, former Director of Jordan Department, Israel Ministry of Foreign Affairs, interview, March 2007).

Eilat and Aqaba gradually reached unprecedented levels of institutionalized CBC. Their mandate was to address immediate problems and coordinate solutions. The committees met alternatively on each side of the border and cooperatively established agendas. Israeli committee participants were municipal figures, national governmental officials of relevant ministries, professionals, private individuals and volunteers, representing local interests. Jordanian participants were ASEZA officials.

Alongside problem-solving, these committees sought to identify fields of mutual interest for realizing the Peace Treaty:

> The recognition that cooperation in areas that know no boundaries, these sub- committees constitute the means to deepening peaceful relationships and trickle them down to everyday life.
> (2006 Eilat-Aqaba Municipal Progress Report)

The scope of ideas documented in the Aqaba-Eilat Coordination Committee Reports cover a range of municipal interests and needs. Although suggestions surpass actual cooperation, they imply developing local perceptions of interdependence, particularly related to the region's environmental uniqueness, climatic challenges and remoteness from both countries' national centres. Table 10.1 outlines some of these suggestions, and varying degrees of their realization.

The committees dedicated to environmental and emergency issues were more active than those focused on social and cultural interaction. Cultural events and educational programs were only episodic and dependent on foreign funding. Most cooperative initiatives focused on solving immediate problems and improving local quality of life rather than on politics, social contact or normalization. Problem-solving has indeed been shown to be a central framework for developing CBC in many borderland municipal contexts (Mikhailova 2018). Problem-solving afforded legitimacy to CBC and shielded it from Jordanian public critique, sustaining it despite many regional crises and setbacks in Israel-Jordan relations (Abu Rashid, Chairman, Amman Center for Peace and Development, interview, May 2009). This functional and needs-based approach reflects the post-conflict context of the committees' work and is a recognized feature of Israel-Jordanian CBC (Arieli 2016).

Table 10.1 Municipal committee-led cooperation

Area	Activity	Realization
Emergency	• Joint early-warning systems for floods, earthquakes, and water pollution; • Expanding potential of local emergency forces by jointly training medical and fire-fighting teams and facilitated equipment connectivity.	Partial fulfilment: • Early warning systems are person-dependent, not institutionalized as regular inter-municipal procedures; • Irregular joint training.
Pest and stray-animal control	Joint animal and pest monitoring.	Ongoing, regular cooperation.
Science, education and culture	• Shared sports and cultural programs, concerts, and underwater photography (e.g., 'Swimming for Peace', *Eilati**, April 2007); • Joint science grants and projects (e.g., Israeli-Jordanian-U.S. air quality monitoring 'Beam of Light', see 'Measuring Air Quality Through Spectral Analysis Of Beam Of Light', *ScienceDaily*, 09.11.2007).	Partial fulfilment: Irregular cooperation due to (Israeli) visa procedures and (Jordanian) opposition to normalization.
Economic and infrastructure complementarity	• Inter-seaport coordination for equipment sharing and increased commercial capacity; • Jordanian employment in Eilat hotels.	Partial fulfilment: • Marginal seaport cooperation due to disparity in size and function; • Significant institutionalization and growth of Jordanian employment in Eilat.
Tourism	Cooperative training, marketing and development of cross-border tourism options.	Partial fulfilment: • No regional marketing, minimal cooperation between individual tourist agents; • Successful private-public initiative expanding Jordanian employment in Eilat.
Marine environment	Increased environmental Gulf security through joint research and monitoring water pollution.	Ongoing cooperation involving national, municipal and academic actors.

* local Eilat newspaper.

The committees have been inactive since 2010, grimly reflecting the state-level political climate, and preference for low-key problem-solving between officials rather than highly visible municipal institutional structures. Communication remains regular but is significantly narrower, engaging only specific municipal individuals and agencies within the two cities. Therefore, an important insight from this case study is the critical importance of individuals, municipal policy entrepreneurs, who build channels of communication and trust and personally facilitate problem-solving and cooperative logistics (Arieli and Cohen 2013). Personal involvement elevates cooperative efforts, overcoming hurdles detrimental to sensitive municipal interactions.

Despite committee inactivity, joint marine-monitoring remained the most regular yet complex and comprehensive expression of Eilat–Aqaba municipal CBC. It involved multi-level governmental, municipal, and interpersonal communication and cooperation regarding shared interests in the Red Sea Gulf environment, as outlined in the Treaty's Article 18. Early installation of joint procedures to safeguard against oil spills in the Upper Gulf of Aqaba reflected awareness of the fragile and diverse natural landscape, the Gulf's collision-prone narrow southern entrance, intense development, proximity of Aqaba and Eilat coastlines and both cities' dependence on tourism. Cooperative procedures included testing decision-making processes, and activating resources, equipment and institutions in both countries. Joint oil-spill control exercises have occurred bi-annually between 1998 and 2014, coordinated by municipal environmental authorities. Despite their significance to both the Gulf environment and the cities' relationship, these exercises were never publicized. Marine scientific cooperation continues, informed by shared interests, developing relationships and interpersonal trust. Yet few joint-scientific publications have followed, due to Jordanian reservations (Shashar, interview, May 2020).

ASEZA too promotes local interests through CBC, while refraining from publicity to minimize public criticism (Al-Moghrabi, interview, 2009). This is reflected in ASEZA Annual Report publications celebrating investment and vacationing opportunities in Aqaba without mentioning cross-border tourism and business opportunities. This policy has indeed safeguarded municipal-level cooperation since 2005, throughout multiple Israel-Jordan national-level crises and continuous Jordanian labour unions' boycott of communication and cooperation with Israelis.

Notwithstanding regression from formal inter-municipal committees to informal interpersonal contact amongst key individuals, strategic municipal cooperation has continued, and even expanded, over the years. A significant display of deepening functional cooperation is the gradually increasing Jordanian employment in Eilat hotels. Until the ongoing COVID-19 crisis which instigated border closure (by mutual consent), 2000 Jordanians worked in Eilat hotels, daily crossing the border through special institutionalized arrangements. Employment opportunities draw candidates from throughout Jordan, seeking employment and minimal Israeli wages which are significantly higher than parallel opportunities in Jordan. This project was coordinated both locally and nationally, involving government agencies in both countries as well as private entrepreneurs of hotels, transportation, and employment agencies. A recent development is the entry of 700 Jordanians from Aqaba to Eilat, to receive the COVID-19 vaccination and resume their work in Eilat hotels, ahead of the Passover holiday season (Jpost 2021).

Appreciation of the value of local cooperation has not spilled into tourism. Beyond individual initiatives by tourist operators in Aqaba and Eilat, there is no cooperation in developing and marketing regional, cross-border tourism, despite obvious potential. This would entail currently unfeasible levels of trust and cooperation between sectors of both cities and countries, plus significantly higher visibility of cooperation and normalized relations.

Logistics of cooperation

Cross-border traffic of people, vehicles and freight are significant expressions of municipal ties. The Wadi Arava-Rabin border crossing served 1.2 million people in 2019, against just 45,000 in 1994. This growing traffic has triggered both Israeli and Jordanian investment in border-crossing infrastructure and facilities. Yet this impressive cross-border traffic growth hardly reflects social or economic inter-municipal interests beyond tourists and (recently) Jordanian day-labourers, comprising the bulk of cross-border traffic. Border crossing is far from routine for Aqaba or Eilat residents. Beyond ethnic-cultural differences, this is explained by logistic and communication barriers, reflecting and enhancing the estranged political climate.

Logistics are the vehicles of inter-municipal cooperative activities, yet they also reveal policy makers' appreciation of CBC. The border crossing is 7 km from Eilat and Aqaba city centres. A bus service from the border-crossing to Aqaba and Eilat was established in 1996 but cancelled soon after. Border-crossing commissions are approximately $34. Border-crossing and transportation costs are significant for people of lower- or middle-class incomes in both cities, deterring casual cross-border touring and shopping which could have generated economic and even social interaction.

Jordanian visas are granted to Israelis at the terminal, while Jordanians must apply at Amman's Israeli Embassy, entailing significant cost, delay and uncertainty. Israel's visa policy stems from prioritizing security over civilian cross-border activity. It has caused frustration, delays and embarrassment for Jordanian officials and businesspeople involved in cross-border initiatives. Israel's visa policy has challenged municipal joint committees' work and remains detrimental to cross-border initiatives. The policy limits the scope of shared educational programs for Israeli and Jordanian students which resort to online communication, hardly realizing the social potential of cross-border neighbours.

Since 2008, Israel and Jordan have maintained lists of key figures from public and private sectors who are granted special visas to facilitate their border crossing. These include directors of land and seaports, and health agencies, alongside government officials active in cross-border interaction. While problems and delays remain, this facilitates somewhat regular cross-border coordination, demonstrating national-governmental sensitivity to local needs. Recently,

Jordan has introduced a policy requiring Israelis to arrange Jordanian tourist operator accompaniment. The formal rationale is the need to guarantee the safety of Israelis in Jordan, although one may assume there is economic motivation for this policy as well.

Conclusion

Geographic proximity and political circumstances frame adjacent communities' perceptions of cross-border opportunities. Aqaba and Eilat, interfacing across a peaceful border in a conflictual region, display only basic interaction. As in many twin-city contexts, problem-solving is a dominant engine of cross-border municipal relations, generating local government structure adjustment in the form of (short-lived) joint-municipal committees and occasional municipal-level action based on information-sharing and local networking (Mikhailova 2018), somewhat reflecting a new dimension of governance. Regular and irregular cross-border networking between municipal officials and figures has sustained time and political crises. Cooperation is resilient and flexible in maintaining channels of cross-border relations and addressing local needs despite political and social challenges. Cooperative initiatives are therefore far more substantial than mere symbolic gestures or opportunistic exercises. This municipal region exhibits many local benefits of border permeability (Sohn 2014):

a. Positional benefits of border-proximity are demonstrated in cooperative pest-control;
b. Transactional benefits of adjacent systems are demonstrated in inter-municipal committees and marine environmental cooperation;
c. Differential benefits exploiting 'factor-cost' differentials are demonstrated in cross-border hotel employment programs;
d. The border as a 'locus of hybridization' producing 'new ways of doing and thinking' is demonstrated in inter-municipal committees and informal joint problem-solving.

Nevertheless, the border as a symbolic resource of place-making and shared identity has not developed and has minimal prospects given the ethnic, social and economic disparities between the cities and the regional context.

Twenty-five years after the 1994 Israel-Jordan peace treaty, Aqaba and Eilat demonstrate both common and unique patterns of cross-border urban twinning. The post-conflict environment rooted in regional politics is the main factor inhibiting significant development of cross-border relations. Distinct socio-economic, religious, cultural and structural disparities and economic competition further amplify the social distance between the cities' populations.

The cities' relationship is mainly institutionalized, lacking spill-over into social realms, a common feature and challenge of the twin-city phenomenon. The cross-border urban arena in this post-conflict context is politically sensitive, yet it faces recognized challenges of inter-municipal service delivery, economic development and environmental, social and cultural interests. The conflict environment poses additional challenges of social distrust and fear, and deep deference by local authorities to national directives. This is the context for identifying subtle twin-city mechanisms and manifestations in Aqaba and Eilat.

Municipal structures in the Aqaba-Eilat context are therefore theoretically insight-producing. The distinctly peripheral location magnifies shared development and environmental interests driving cooperation, despite imbalance between local government agencies. Municipal structures, challenged by national-level authorities, have proven effective in facilitating direct CBC in local problem-solving. Neither city experiences major conflict with its national government despite the element of foreign policy of these interactions, demonstrating how securitization common to post-conflict environments can accommodate official, semi-official and non-official local municipal cross-border initiatives. Securitization may even be pre-conditional for stable CBC implementation.

Cross-border twin-city development in conflict and post-conflict environments is slow, subtle, even reversible. Notwithstanding, Aqaba and Eilat demonstrate surprising compliance to patterns recognized in non-conflict contexts, of conflicting interests and practices regarding desirable degrees of cross-border municipal cooperation and interdependence.

Notes

1 In 2007 the project was adopted by the Instituto per la Cooperazione Universitaria (ICU) and became AETP: Aqaba-Eilat: One More Step Toward Peace.

References

Arieli, T., 2012. Borders of peace in policy and practice: National and local perspectives of Israel-Jordan border management. *Geopolitics*, 17(3), 658–680. doi.org/10.1080/146 50045.2011.638015

Arieli, T., 2016. Municipal cooperation across securitized borders in the post-conflict environment: The Gulf of Aqaba. *Territory, Politics, Governance*, 4(3), 319–336. doi. org/10.1080/21622671.2015.1042026

Arieli, T. & Cohen, N., 2013. Policy entrepreneurs and post-conflict cross-border cooperation: A conceptual framework and the Israeli Jordanian case. *Policy Sciences*, 46(3), 237–256. doi.org/10.1007/s11077-012-9171-9

Bar-Tal, D., 2007. Sociopsychological foundations of intractable conflicts. *American Behavioral Scientist*, 50(11), 1430–1453. doi:10.1177/0002764207302462

Cohen, N. & Ben-Porat, G., 2008. Business communities and peace: The cost-benefit calculations of political involvement. *Peace and Change*, 33(3), 426–446. doi:10.1111/j.1468-0130.2008.00505.x

Crocker, C. A., Hampson, F.O., Aall, P., 2020. Intractable conflicts and the challenge of mediation, in Crocker, C. A., Hampson, F.O., Aall, P. (Eds.) *International Negotiation and Mediation in Violent Conflict: The Changing Context of Peacemaking*. Chapter 3, Routledge.

Ersan, M. 2020. Israel-UAE deal: Will Jordan's custodianship of the Al-Aqsa Mosque be affected?, *Middle East Eye*, September 9. Available: https://www.middleeasteye.net/news/israel-uae-deal-jordan-jerusalem-al-aqsa-mosque-custodianship Accessed 01.03.2021.

Ganster, P. & Collins, K., 2016. Binational cooperation and twinning: A view from the US–Mexican Border, San Diego, California, and Tijuana, Baja California. *Journal of Borderlands Studies*, 32(4), 497–511. doi.org/10.1080/08865655.2016.1198582

Garrard, J. & Mikhailova, E., 2019. (eds). *Twin Cities: Urban Communities, Borders and Relationships Over Time*, Routledge, Taylor and Francis Group.

Gilles, P. et al., 2013. *Theorizing Borders through Analyses of Power Relationships*, P.I.E. Peter Lang.

Henrikson, A.K., 2000. Facing across borders: The diplomacy of bon voisinage. *International Political Science Review*, 21(2), 121–147. doi:org/10.1177/0192512100212002

House, J. 1980. The frontier zone. *International Political Science Review*, 1(4), 456–477. doi.org/10.1177/019251218000100403

Houtum, H.V., 2000. III European perspectives on borderlands. *Journal of Borderlands Studies*, 15(1), 56–83. doi:10.1080/08865655.2000.9695542

Jańczak, J., 2013. Revised boundaries and re-frontierization border twin towns in Central Europe. *Revue d'études comparatives Est-Ouest*, 44(04), 53–92.

Joenniemi, P., 2014. City-twinning as local foreign policy: The case of Kirkenes-Nickel. *CEURUS EU-Russia paper*, (15). http://ceurus.ut.ee/wp-content/uploads/2011/06/EU-Russian-paper-15_Joenniemi.pdf

Joenniemi, P. & Jańczak, J., 2017. Theorizing town twinning – towards a global perspective. *Journal of Borderlands Studies*, 32(4), 423–428. doi:10.1080/08865655.201 6.1267583

Jpost, 2021. *Israel to Vaccinate, Allow Entry to Jordanian Laborers for Passover*. Available: https://bit.ly/3v3uuVL Accessed 09.03.2021.

Kardoosh, M. 2019. Jordanians now see Israel as an implacable enemy, despite 25 years of peace, *Haaretz*, March 11. Available: https://www.haaretz.com/middle-east-news/.premium-25-years-after-peace-with-israel-jordanians-see-little-benefit-1.8067457 Accessed 01.03.2021.

Keating, M., 2014. Introduction: Rescaling interests. *Territory, Politics, Governance*, 2(3), 239–248. doi:10.1080/21622671.2014.954604

Kriesberg, L., 1993. Intractable conflicts, *Peace Review*, 5(4), 417–421.

Martínez O. J., 2005. *Border People: Life and Society in the U.S.-Mexico Borderlands*, University of Arizona Press.

Michael, K., 2007. The Israel Defense Forces as an epistemic authority: An intellectual challenge in the reality of the Israeli–Palestinian conflict. *Journal of Strategic Studies*, 30(3), 421–446. doi:10.1016/j.geoforum.2009.06.006

Mikhailova, E., 2018. Collaborative problem-solving in the cross-border context: Learning from paired local communities along the Russian border, *Journal of Borderlands Studies*, 33:3, 445–464. doi.org/10.1080/08865655.2016.1195702

Minghi, J.V., 1991. From conflict to harmony in border landscapes, in Rumley, D. & Minghi, J.V., *The Geography of Border Landscapes*, Routledge, 43–62.

Newman, D., 2003. On borders and power: A theoretical framework. *Journal of Borderlands Studies*, 18(1), 13–25. doi.org/10.1080/08865655.2003.9695598

Nugent, P., 2012. Border Towns and cities in comparative perspective, in Wilson, T.M. & Donnan, H., 2016. *A Companion to Border Studies*, Wiley Blackwell.

Peck, J. & Theodore, N., 2012. Follow the policy: A distended case approach. *Environment and Planning A: Economy and Space*, 44(1), 21–30. doi.org/10.1068/a44179

Portman, M. E., & Teff-Seker, Y., 2017. Factors of success and failure for transboundary environmental cooperation: Projects in the Gulf of Aqaba. *Journal of Environmental Policy & Planning*, 19(6), 810–826. doi:10.1080/1523908x.2017.1292873

Press-Barnathan, G., 2006. The neglected dimension of commercial liberalism: Economic cooperation and transition to peace. *Journal of Peace Research*, 43(3), 261–278. doi.org/10.1177/0022343306063931

Scott, J.W., 1999. European and North American contexts for cross-border regionalism. *Regional Studies*, 33(7), 605–617. doi.org/10.1080/00343409950078657

Shamir, S., 2004. Overview of economic relations and projects – Introduction. *In Israel-Jordan Relations: Projects, Economics, Business*, 19–21. Ramot and Tel Aviv University.

Soberon, J.R.X. & Rofé, Y., 2016. Modeling the bi-national city process of Eilat–Aqaba, a bi-national city in a cold peace setting. *GeoJournal*, 82(6), 1263–1274. doi:10.1007/s10708-016-9748-5

Sohn, C., 2014. The border as a resource in the global urban space: A contribution to the cross-border metropolis hypothesis. *International Journal of Urban and Regional Research*, 38(5), 1697–1711. doi.org/10.1111/1468-2427.12071

11 Lomé and Aflao

Ambivalent affinity at the Togo-Ghana border[1]

Paul Nugent

The phenomenon of urban centres located on either side of international borders is familiar from the Americas and Europe, but is also a recurring feature
across Africa (Nugent 2012). What is distinctive is that so many capital cities
rest on, or very close to, a border (Soi and Nugent 2017, 537–538). Thus, Kinshasa and Brazzaville, two capitals, interface across the Congo River. However,
more common is where a capital city co-exists with a smaller town. Typically, a
river seems like a natural border. By contrast, Lomé, Togo's capital, and Aflao,
the adjacent town in Ghana, do not even have a separating no-man's land on
a stretch of border also characterised by straight lines. This imparts particular
intensity to their relationship, as I wish to demonstrate.

Terms commonly depicting the relationship between border towns vary from
neighbours to pairs to twins. Each invokes a potentially misleading metaphor.
To suggest such towns are neighbours implies an almost accidental proximity,
whereas their trajectories have often been closely intertwined (Buursink 2001).
Nor are they necessarily pairs because they are often vastly different in population size, physical dimensions and economic profile. The metaphor of 'twin
towns' has the virtue of signalling intimate connection, without presuming
relationships are equal or exude great warmth. A degree of ambivalence is very
common amongst twin cities, deriving partly from human flows. It is manifest on
asymmetrical borders, like that between the USA and Mexico, where mobility in
either direction is shaped by underlying patterns of inequality (Chapter 21, this
volume; and, with somewhat different implications, the first volume of twin cities (Garrard & Mikhailova 2019, chapter 11). What African cases reveal is that,
even where borders separate countries with broadly similar profiles, border towns
still exhibit considerable diversity based on high in-migration levels (Nugent
2020). I refer to ambiguous affinity because, while Aflao and Lomé exhibit a
close relationship – reflected in trade relations and migratory flows – there is also
dissonance. In particular, there are different interpretations about who really
'owns' these spaces historically, and with what consequences. Moreover, tensions abound as national institutions meet at the border. At a mundane level,
this is reflected in interactions between border officials struggling to communicate, both literally and more figuratively. While Ewe is this border's lingua
franca, the official language is French in Togo and English in Ghana. Because

DOI: 10.4324/9781003102526-14

officials typically originate elsewhere, they are often unfamiliar with the other administrative language, the lingua franca or any of the other languages present. Meanwhile, administrative structures operate according to very different logics. This colours daily experiences for those whose livelihoods depend on negotiating bureaucratic obstacles day-by-day. I will probe three dimensions of the relationship between Lomé and Aflao: namely, their relationship's historical foundations; their distinct demographic trajectories, and the manner wherein the border affords sites of both interaction and friction.

Aflao and Lomé's historical emergence

To the naked eye, Lomé and Aflao appear a single conurbation with only an incomplete separating fence. One might reasonably conclude either that this was once a single town or that Aflao represents a spill-over from the Togolese capital. Neither interpretation is accurate. Aflao is a very old settlement, featuring in European traders' descriptions of the coast from the 17th century and quite likely emerging over a century before (Justesen 2005). In the mid-19th century, the area immediately east of Aflao was hardly populated, excepting some fishing villages around Bè. Bè inhabitants were apparently unrelated to Aflao's and, unlike the latter, faced the lagoon rather than the ocean (Marguerat 1992, 7). Hence, Lomé was very much a product of colonialism – a clue to which lies in its administrative district's grid layout, contrasting starkly with Aflao's irregular settlement pattern. We start, therefore, by exploring how this contrasting configuration emerged through successive migration waves and iterations of space-making in the colonial era and since independence.

During the centuries of the trans-Atlantic slave trade, Aflao was shaped by its intermediary position between the polities of Anlo to the west and Little Popo (or Anécho) in present-day eastern Togo. These polities traded most of the slaves on this coastal stretch, although the Danes acquired some directly through Aflao. In 1850, the Danish withdrew from the coast, after which the British slowly became more involved in what was now 'legitimate' trade, eventually loosely asserting a sphere of influence over the trans-Volta. The breaking of the Asante kingdom's power following the 1874 British invasion produced a shifting of trade routes east of the Volta River – to Aflao's potential benefit. Britain's decision to declare a colony in 1874 was largely intended to assert control over that trade. However, the initial Gold Coast border did not include Aflao, and consequently Denu (an Ewe name connoting 'by the border') flourished as a smugglers' haven a couple of kilometres west of the town. In 1879 Britain redrew the border to incorporate Denu and Aflao proper. Repeating the earlier pattern, a motley collection of traders – Sierra Leoneans, Anlos, Minas from Anécho, 'Hausa'[2] from the north, alongside two European trading firms (Marguerat 1993, 27) – settled just beyond the Gold Coast Customs cordon. Lomé therefore owes its origins to a particular moment when a border's imposition created ideal conditions for a flourishing contraband trade. Crucially, the medley of people who settled Lomé had no particular relationship with Aflao, although the latter claimed farming lands well beyond the border. Hence, the *quartier* of Kodjoviakopé, where the main border-crossing was located, was settled both by Anlos and Minas. The Anlo are an Ewe subset, but they have only modest attachment to the Aflao – who, besides providing Aflao with its name, are also another sub-set. Anécho's peoples speak Mina, which is close to Ewe, but trace their more distant origins to Elmina along contemporary Ghana's central coastline. They do not regard themselves as close ethnic kinsmen of the Aflao or any other Ewe subgroup. Significantly, Kodjoviakopé (literally meaning 'small Kodzo's village') derives from the name of the notional founder, Joseph Kodzovia Anthonio da Souza. He was a Brazilian merchant's son who had initially settled in Anécho before establishing himself at the newly defined border (Spire 2007, 193). Although da Souza was unrelated to the people of Aflao, he and his associates had close links with the settlers of Denu, a relationship that is valued today.

In 1884, the Germans announced their arrival on this coastal stretch, formally agreeing a colonial border with the British in 1890. This comprised a straight line running northwards from the coast to 6° 10′ north and then another running westwards until the Aka River – creating a dog's leg. The Germans moved the capital to Lomé in 1897, constructing an administrative centre, whilst having to accept that particular sections were already settled. They also constructed a wharf and three railway lines to funnel trade from the interior through Lomé. After 1900, the colonial capital attracted additional settlers searching for work and trading opportunities. Despite a period of Customs alignment, the fact that the British and German authorities could not reconcile their duties left ample scope for smuggling.

There was a brief moment after World War I when the practical significance of the border between Lomé and Aflao was effaced. After Britain and France invaded Togo, they provisionally divided the colony in a manner placing Lomé under British control. But the French argued hard for the railways and Lomé itself, claiming it provided a more natural port than Cotonou in Dahomey (Nugent 2002, 26–35). Britain conceded, entailing the former colony's physical partition. Whereas the interior was divided to place the Kpalimé-Lomé railway in French Togoland, the border's southern stretch remained as under the 1890 agreement. In the interwar period, a somewhat paradoxical situation arose whereby British trading firms like John Holt controlled much of Lomé's import-export trade, while the Gold Coast Customs Preventive Service sought to maintain a hard border. The reason was that Gold Coast/British Togoland did not levy direct taxation, meaning that customs duties were the principal source of government revenue (Nugent 2019, 174–176). Duties on Dutch gin accounted for much of those revenues. Gin imported by British firms was often purchased by smaller traders who then smuggled it into British territory. The terrain around Aflao was ideal for contraband trade. Some was transacted by sea under cover of darkness. The lagoons immediately north of Lomé's administrative district were particularly difficult to patrol. There were also many places where smugglers could headload goods across the border on foot.

After World War II, Ewe/Togoland unificationists sought to erase the border a second time, but there was little enthusiasm for this political project. In the 1960s and 1970s, successive civilian and military authorities in Ghana periodically closed the border to pressurise their Togolese counterparts – initially to force them to accept a territorial merger and increasingly to elicit cooperation in eradicating smuggling (Nugent 2019, 385–390). These closures were symbolically enacted at Lomé-Aflao where most traffic and people crossed. In fact, these measures signified the weakness of Ghana, where national economic implosion and mounting consumer shortages became increasingly evident from the mid-1960s. In Togo, a very particular social contract emerged, binding together a northern government and its base in the military, Lomé's women cloth merchants and the many cross-border traders (Nugent 2019, 449–452). The Eyadéma regime expanded the port of Lomé after 1968 which was consecrated as a free port serving the countries of the Sahel. But a generous tax regime also enabled Dutch textiles to be imported relatively cheaply, thereby cementing Lomé's dominant position within sub-regional commerce. The Asigamé textile market in downtown Lomé was controlled by the so-called 'Nana Benz' (Sylvanus 2016) – wealthy and well-connected female merchants, largely of Mina origin, with an acquired taste for Mercedes-Benz cars. They, in turn, supplied countless smaller traders transporting textiles to the interior or smuggling them across the border. But Ghana's endemic shortages meant all manner of consumer goods were traded across the border. Aflao's location meant it continued to occupy a leading position within contraband trading.

During the 1980s and 1990s, the Lomé-Aflao border became a site of recurrent tension. At the height of Ghana's 1982–1983 revolution, Aflao's population was subjected to constant surveillance as border guards and cadres attempted to enforce a curfew, border closures and price control. Although shortages in Ghana eased after 1985, Aflao continued to account for many smuggling cases brought before the public tribunals – with petroleum products and textiles topping the list. Complicating the local dynamic, both governments accused each other of subversion. The Togolese authorities felt especially vulnerable because Lomé was located hard against the border. In 1986–1987, after armed insurgents infiltrated from Ghana, the Togolese authorities erected a fence along the border's southern stretch. Work also began digging a trench, although this was eventually abandoned. During the early 1990s, tensions escalated once again as the Togolese regime sought to repress demands for a return to democratic rule. The violent suppression of demonstrations in Lomé, reflected in the many corpses washing up in the lagoon, produced a refugee exodus in 1993–1994. Only 10% were accommodated in the designated refugee camp at Klikor, most simply blending into Aflao and surrounding towns. This pattern recurred in 2005 during post-election violence in Lomé (Anon. 2005). These paroxysms had two important consequences: firstly, the instability killed much of Lomé's economic vitality; secondly, in the wake of taxi strikes, it prepared the way for a new pattern of urban transportation based on the *zemidjan* or motorcycle taxi. This also revolutionised cross-border trade.

What emerges from this brief account is that Aflao and Lomé's relationship since 1874 has been fundamentally shaped by trading opportunities, thereby attracting settlers from afar. Aflao's people have construed themselves as the landowners and most of Lomé's population as 'strangers'. A discourse of autochthony (i.e. about who had claim to be the 'original' peoples) has little traction in Lomé, where the meaningful contrast is between the city and its peripheries, of which Aflao is a very particular one. Nevertheless, people's economic fortunes on either side have been thoroughly entwined, while there has been sympathy in Aflao for opposition to a government construed as *nordiste* since the 1960s. Hence, distancing in a socio-cultural context contrasts with manifestations of empathy in the political domain.

Demographic patterns

With the historical context established, I now consider underlying demographic patterns. The reworked urban data of the OECD/Sahel and West Africa Club (SWAC) indicates that the population of border towns and cities across the region has grown considerably faster than national averages since 1950 (OECD/SWAC 2019, 21–33). Lomé and Aflao, located on a belt of virtually continuous urbanism running roughly between Lagos and Abidjan, are prime examples. Lomé's demographic growth has been driven, firstly, by the capital-city effect, reflected in proliferating civil-service positions and demands from urban transport and construction. Secondly, it has been spurred by commercial dynamics arising from the port's existence and cross-border trading opportunities. The city's cosmopolitanism arises from successive waves of settlers arriving from other Togo regions, especially the poorer north, and neighbouring Sahelian countries, like Mali, Niger and Nigeria. By contrast, Aflao is a provincial town at Ghana's geographical margin. Yet its demographic trajectory follows that of its larger neighbour and has been shaped by it. People have converged on Aflao from elsewhere in Ghana; also from the countries to the north.

Overall, population growth in Lomé since the 1950s has been sustained and impressive. In 1950, Lomé was a very modest centre, with fewer than 40,000 people; within 20 years it had become a city of 228,179 (Table 11.1). In 2015, Lomé's population was estimated at 1,700,000, meaning it had become 44.6 times larger. This was much faster than Accra, which had grown a 'mere' 28.1 times over the same period. Lomé's more rapid expansion clearly owed much to

Table 11.1 Comparison of urban growth patterns, 1950–2015.

City	1950	1970	1990	2010	2015	2015 pop. density per square km
Lomé	38,904	228,179	609,223	1,499,284	1,733,330	4,913
Accra	158,196	624,091	1,185,614	3,269,813	4,452,483	3,718
Aflao	2,996	11,397	26,393	50,699	84,649	6,927
Elmina	6,278	11,401	18,415	28,144	31,884	5,673

Source: OECD, Africapolis, Agglomeration level, https://stats.oecd.org/Index.aspx?QueryId=85686

its position at the epicentre of a well-oiled entrepôt state, supplying goods (especially textiles) to the entire West African sub-region and as far afield as Zaire. Whereas Accra entered severe recession during the 1970s, Lomé boomed as a commercial and financial centre. Entrepreneurs of different nationalities established themselves there. Unlike Lagos, whose prolific growth was oil-propelled, Lomé enjoyed a reputation as a physically secure city, with low taxes and functioning institutions – burnishing Togo's little-Switzerland image. Employment prospects attracted populations from the poorer north and across the Sahel, seeking work in the informal sector – especially around the markets – but also in the port, urban transport and in the construction industry, stimulated by a building boom. Aflao's growth was less spectacular, but it was nevertheless 28.3 times larger than in 1950, slightly above Accra's. By 2015, Aflao was over double Lomé's size back in 1950. A more meaningful comparison perhaps is with a non-border Ghanaian town. Aflao and Elmina were almost exactly the same size in 1970, but over the next two decades, the former increased its population by 230% whereas Elmina grew more steadily at 60%. Today, Aflao is 2.7 times larger than Elmina. We should also note that, whereas Elmina has both a fishing industry and a flourishing tourism sector, Aflao enjoys neither advantage. Its expansion is explicable only by Lomé's looming presence next door.

However, Lomé and Aflao's population growth shows significantly different spatial patterns. Lomé's expansion has been blocked by the sea to the south, wetlands to the north-east, and the border with Aflao to the west. The city has therefore snaked northwards, with one spur leading due north towards Agoué-Nyive, and the other following the contours of the dog-leg border, then broadly heading north-westerly towards Sanguera when it meets Ghana's border again (Figure 11.1). The relentless push northwards continues today. In the process, the city has progressively swallowed the countryside, absorbing rural areas once used by Aflao farmers. Before being engulfed, people from Aflao were selling produce to Togolese buyers – often northern civil servants and businessmen. Once the latter built houses, the city authorities caught up and inserted supporting infrastructure.

Aflao's population has generally concentrated in the narrow land-strip between lagoon and sea. This is reflected today in very high population density (6,927 people per square kilometre). A 1974 map, based on an aerial survey, shows most people bunched around Aflao's centre, with largely empty space between the latter and Denu (Survey Department, Map of Ghana 1974, 1:50,000, Sheet E0601C3). Today, satellite maps and direct observation reveal Aflao and Denu as a single conurbation. North of the lagoon, there had been a sharp distinction between the border's two sides. Whereas Togolese settlements were pushed hard against the line, the same 1974 map revealed a scattering of very small settlements on Ghana's side. However, satellite images today reveal significant population clustering also on Ghana's side, especially at the point where the border takes a sharp deviation. Hence the *quartiers* of Akosombo and Casablanca now have counterpart settlements on Ghana's side. This reflects two distinct processes. First, Aflao residents owning sufficient land, or being able to buy, have built houses in former green spaces. Constructing feeder roads, like

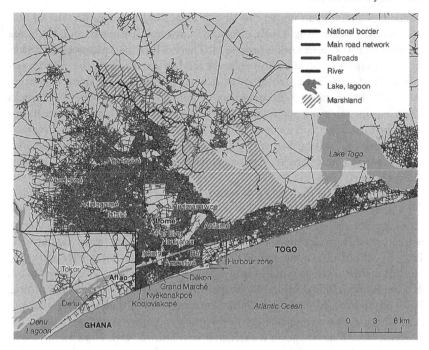

Figure 11.1 Lomé and Aflao.

Source: OECD/SWAC (2019), 'Population and Morphology of Border Cities', West African Papers, No. 21, OECD Publishing, Paris, https://doi.org/10.1787/80dfd9d8-en; by permission.

that servicing the Diamond Cement Factory, has partly fuelled this exodus to what was hitherto considered 'the bush'. Second, the pattern reflects an increasingly visible phenomenon of cross-border commuting. In 2019, the cost of rented accommodation in Lomé – CFA 9000–10,000 per room per month when a middle-range primary school teacher was earning around CFA 40,000 – was nearly twice that on Aflao's peripheries (interview, Nyekonakpoé, 20 March 2019). Hence, traders selling goods in Asigamé often prefer living on Ghana's side in villages like Kpakakope and Venavikope. Each evening as the markets close, one can witness a steady stream of people, walking or riding on the backs of *zemidjans*, as they return to their sleeping quarters. Many of these commuters originate from the Sahel, their presence signalled by mosques dotted around the border. At the crossing points, there are also stores – some surprisingly large – catering to customers living on either side.

Commercial ties

This draws attention to the economic connections shaping the larger demographic patterns. Today, there are four sets of commercial flows co-existing within the same spaces. Firstly, there is traffic passing along the Abidjan-Lagos Corridor. Heavy goods vehicles and buses currently complete their customs formalities at the main border-crossing at Aflao/Kodjviakopé. Because of limited progress in creating a fully electronic customs-clearing system, there are lengthy delays as paperwork is processed. There is little parking space for trucks around Kodjoviakopé, meaning vehicles tend to queue along the Aflao roadside – creating persistent congestion at what is effectively a narrow aperture used by both commercial and passenger vehicles. The slow procession of trucks has an upside, creating employment for those assisting with customs clearance – freight-forwarders and fixers of various kinds – and for others preparing food and buying and selling basics like cigarettes. There are also many currency changers and phone-credit vendors catering to passenger vehicles moving along the corridor.

Secondly, there is the trade still centring around Asigamé's market complex. This is where trade in imported textiles still continues, but other items like shoes, handbags, T-shirts and electronics are also sold on the fringes. This market is frequented by traders from across Togo, but it also retains wider significance. Ivoirien cloth traders come to buy in Lomé, maintaining that quality and range are better than cloth imported through the port of Abidjan. These days, much of it comes from China, which is cheaper than Dutch wax print. Each morning, buses arrive from Abidjan and Kumasi carrying women traders. They alight in Aflao sharing taxis to Asigamé, or walk across the border taking a vehicle on the Togolese side. This operation is extremely streamlined, with Lomé hotels catering purely to Ivoirien traders staying overnight, and companies packaging goods and returning them on trucks for modest fees. The following day, the women return to Aflao, boarding buses home. While waiting, they provide a captive market for myriad street-sellers frequenting the transport yard.

The third type of commerce comprises formal markets on both sides of the border opening on specific days. Although Aflao has a market, it apparently lost its lustre after the border was fenced. But Denu has a large market every four days frequented by Ghanaian and Togolese traders. Equally, Aflao traders sell in Lomé's various markets. Fish sellers come even from Ghana's western borderlands, whereas fruit and vegetable sellers originate closer to the border. Revolving markets within the borderlands provide a means whereby goods circulate over much larger distances.

Finally, there is local trade outside a formal market setting. Aflao traders procure goods from Accra, but also from more adjacent Asigamé. Conversely, traders with stores in Lomé often buy specific items from Aflao. For example, the steady trade in Ghanaian soft drinks which are considerably cheaper, and in the two main brands of alcoholic bitters for which there is significant demand. By contrast, very little bottled beer is traded either way. This reveals strikingly how some very specific consumer preferences have moulded themselves to the international border. Despite falling government subsidies, petroleum is considerably cheaper in Ghana, fuelling clandestine trade towards Lomé. There are nearly 20 filling stations on the road-section between Denu and Aflao – far beyond local demand or what passing trucks need. Much of the petrol is purchased by *zemidjan* riders who fill large plastic demijohns, transporting them to Lomé through unofficial routes.

Underpinning these different types of trade are distinct modes of cross-border transport. As indicated, those coming to Aflao from afar generally use taxis when crossing the main border. Market women carrying bulky consignments like fish do the same. Local traders often employ young men to headload their goods or transport them by handcart. But not all traders choose the official crossing. Those transporting smaller quantities of goods, to evade payment of duty, or avoid downtown Lomé, often find it convenient to employ a *zemidjan* to take them through an unofficial crossing. This typically involves negotiation with border officials on either side, and there are individuals making it their business to act as intermediaries in return for fees. Although many believe there is greater risk in using unofficial crossings – like robbery – lower cost means they will often take it.

Underlying tensions

Finally, we must note how cross-border interactions entail a complex mixture of mutual cooperation and tension. Many people have family members on either side of the border whom they visit and with whom they share lifecycle events like marriages and funerals. The concentration of people in Lomé claiming Aflao identity remains greatest in the northern suburbs nearest the border, like Adidogome, Totsi and Sagbado (Nyassogbo 2007, 213–215) – even though they are now a minority overall and constitute only 20% at Adidogome itself. And of course, traders and *zemidjan* riders constantly shuttle back and forth. During the 2020 border closure induced by COVID-19, there was a surge in *zemidjan* activity as residents sought to remain mobile. Ghana's immigration authorities increased border controls, and made many arrests, claiming arrestees were vectors for the disease.

While municipal authorities maintain arm's-length relationships, traditional authorities still communicate closely. Aflao's paramount chief, Togbe Amenya Fiti V, notes that some of his senior sub-chiefs are in Lomé, and he remains in close contact by phone. Chiefs and people from both sides participate in shared events, most notably during Aflao's biannual Godigbe festival. But Lomé is also a city of incomers, and many city dwellers know little about Aflao, normally only venturing there to catch transport to Accra. By contrast, everybody in Aflao is well-acquainted with the city next door because of its markets and urban amenities. Even around chieftaincy, there are marked differences. In Ghana, chiefs are autonomous from the state and expected to remain above partisan politics. In Togo, chiefs are part of the state apparatus, expected to align themselves with the government stance. Hence, the *chef de canton* of Djidjolé, whose origins lie in Aflao, is much more constrained than Togbe Fiti himself.

In Lomé, city politics are played out across urban space. Opposition to the government has been most pronounced in the city's southern *quartiers* – especially at Kodjoviakopé, Nyékonakpoé, Bè and, to some extent, Tokoin – areas mostly populated by Ewes and Minas. By contrast, northerners have mostly clustered in newer northern *quartiers* – like Agbalepedo, Totsi, Avedzi, Agoué and Djidjolé (Nyassogbo 2007). During acute political tension, protestors have managed to paralyze the commercial centre, whereas the northern suburbs generally remain quiet. In Aflao, different tensions manifest themselves. The town prides itself on its cosmopolitanism, and Togbe Fiti speaks especially highly of the Malians who have their own association and are highly organised. When Lomé experiences violence, Aflao people's sympathies are generally with the southern *quartiers*. Within Ghana, Aflao is reputedly solidly loyal to the National Democratic Congress (NDC) which refers to the area as its 'World Bank'. The New Patriotic Party (NPP) has repeatedly claimed that Togolese cross to register and to vote. This has produced the practice of closing the border during elections, but it is also reflected in ongoing efforts to purge the voters' roll of 'alien' voters. Although Aflao is well-positioned when the NDC is in power, whenever an NPP government takes over, it is excluded from the circles of influence and must deal with the reality of centrally appointed regional ministers.

Although there remains much cross-border daily interaction, residents also face several practical difficulties. The border fence the Togolese authorities unilaterally erected in 1986–1987 was always deeply resented as a barrier to cross-border movement. During the early 1990s protests, it was pulled down in many places – and never replaced. The two sets of authorities made some concession to local sensibilities, introducing a state of exception via a pedestrian portal at a place called Beat 9 (Figure 11.2). People living nearby could cross without normal border formalities. Although the system worked well for a time, officials now allegedly demand money for the privilege of crossing. Hence, one Lomé trader explained she had been sending her children to school in Aflao because she was Nigerian and wanted them to learn English. But, after detention for some hours while Ghana officials demanded significant sums of money, she

Figure 11.2. Beat 9 Pedestrian Portal.

Source: Photo by Paul Nugent.

had to withdraw her children from school (interview, Kodjoviakopé, 29 March 2019). The money demanded means many people prefer using the unofficial crossings assisted by the *zemidjans*. The fence is currently in poor repair, and one can view convoys of *zemidjans* and motor-tri-cycles lining up to accelerate through when it is suddenly pulled back. But this only makes sense for traders carrying goods in commercial quantities because a payment of CFA 5000 (around US $9) must be made to Togolese soldiers sitting nearby. At the time of writing, Ghana's government is reputedly exploring whether to construct a new border fence with surveillance cameras, which may well have dramatic effects on border interactions.

Even more worrying for Aflao people has been the opening of a One-Stop Border Post (OSBP) at Noepe, north-west of Lomé. This is designed to facilitate traffic flow along the corridor, by supposedly reducing congestion at Aflao. But it also reduces the possibility that protestors in downtown Lomé could interfere with traffic along the corridor. It is perceived as a particular threat in Aflao itself, whose commercial life revolves around what stops at the border. The decision to open the OSBP is widely interpreted as an attempt to punish Aflao citizens for their voting habits and conversely reward the town of Dzodze (the closest Ghanaian town to the OSBP) which has provided successive NPP Regional Ministers. Prior to the extended COVID-induced border closure of 2020, those involved in transporting goods still preferred using Aflao, leaving the OSBP mostly unused. But if the two governments ever insisted on closing the Aflao-Kodjoviakopé crossing to vehicles, the impact on Aflao particularly would be very serious.

Conclusion

Lomé and Aflao's relationship is clearly complex. Their destinies have been intertwined since the 19th century, and Aflao's rapid growth today is heavily influenced by proximity to the capital city. But while there are *quartiers* beyond the administrative centre with strong links to Aflao, most of Lomé's population traces its origins to other places. The relationships that have emerged are based on mutual advantage arising from trade and associated livelihoods. But a sense of national difference also plays itself out at the border. In Aflao, there is a profound sense of being Ghanaian, and the difference from being Togolese. Meanwhile, there is considerable acceptance of strangers from neighbouring countries who have helped enhance the town's reputation. Indeed, this receptivity is taken as quintessentially Ghanaian. In Togo, the conscious promotion of national identity is deeply bound up with the symbols pervading the capital city. Yet Lomé is also a city deeply divided ethnically and politically. Living in the borderlands therefore involves negotiating multiple levels of affinity and difference.

Notes

1 This article's underpinning research was conducted under a European Research Council (ERC) Advanced Grant for the project *African Governance and Space: Transport Corridors, Border Towns and Port Cities in Transition* (AFRIGOS) [ADG-2014-670851]. I would like to thank Kpatogbé Agossou for assistance during fieldwork

and additional input on Lomé's spatial layout; also Winston Stevens, Basil Sennor and Gabriel Shiador of the Ghana Shippers Authority for facilitating access in Aflao; and finally, Togbe Amenya Fiti V. I would also like to acknowledge the support of the Direction de la Recherche Scientifique et Technique in Lomé.

2 They may have been Yoruba Muslims rather than ethnic Hausa because where they settled was called Anagokope. Anago was the common name for Yoruba.

References

Anon. (2005), '15,000 refugees but no camps needed', *New Humanitarian* 26 May, available: https://www.thenewhumanitarian.org/fr/node/222238 (accessed 07.06.2020).

Buursink, J. (2001) 'The binational reality of border-crossing cities'. *GeoJournal* 54, (2001), 7–19. doi.org/10.1023/A:1021180329607

Justesen, O. (ed.) (2005) *Danish sources for the history of Ghana, 1657–1754, Volumes 1 & 2*. Copenhagen: Royal Danish Academy of Science.

Garrard, J., & Mikhailova, E. (2019). *Twin Cities: Urban Communities, Borders and Relationships over Time*. New York: Routledge.

Marguerat, Y. (1992) *Lomé: une brève histoire de la capital du Togo*. Lomé, Paris: Editions Haho & Karthala.

Marguerat, Y. (1993) *Dynamique urbaine, jeunesse et histoire au Togo: articles et documents (1984–1993)*. Lomé: Presses de l'UB.

Nugent, P. (2002) *Smugglers, secessionists and loyal citizens on the Ghana-Togo frontier: The lie of the borderlands*. Oxford: James Currey.

Nugent, P. (2012) 'Border towns and cities in comparative perspective", in Wilson, T. and Donnan, H. (eds.), *A companion to border studies*. Chichester: Wiley-Blackwell, 557–572.

Nugent, P. (2019) *Boundaries, communities and state-making in West Africa: The centrality of the margins*. Cambridge: Cambridge University Press, doi.org/10.1017/9781139105828

Nugent, P. (2020) 'Symmetry and affinity: comparing borders and border-making processes in Africa, in Fassin, D. (ed.), *Deepening divides: How territorial borders and social boundaries delineate our world*. London: Pluto,233–255.

Nyassogbo, G.K. (2007) 'Intégration or segregation ethnique: le cas du quartier d'Adidogomé', in Gervais-Lambony, P and Nyassogbo, G.K. (eds.), *Lomé: dynamiques d'une ville Africaine*. Paris: Karthala, 211–231.

OECD/SWAC (2019) 'Population and morphology of border cities,' *West African Papers*, No.21. Paris: OECD Publishing, doi.org/10.1787/80dfd9d8-en.

Soi, I. and Nugent, P. (2017) 'Peripheral urbanism in Africa: border towns and twin towns in Africa.' *Journal of Borderlands Studies* 32(4): 535–556, doi.org/10.1080/17531055.2020.1768468

Spire, A. (2007), 'Kodjoviakopé à Lomé: le temps et la constitution d'un terroir urbain', in Gervais-Lambony, P and Nyassogbo, G.K. (eds.), *Lomé: dynamiques d'une ville Africaine*. Paris: Karthala, 189–210.

Sylvanus, N. (2016) *Patterns in circulation: cloth, gender, and materiality in West Africa*. Chicago: University of Chicago Press.

12 Ketu and Imeko

Yoruba twin cities astride the Bénin-Nigeria border in West Africa

Anthony Asiwaju

Global African overview

Archetypical examples of Yoruba urbanization and replication (Mabogunje 1962; Igue 1979), Ketu and Imeko illustrate its manifestations in a typical African Cross-Border Area (CBA). The Yoruba are arguably Africa's most studied and best-known ancient city-dwelling people and civilization (Baldwin and Baldwin 1976; Akintoye 2010). This case study, drawn from this historic culture-area astride the Nigeria-Bénin border, is set within the emerging literature on African border towns (Nugent 2012; ABORNE 2019); and longer-established comparative borderlands research (Martinez 1986; Miles 2014).

Under 30 km apart in West Africa's Nigeria-Bénin CBA, Ketu and Imeko typify intertwined replica Yoruba ancestral cities and African border-twin cities. Although mutually distinct, the founders of Imeko are traditionally believed to have come from Ketu. The two Yoruba ancestral cities began and remained within the territorial jurisdiction of the ancient Yoruba Kingdom of Ketu (Parrinder 1956) until bitter and protracted rivalry arose in the late-18th century, becoming deadly conflict in the 19th. This ultimately weakened the kingdom, exposing its two leading cities to fatal invasion by Dahomey, the belligerent Fon kingdom (Akinjogbin 1967) to their west, reducing both cities to ruins (Imeko 1882 and Ketu 1886) until resettlement at the dawn of European colonial rule.

Both city histories recommenced in 1893. Their rebirths utilized the pre-existing sites and foundations, registering their traditional and inseparable connection to shared Ketu-Yoruba culture, ethnicity, lineage and kinship. However, their locations on different sides of the new Franco-British colonial border separating French Dahomey from British Nigeria, and thereby in two divergent colonial, later national, sovereign territories, meant they developed separately.

The discussion that follows is set within several interrelated contexts: first, wider contemporary African policy concerns about cross-border towns, promoting cross-border cooperation and deeper regional integration; second, Yoruba history as one of Africa's urban-dwelling cultures and civilizations; third, more specific conceptual issues around separate development, interdependence, inequality and permanence; finally, concerns about formalizing the largely informal cross-border relations. It draws heavily on many publications by myself and

DOI: 10.4324/9781003102526-15

others (notably Parrinder 1956; Asiwaju 2017), updated by my lived experience of residence in Imeko and frequently visiting Ketu across the border.

While some scholars contest the notion of 'twin' in 'border-twin cities' (Botia and Motta 2019; Jańczak 2019), it blends nicely with cultural nuances of replication in Yoruba urban culture and traditional religion which embrace multiple births, especially twins, and include *OrisaIbeji*, the god/goddess of twins, amongst their deities. Furthermore, Yoruba beliefs accommodate co-existing ideas about similarity, harmony, interdependence and convergence alongside realities about separate identity and divergent development. Thus, Yoruba ancestral-city evolution consists of successive generations (Igue and Iroko 1976; Igue 1979). Ketu and Imeko are not just twins; they exemplify Yoruba urban replications of considerable antiquity, embracing foundation and second-generation phases. Transformation from primordial to modern border-twins occurred in the later era, starting with the European-imperialist partition and boundary-delimitation of West Africa in the 1889 Anglo-French Treaty. This delimitation produced the inter-colonial, now international, boundary between former British Nigeria and French Dahomey, now the Republic of Bénin.

Primordial replication to modern border-twin cities

What follows are four historically and thematically overlapping sections exploring the themes of separation and togetherness distinguishing the history and experience of these cities, and suggesting that togetherness has become increasingly predominant, impacted especially by tribal, regional, even international, factors.

We begin with a sketch of the transition from primordial replication to modern border-twin cities. Like many other first-generation Yoruba cities, Ketu's foundation is usually attributed to a big-bang dispersion of 'direct descendants' of Oduduwa, acclaimed cultural hero and founding ruler of the Yoruba primordial 'mother' city and city-state metropolis of Ile-Ife. Imeko subsequently originated from Ketu (Parrinder 1956; Asiwaju 2017). It was first permanently settled on an abandoned homestead, now cultivated forest farmland, in the rift valley east of its present plateau location. Oral traditions hold the re-location (probably in the mid-17th century) was preceded by another century of shifting settlements amidst forest-savanna wilderness north and north-east of the present site. Now separated by the north-south Nigeria-Bénin border, Ketu and Imeko, the Yoruba kingdom's two major cities, are morphological replica border-twin cities. Border separation is mitigated by inseparability of more enduring kinds: inter-city social, economic and cultural communications; mediated by numerous overlapping satellite villages and small towns. Most prominent is Ilara, the spectacular binational town astride the border, the vital organ inseparably joining the Siamese-twin border cities (Gonsallo 1986).

Ketu and Imeko mirror each other, notwithstanding each evolving its own distinctive style and fashions. They both exhibit the four typical characteristics of Yoruba ancestral cities: (1) an imposing royal palace housing an *Oba* (King), uniquely distinguished by the customary beaded crown; (2) a foundation central

inner-city royal market, traditionally close to the palace; (3) fortifications, usually an encircling wall and moat; (4) a central city gate (*Akaba Ilu*), the wall's
only opening, in ancient times permitting entry and exit.

In Ketu, these irreducible minimums of Yoruba city-dom (Igue 1979) are manifest, making it not just typically Yoruba but also distinctively African. To start
with, the qualifying criterion of a prominent palace was met by one built on an
enclosed three-hectare area, only about one-third the size of the more elaborate
pre-colonial space before ancient Dahomey destroyed the city in 1886. Following the 1893 resettlement, the palace underwent several reconstructions.

The inner-city central market, typically close by the royal palace, functioned
in two conjoining spaces: one large (*Ojanla*), one small (*Ojakere*). Over the years,
other support-markets emerged, boosting Ketu's ever-expanding socio-economic
activities, connected to surrounding agriculturally productive rural communities. Most important was the Asena regional market outside the fortifications,
east of the inner city.

But of ancient Yoruba Ketu's historic structural features, none surpassed its fortifications. These included the massive encircling wall and accompanying deep
moat that 'at their greatest extent were over two miles long, encompassing an
area of some eighty-five hectares'. There was also 'the fortified gate … still preserved and … the best example of military architecture in Dahomey' (Parrinder
1956, 88) and wider Yorubaland. Outside observers were astonished, including
the Christian Missionary Society's Samuel Ajayi Crowther visiting in 1853 who
saw the fortifications as 'the most considerable walls in Yoruba country'. He felt,
following the Euro-centric mentality of the time, that 'a European must be the
giant whom tradition speaks of as their builder' (quoted Parrinder 1956, 88).
Significantly, these works, reconstructed and preserved at resettlement in the
early 1890s, recently became a UNESCO monument.

Imeko morphologically replicated 'parent' Ketu (Asiwaju 2017, 15–16). There
was no imposing palace until 1927, when Imeko's first Onimeko, Oba Durosinmi
Oyekan (1926–1937), was crowned, giving it traditional Yoruba 'chartered' status. Thereupon, a Yoruba palace structure duly emerged using the original ruling
house's Ijumu residence alongside two pre-existing ancient inner-city markets:
OjaLaku, big market, like Ketu's *OjaNla*, to the left, and *OjaKekere*, small market,
equivalent of Ketu's *OjaKere*. Also identical is the bigger space where traditional
foundation shrines are concentrated, providing the arena for major traditional
rituals, notably the annual *Oro* festival and popular *Gelede* in dance. With rapid
city growth, other markets emerged, especially the *Ajegunle*, whose weekly openings mirror Ketu's *Asena*. Imeko also acquired massive fortifications comprising
a semi-circular defence wall and deep moat, about 7 km, modelled on ancient
Ketu's and attributed to similar tradition of origin (Asiwaju 2017).

Other replications included Imeko's historic quarters and compounds, mostly
copying Ketu's, many identically named. More important here is that all these
vital cultural, ethnic and kinship ties have strongly enhanced the twins' mutual
attraction, notwithstanding occasional disagreements.

This context of solidarity overtopping separation allows us to understand a
statement in 1960 by the 48th Alaketu (King) of Ketu, Oba Adegbite Adewori

(1937–1963), who reportedly maintained that the border 'separated the English and the French, not the Yoruba' (Anene 1970, 18). This is now a dictum: notwithstanding the international boundary, every Alaketu must perform a pre-coronation ritual pilgrimage to certain historic Ketu-related towns on both sides beginning with Imeko, Ketu scion city in Nigeria (Parrinder 1956, 83).

Today these rituals have magnified into unprecedented big-time and elaborate visit exchanges by Yoruba Oba on both sides of the border, beginning in 1983. Aside from enhancing both cities' images, and Yoruba urbanism generally, the grassroots-level transborder solidarity by Yoruba royalties boosts bilateral policy initiatives for cross-border cooperation, strengthening wider integration drives at both West African sub-regional and continental levels. This official policy implication explains why local-level interaction between Yoruba royalties has been tacitly encouraged by territorial authorities on both sides. We will return to this later.

Paradox of separate development

We now focus on the intricacies of separate development within realities of interconnectedness. The problem is the effects of Anglo-French partition on two sovereign-state territories differing markedly in colonial and post-colonial background, alongside the ever-present inseparability of closely shared culture around the Ketu-Yoruba mother tongue, kinship and local economy.

At one level, the pulls have been in opposing directions of socialization into parallel colonization and nationalization processes. At another, important undercurrents of interconnection have flowed vigorously, hampering serious official engagement with sensitive issues of territorial distinctiveness of citizenship and national identity. The intricacies become demonstrable in Ketu and Imeko's interconnected development, each simultaneously a city within its own colonial and later national-sovereign territory, and a twin within a distinct trans-frontier region which 'although traversed by [an] international [boundary] nevertheless constitute[s] a unit' (Vedovato 1995).

Here, we can repeat that, aside their location some 30km from each other and now separated by an international boundary, Ketu and Imeko were founded at least three centuries apart. Subsequently, one has also steadily lost population to the other, due to sharply differentiated colonial regimes.

Ketu

Ketu, indisputably the older city, emerged not later than the 11th century. It has the longest-established king list of any Yoruba monarchy. However, it was founded in a uniquely difficult environment on an extensive red-laterite arid plateau about 100 m above sea level. While useful for defence, its location completely lacked drinkable water, had poor soil and thus needed to get its food supply from far-flung outlying farms and farming villages to its south, west and, especially, north and north-east. Thus, the walled city could never sustain a large urban population. Estimates are difficult for any pre-literate society, but varied between 15,000 and 30,000, with more outward than inward migration, with many groups and settlements, especially on the Nigerian side, able to trace their origins back to Ketu. The overall result for Ketu has been a generally austere if not plainly poverty-stricken traditional urban life.

Furthermore, the city's location in Yorubaland's western fringes put it on the frontier between Yoruba and Adja cultures, and thus close to hostile groups who eventually destroyed Ketu in the mid-1880s, enslaving many in ancient Dahomey and slaughtering many others. Ketu's population rapidly decreased and then totally dispersed – many Ketu Yoruba fleeing elsewhere in West Africa and founding descendant settlements as far away as Latin America and the West Indies.

It was those Ketu in Danhomean captivity, liberated when France conquered Danhomey in 1892, who regrouped and returned to reconstruct Ketu in 1893. They and other returnees unanimously crowned a new Alaketu of Ketu, Odu of the Mesa Ruling House. It was he who signed the Protectorate Treaty with France in January 1894.

However, out-migration continued in the new colonial period and beyond, due largely to the social destabilization of repressive French colonial administration, featuring, especially, military conscription, forced labour, burdensome taxation and other police-state practices, such as the *indigénat* (Asiwaju 1976). This produced widespread revolt by French West African subject populations, including massive protest migrations and walkouts by several groups along the border with British West Africa where autocracy was relatively mild. Consequently, the 'masses in flight' from French Dahomey always sought and found ready asylum among kith and kin on the British Nigerian side (Asiwaju 1976). All Ketu's huge population losses during the colonial period, especially until World War II, were direct gains for Imeko and the wider colonial 'Meko District' (Asiwaju 1976).

In the post-colonial period, most colonial-period migrants remained to permanently swell the population of Imeko and its satellite towns and expand rapidly growing refugee settlements in Oke-Agbede/Moriwi and Alagbe. In the 1970s and 1980s, the Marxist-Leninist Socialist Revolution, converting the Dahomey Republic into today's Bénin Republic, was also rather repressive, re-igniting mass migration from Ketu into Nigeria, especially Imeko. Local socialist revolutionaries were often antagonistic to indigenous cultural practices and institutions, deemed conservative and hostile to new revolutionary change.

Besides long-continuing and traditional population loss, Ketu's development as a leading Yoruba ancestral city was further compromised by systematic shrinkage in physical size and traditional prestige created by French colonial territorial administration.

Unlike most Nigerian Yoruba ancestral cities, many of which, like Imeko, became influential civil-administration centres, Ketu, until recently, enjoyed no such status, being tossed around in frequently shifting attachments to other French-colonial centres (Asiwaju 1976). Starting in January 1894 as a 'protectorate', it soon became an attachment to the 'Région de Sagon' – a remote and previously unimportant Fon village to its west. By 1910 it had been designated a 'canton' in the Cercle de Zagnanado, another unimportant Fon village further west. In 1915, following anti-French armed revolts in Ohori-Ije and Ketu itself, Ketu Canton became a military post attached to the newly-constituted 'Cercle de Holli-Kétou', headquartered in Ipobe, a smaller Yoruba town formerly sub-ordinate to Ketu. In 1934, Ketu Canton was re-attached to a new *Sub-Division de Zagnanado*.

Thus it remained until 1947 when, most politically detestably, Ketu Canton was attached to the 'Cercle d'Abomey', capital of the ancient Fon kingdom of Dahomey that had destroyed Ketu in 1886. In 1958 just before French Dahomey's independence, Ketu Canton became a *Sous-Préfécture* in the new *Préfécture de l'Ouémé*, headquartered in Porto Novo, French Dahomey's capital. It was the *Sous-Prefecture de Kétou* that has evolved into today's *Commune de Kétou*, attached to the new *Département de Plateau*, headquartered in Ipobe, not Ketu unfortunately. This melancholy story capped Ketu's long-continuing isolation, particularly from elsewhere in Dahomey, and now the Bénin Republic: there were no all-season roads in any direction, even eastwards towards Nigeria.

These historical vicissitudes meant Ketu failed to benefit from developments amongst most other ancient Yoruba capitals in Nigeria and, indeed, wider British Africa. When in Ashanti a place became a District headquarters, this ensured that 'its trade becomes brisker; it attracts immigrants ... receives more attention from the District Commissioner; and its prestige rises' (Busia 1968, 104). Ketu's history has apparently not borne out the enthusiastic hopes of its 1893 re-founders – that it would regain its status as a premier Yoruba ancestral city.

However, the situation has now significantly improved, thanks largely to increased political involvement by its Western-educated elite in national politics. This became especially marked with the democratic restoration following the ending of the former Marxist-Leninist state in 1989. Ketu has now become a stable local-government headquarters, comparable to contemporary Yoruba ancestral cities in Nigeria. Ketu's isolation has also been significantly reversed by an all-season network of motor roads, restoring its pre-colonial nodal location on historic trade routes.

Alongside planned roads within the old inner city, the last 20 years have seen massive and transformative urban renewal, funded by national government. This involves reconstructing Ketu's drainage system and modern road networks connecting the inner city with better-planned new outskirts: northwards along the Adaplamè motor road; westwards towards Ilara on the border and Imeko; and especially south-westwards to Ipobe, regional-administrative headquarters. This axis hosts cardinal state structures like the regional Customs and Excise Complex, the regional headquarters of the National-Gendarmerie and the central administrative offices for Bénin's recently established National University of Agriculture in Ketu. Here also is the largest concentration of elite-housing estates, high-star hotels plus many cafés and restaurants, servicing a growing modern urban population and expanding tourist industry.

The aforementioned historic *Asena* Market has been expanded and modernized. Its traditional weekly sessions conjure memories of its mid-19th-century heyday. It remains the ancient city's main regional market, also now linking the old inner city with the new expanding suburbs.

Overall, Ketu now boasts a good range of essential urban infrastructure: reliable pipe-born drinking water, stable public electricity and health services provided by a state general hospital and well-stocked pharmacies, all inter-linked with accessible more advanced medical services elsewhere including the capital, Porto Novo. As will be evident shortly, this urban infrastructure, especially in the health sector, is a leading factor pulling people from Imeko and related Nigerian towns into Ketu and its Bénin hinterlands. Ketu is no longer a 'shadow of its former self'.

Imeko

Imeko's contemporary history significantly diverges from Ketu's, while interconnecting at very important points buttressing the similarities. Both are currently headquarters for important local-administrative units in their respective countries: Imeko for the Imeko/Afon Local-Government Area of Nigeria's Ogun State; Ketu for the Ketu Commune of Bénin's Département of Plateau. Other similarities include the historical, kinship, ethnic and physical layout already noted, all set within sprawling modern suburbs. Both now have modern urban infrastructure, though quality and range appear higher on Bénin's side. Both are headed by autonomous traditional Yoruba Oba, honouring their expanding urbanity.

However, closer scrutiny suggests most similarities, even replications, disguise real differences of *type*, consequent upon divergent historical trajectories. Both share plateau locations, and like several other Yoruba urban replications (Igue and Iroko 1976), Imeko's plateau location partly flowed from a nostalgia about ancestral Ketu. Nevertheless, distanced by space and time, Imeko's early-17th century founders probably sought a less naturally hostile plateau.

Different the plateaus certainly were. In 1906, Imeko District's resident Commissioner, substantively a senior customs officer, Captain Butterworth, wrote 'Imeko is situated on the two edges of a plateau ... an area of 5 square miles about 580 feet above sea level, surrounding country sparsely timbered grassland'. He also noted extensive areas north and south of the ancient city, comprising 'successive plateaus, valleys and slight undulations'; and 'Good water from native springs ... all the year round' and 'the soil is rich' (Butterworth 1906). Furthermore, extensive fortifications, modelled on Ketu's, plus strong natural defences provided by deep rift valleys, enabled a flourishing mixed economy: initially subsistence agriculture, later cash-crop production, plus ample wild-game possibilities.

Other economic interests and activities soon followed from within and outside, supplying growing demands for basic needs like clothing and shelter, and specialized agricultural and hunting equipment. Imeko's site also evidences iron smelting and related metalworks of considerable antiquity (Asiwaju 2017). Also evident were material cultural productions: household pottery and woodworks, alongside leisure and social entertainments. Imeko became famous for cotton production and textiles. Its nodal position on criss-crossing ancient trade routes, especially east-west through Yorubaland from Bénin City to Ketu and beyond, fuelled early development.

All this made Imeko a Mecca for multi-directional inward migration including from Ketu. Indeed, by the 1760s, the Onimeko, 'chief of Meko began to feel himself more important than the chiefs of other villages under the dominion of Ketu ... (and) thought the time had come for him to wear a royal crown', sending envoys to Ketu to request the Alaketu's authorization (Parrinder 1956, 56ff). Refusal sparked the dispute eventually producing Dahomey's destruction of both cities in the 1880s.

However, once both cities were re-founded, Imeko's rulers resumed their claims, seeing them granted in 1927. The two cities were now separated by a colonial boundary, easing Ketu's consent needed for confirmation by the British Nigerian authorities. By now, Imeko was becoming a Nigerian, indeed 'African border boom town' (Asiwaju 2017). This has continued; it now has at least 100,000 inhabitants, spread over 1000 hectares of built-up area, with the historic inner city located in the core centre.[1]

This phenomenal post-resettlement development resulted largely from Imeko being made a District administrative headquarters by the British Nigerian colonial authorities. Unlike Ketu, where, as noted, the vagaries of French administrative and territorial administrations caused much local instability and disruption, right from the start of British colonial rule in 1893, Imeko was recognized as a very important town, headquarters of 'Meko District', one of 11 for administering the far-flung Western District of Lagos Protectorate, extending along 200 km of the Nigeria-Dahomey border. The regional headquarters was initially located in Badagry on the Atlantic coast. From there, British Travelling Commissioners conducted regular duty tours of this extensive area. Imeko retained this status throughout British Nigerian rule. Indeed, from 1900 to 1914, it doubled as headquarters of the wider regional administration, following transfer of the territorial headquarters from mosquito-infested Badagry to higher and healthier Imeko, much better for malaria-prone European Travelling Commissioners.

In 1914 Imeko lost this additional role to more central Ilaro. However, it retained its Meko-District Headquarters status. Indeed, it emerged as one of 11 local administrations eventually styled 'Native Authorities', each controlled by the Oba, the headquarter-town's traditional ruler combining executive, legislative and judicial responsibilities for its district.

All this impacted positively on Imeko's development and status, especially due to the widely acknowledged prominence of successive Onimeko and the esteem they commanded in assemblies of other Yoruba Obas at district, zonal and regional levels on the Nigerian side of the border. As 'sole native authority' for 'Meko District' in the heyday of British colonial over-rule, the Onimeko's awesome authority and influence largely accounted for Imeko's deliberate long-term open-door policy attracting those fleeing French colonial excesses in and around Ketu, and tempting them to join kith and kin across the border. His policy impacted beyond Imeko, drawing people from the whole of French-Dahomey's Ketu Canton into Meko-District's northern sector, producing numbers of new and ultimately sizeable Yoruba urban settlements, now satellite towns linked by history, culture and kinship.

Imeko's image has continued growing, partly due to contributions from certain key citizens, whose efforts complemented the traditional Onimekos during the colonial and post-colonial periods. They contributed schools, large churches and other institutions with regional, even national reputations. Most notable have been Prophet Frederick Akanbi Adeoye, founder of Nazareth High School, the first secondary school, and Reverend Samuel Bilewu Oshofa, founder of the Celestial Church of Christ, indisputably Africa's biggest white-garment church; also the famous Celestial City in the early 1980s, now International Headquarters for the Nigeria-Bénin cross-border indigenous African Church, making Imeko a Jerusalem of sorts, especially at Christmas when it becomes thronged with millions of pilgrims from across the world. The church is now planning its own Celestial International University in Imeko.

Interconnectivity and interdependence

As already evident, separate development under contrasting colonial, then national, regimes has not severed more enduring cultural, ethnic, kinship, lineage and nuclear-family knots joining Ketu and Imeko. If anything, enduring socio-cultural networks of intricate ties have been reinforced by equally binding nexuses of ever-thriving informal cross-border business transactions on an enormous scale, based on mutual trust. This has glued the two Yoruba border-twin cities, doubly so because, Siamese twin-like, they are inseparably co-joined by Ilara, the binational Ketu satellite town straddling the border mid-way (Gonsallo 1986).

Previous sections have begun exploring these themes. We now take them further. The important fact of Imeko's founding population primarily deriving from Ketu shows in commonplace encounters embracing many people. This entails members of many historic compounds, quarters and lineages in Imeko responding to identical lineage praise-names as kith and kin of more established Ketu homesteads. These long-enduring kinship ties are the key factors permanently linking the two cities since antiquity – borders, separate regimes and developments notwithstanding. It also explains why Imeko and District, before any other town or region on the Nigerian side, constituted the chosen alternative home base for kinsmen fleeing repressive French colonial administration. Such ties are continuously strengthened by unceasing inter-city cross-border marriages and child bearing, further enlarging family connections linking both cities, as elsewhere in cross-border Africa (Asiwaju 1985).

Reinforcing kinship is over-arching Ketu-Yoruba culture, notably strong commonalities of mother-tongue Ketu dialect, evenly-shared religious traditions and inter-faith harmony. Yoruba loyalty overlays differences between Muslims, Christians and indigenous beliefs. This expresses in shared respect for, if not active participation in, ancestral-city festivals, religious affiliation notwithstanding. Ketu and Imeko people commonly, if separately, celebrate the annual ancestral Oro festival. Ketu dialect has also produced distinguished oral artistry in widely respected traditions of indigenous poetry and diverse musical compositions. Most notable is the *Gelede* genre, the renowned multiple art form widely adopted by almost all Yoruba sub-groups. One expression of Ketu cultural excellence is *Gelede*'s recognition as one of man's 'Intangible Heritages' by UNESCO (Figure 12.1). Though it was artists from Kilibo in Sabe (Savè) north of Ketu

Figure 12.1 Gelede Mask, carved wooden headwear for the uniquely Ketu-Yoruba multiple art form proclaimed in 2001 as a UNESCO Intangible Cultural Heritage. This particular mask, showing a colonial administrator in a man-carried hammock, is the carver's impression of European colonialism as a blackman's burden. Artist: Sekoni Doga, Imeko, 1974.

Source: Photo by Adixphotography, Abeokuta, Ogun State, Nigeria, by permission.

in Bénin, that won the UNESCO competition, it remains the widely acknowledged historical fact that *Gelede* art originated in Ketu, enhancing deep cultural solidarity between Ketu and Imeko as leading diffusion centres.

Overall, despite separate development, certain underlying, wide-ranging and deep-laid inseparabilities have actively linked Ketu to Imeko. All this underpins vast cross-border business transactions and interconnected regional flows and mobility. This is based mainly on mutual trust between local operators on both sides, relying mainly on affinity and bonding networks. No reliable statistics exist, exchanges being mostly informal, unofficial, unrecorded, clandestine, contraband or smuggled, depending on who wears the lenses (Igue and Soulé 2005; Nugent 2002).

However labelled, these economic exchanges are enormous, impacting both towns' urban growth. Furthermore, this mushrooming informal-exchange system directly results from the border itself, notwithstanding border-control efforts (Amdii 1991). The impacts on Ketu's and Imeko's development and abiding bonds are best gauged by the phenomenal growth in physical size and social status of Ilara, the main hub and clearinghouse for cross-border business transactions and these twins' connection point. This is evident on both sides of this binational town: on Bénin's side, called Ilara Kanga on account of a deep well (kanga) dug there in the mid-1950s, and now hosting the warehouses of traders engaged in re-exporting mainly Asian rice and vegetable oil into Nigeria; and the Nigerian side accommodating multiple petrol-filling stations, 41 at the last research count in late 2016, far beyond local needs and thus obviously targeting the booming market on the Bénin side. Ilara has grown rapidly from a hamlet when border demarcation occurred in 1912 (Moulero, n.d.) to a full-grown Yoruba town with its own Oba since 1996.

Thus, while agriculture remains people's main occupation in and around Ketu and Imeko, cross-border business has long been the main wealth source for the border-twins' business elite. This will remain so while vast profits continue and territorial authorities on both sides continue looking the other way rather than squarely facing the task of harmonizing policy within wider regional-integration frameworks like the ECOWAS, African Union (AU) and even the Paris-headquartered international United Towns Organisation. Currently, the differences in the policy regimes mean that what is forbidden on one side of the border may well be what is permitted on the other.

This situation of parallel economies and asymmetric markets has been extremely profitable to cross-border exchanges and for practitioners and their partner corrupt border-enforcement officials on both sides. However, it tends to bring chaos and fragility to Ketu and Imeko's otherwise-strong inseparability. This sometimes surfaces, threatening the sustainability of the informal cross-border economy when one or other government decides to enforce the rules, producing colossal losses to the operators.

As the foregoing implies, there are dysfunctional and negative as well as functional and positive sides to inseparability. Obvious preference for durability raises the issue of formalization and institutionalization. African cross-border relationships, including and especially between border-twin cities, need to be purposively formalized. Hence, the policy-advocacy work over the years aimed at positively reconceptualizing borders between African states and interrelated border twin cities. Also needed is their systematic conversion from inherited negative colonial postures that see borders as barriers into positive proactive roles and functions as bridges (as in post-1945 Europe). This accords with Africa's post-colonial states' overall policy commitments to regional and continental integration. Accordingly, CBAs, including the twin border cities astride them, structurally marginalized and infrastructurally disconnected from core centres on either side, are required to receive a special development focus. This should enable the CBAs and constituent border twin cities to become cornerstones for much-needed deepening of African integration processes, as in the European Union where many such areas have become 'power houses' and border twin cities within them 'laboratories' of European integration (Vedovato 1995; Jańczak 2019).

Notes

1 Interview conducted by the author with Onimeko of Imeko Oba Benjamin Alabi Olanite in 2020.

References

ABORNE (2019) Annual Conference in Lome, Togo, on African Border Towns.

Akinjogbin, I.A. (1967) *Dahomey and Its Neighbours*. Cambridge: Cambridge University Press.

Akintoye, S.A. (2010) *A History of the Yoruba People*. Dakar: Amalion Publishing.

Amdii, I.E.S. (ed.) (1991) *100 Years of Nigerian Customs and Excise, 1891–1991*. Zaria: Ahmadu Bello University Press.

Anene, J.C. (1970) *The International Boundaries of Nigeria: The Framework of an Emergent African Nation*. London: Longman.

Asiwaju, A.I. (1976) *Western Yorubaland Under European Rule, 1889–1945: A Comparative Analysis of French and British Colonialism*. London: Longman.

Asiwaju, A.I. Ed. (1985) *Partitioned Africans: Ethnic Relations across Africa's International Boundaries, 1884–1984*. London: C. Hurst & Co. Publishers.

Asiwaju, A.I. (2017) *African Border Boom Town: Imeko Since c. 1780*. Ibadan: BookBuilders, Editions Africa.

Baldwin, D.E. and C.M. Baldwin (1976) *The Yoruba of Southwestern Nigeria: An Indexed Bibliography*. Boston: G.K. Hall.

Botia, C.Z. and J.A. Motta (2019) Tabatinga, Leticia and Santa Rosa: Emergence, Transformation and Merger of Paired and Triple Cities in the Amazon, in Garrard, J. and Mikhailova, E. (eds.), *Twin Cities: Urban Communities, Borders and Relationship over Time*. London: Routledge, 149–162.

Busia, K.A. (1968) *The Position of the Chiefs in Modern Political System of Ashanti*. London: Oxford University Press for International Africa Institute.

Butterworth, I.W. (1906) *Meko District Travelling Commissioners Diary Entry, 20/10/1906*. Ibadan: National Archives, District Commissioners Office Records, Meko, 1906–1909.

Gonsallo, G. (1986) *Le Rôle de Kétou, d' Ilara et de Meko dans la Dynamique de L'Espace Frontaliére Bénin-Nigeria, Maitrice, Université nationale du Bénin, Abomey-Calavi*, République du Bénin.

Igue, O.J. (1979) Sur L'Origine des Villes Yoruba, Bulletin de l' IFAN. Institut fundamental d'Afrique Noire, Dakar, Sénégal, Tome 41, Series B No. 2.

Igue, O.J. and F. Iroko (1976) *Les Villes Yoruba du Dahomey: l'Example de Kétou*. Universite du Dahomey.

Igue, O.J. and Bio G. Soulé (2005) *L'Etat Entrepot au Bénin: Commerce Informel Ou Solution a la Crise*. Paris: Karthala.

Jańczak, J. (2019) Central European Cross-Border Towns: An Overview, in Garrard, J. and Mikhailova, E. (eds.), *Twin Cities: Urban Communities, Borders and Relationships over Time*, London: Routledge, 203–216.

Mabogunje, A.L. (1962) *Yoruba Towns*. Ibadan: Ibadan University Press.

Martinez, Oscar J. Ed. (1986) *Across Boundaries: Transborder Interaction in Comparative Perspective*. El Paso, Texas: Texas Western Press.

Miles, W.F.S. (2014) *Scars of Partition: Postcolonial Legacies in French and British Borderlands*. Lincoln: University of Nebraska Press.

Moulero, T. (n.d.) Histoire des Villages des Environs de Kétou (Unpublished Manuscript).

Nugent, P. (2002), *Smugglers, Seccessionists and Loyal Citizens of the Ghana-Togo Frontier: The Lie of the Borderlands since 1914*. James Currey, Ohio University Press, and Sub-Saharan Publishers.

Nugent, P. (2012), Border Towns and Cities in Comparative Perspective, in Wilson, T.M. and Donnan, H. (eds.), *A Companion to Border Studies*. New Jersey: Willey-Blackwell.

Parrinder, E.G. (1956) *The Story of Ketu, An Ancient Yoruba Kingdom*. Ibadan: Ibadan University Press.

Vedovato, G. (1995) *Transfrontier Cooperation and the Europe of Tomorrow*. Strasbourg: Council of Europe.

Part II

International Twin Cities

C. In Asia

13 Dandong and Sinuiju

Twin towns on a fragile border

Tony Michell

Introduction

Guo and Freeman (2010) regard borders as regions potentially fragile in many ways: politically, economically, socially, culturally and ethnically. When a border divides twin cities, economic fragility can embrace one or both, whenever either country changes policy. Over the years, both China and Korea have fenced off their twin and limited contact, fearing too much interaction. The Ming dynasty once demolished all settlements within 10 km of the border. The Japanese taxed Korean products flowing into Manchukuo, while more expensive Japanese goods entered untaxed. The DPRK has failed to complete the modern four-lane cross-river bridge built by the Chinese in 2014, fearing complete economic penetration.

The China-DPRK border has attracted many commentators, it being relatively easy to visit the Chinese side and very difficult to access the DPRK's, making quick articles from the perspective of Chinese-side visitors journalistically easy. Some earlier visitors were similarly impressionistic. More scholarly articles include Lankov (2007), Lankov (2015) and Pardo (2013).

The Yalu River (reputedly a Manchu word meaning 'boundary between two countries') winds 795 km inland between North Korea and China. In 1018, the Koryo dynasty (935–1392) signed a treaty with the Liao, Manchuria's then dominant tribe, reconfirming the river-based frontier. This it still remains, through multiple regime changes on either side, with attitudes to trade weakening or strengthening the border area. In the everchanging relations between Han, Mongols, Jurchen, Manchu and Koreans, there were naturally schemes to commandeer the border territories by different actors. Each would have enhanced border fragility and was abandoned. On the south side, Uiju was the gateway city; on the north side, a group of five walled towns. In 1905, when the Japanese-built railway reached the Yalu, the engineers chose a crossing 15 km closer to the sea than Uiju, between a settlement named Sinuiju (New Uiju) and Antung/Andong (renamed Dandong in1965), one of the five Manchurian cities, hardening the crossing to one fixed point, allowing the twins to grow.

DOI: 10.4324/9781003102526-17

Regional context, development history and current situation

From 1018 to 1895, dynasties and regimes rose and fell, and while the border normally remained unchanged geographically, attitudes to its porousness changed with each Chinese regime. In 1894–1895 the Japanese, at war with Qing China, marched through the independent Kingdom of Korea and fought at the border crossing just south of Uiju. The western powers negated the subsequent treaty which would have given the Japanese the Chinese littoral as far as Dalian. Instead, Russia renamed Dalian, Port Arthur; and Antung was declared a treaty port in 1903 with British and US consular districts and a Japanese district with 7000 inhabitants by 1919 (Arnold 1918–1919, Vol. 1, 390). Ocean steamers could reach here at high tide, but many off-loaded into lighters delivering to both sides of the Yalu 25 km downstream. Massive rafts of timber floated down from the mountain areas.

In 1905 the Japanese returned, proclaiming Korea a protectorate, defeating the Russians close to Antung, and besieging Port Arthur. Now Japan acquired Port Arthur and control of what would become the South Manchurian Railway (SMR) from Port Arthur (renamed Dalian) to Shenyang and Changchun. A narrow-gauge railway, becoming standard gauge in 1911, linked Antung with the main line. The 'iron bridge', completed in 1911, defined the growing urban areas on both sides of the Yalu down to the present day, creating the connected twin cities which would share a joint economic destiny through changing political scenarios down to the current severe sanction-regime on the DPRK (starting in 2016).

Under the Treaty of Portsmouth, the Japanese controlled the SMR, but the territory around Antung and Shenyang was governed autonomously as Fengtian by a Chinese warlord. The SMR also controlled Korean railways until 1927, Chosun (Korea) having become a full Japanese colony in 1910. Japan developed docks and warehouses on the Chinese riverside, creating a thriving silk-weaving industry alongside a bean-milling industry feeding Korea and Japan. The most modern hotel in 1920 was a '25 minute jinriksha ride across the bridge to Sinuiju' (Arnold 1918–1919, 398). In 1928, Japan's assassination of General Zhang brought instability to the north bank, ending with the 10 September 1931 Mukden incident whereby Japan created Manchukuo, a puppet state. Although Japan now ruled Chosun as a colony and Manchukuo as a protectorate, differential policies produced tariffs at the Yalu border for Korean-manufactured goods.

In 1926, the year after the SMR returned control of the Korean railways to Japan's colonial government, Sinuiju's population was 23,893. By 1937, when work on the second (road) bridge began, it was 51,347 (approximately 16% Japanese and 12% Chinese) (Lautensach 1945, 256). As industries emerged after the Sup'ong Dam power station's completion in 1940, population grew from 61,143 in 1940 to 114,314 in 1941 and 127,706 in 1942. This massive surge produced the Majon-dong slums. The province's total population reached 1,800,800 in 1944, compared with 1,500,000 in 1930 (Bank of Chosun 1948). The Sup'ong Dam, supplying hydroelectricity to both sides of the Yalu, was then the world's second biggest, but ended timber flows on the Yalu, and other navigation along its length (Moore 2018).

Andong's (later Dandong) demographic history is less well-documented. The city's population in the early-1930s was probably around 45,000–70,000, with a mix of Japanese, Korean, Manchu and Han Chinese, but grew more slowly than Sinuiju's in the 1930s because Japan focused upon Dalian, Shenyang and Chang-chun (Arnold 1918–1919).

Perhaps 7000–8000 visitors passed from Pusan up the rail system in Man-chukuo in 1936 (Han 2005, 96). Train timetables were accelerated to coincide with steamers from Japan to Busan. Railway stock was remodelled with stream-lined locomotives and named trains (ibid., 97). The slogan 'breakfast in Pusan, dinner in Andong' appeared (ibid., 97). As Japan's control extended across north-ern China after 1937, trains left Busan and Seoul for Beijing and other major cities (ibid.). Japanese goods flowed freely along the line, while colonial Korean merchandise received less easy passage. The same process worked in reverse. But from 1941 all barriers between colony and protectorate were removed as part of the Japanese war effort, and further cooperation coordinated.

On 15 August 1945 Korea was split at the 38th parallel: within months eco-nomic reality was split, with major enterprises being nationalised in what would become the DPRK on 10 August 1946. Many were Japanese-owned, so already seized by both North and South. Equally Japanese living and working in North Korea including Sinuiju were rounded up, once the jobs they did could be done by Koreans, and deported to Japan, totalling some 700,000 across North Korea. Accordingly, Sinuiju's population dropped by around 30,000 to about 100,000. 1,500,000 Japanese were in Manchukuo at war's end, including in Andong. They were repatriated or self-repatriated in about three years (Sung 2011a; Maruy-ama 2016, passim). Sinuiju's urban economy shifted from 'colonial mercantil-ism' to state-planned national self-sufficiency over 2–3 years encountering total supply-chain disruption and subsequent shrinking of the private micro-business sphere. Some sense of this disruption can be seen in Kim Il Sung's speeches dur-ing a November 1945 visit where he ordered the use of state funds to stabilise the urban economy and trading with China. (Sung 2011b).

While in Sinuiju, Kim Il Sung talked to the secretary of the North Pyongan Provincial Committee about organising border guards amidst fears the Chinese Nationalist Army 'might advance to the borderlands of our country'. He decreed a strict-traffic order over the bridge and 'exercise(ing) control over all the peo-ple who cross it' (Sung 2011c). On China's side, no such changes occurred. In August 1945, Chinese Nationalist defeat was uncertain, but the Soviet Army and rural communist guerrillas occupied the area, and the Koreans sent a guer-rilla force to assist Chinese-Communist forces. The regional economy was in chaos with monetary inflation and confusion. The settlement of Andong, as distinct from its transit traffic, was far from the centre of interest: apart from Japanese-civilian departure and irregular cross-river contact, it was apparently left in gentle decay. When the People's Republic of China (PRC) was declared in October 1949, Manchuria still had a different currency from the rest of the country (until 1956) and was virtually a separate administration.

Shortly after order was established on the Chinese side, Korea was plunged into civil war. By November 1950, UN forces came far into Pyongan Bukdo, though not to the Yalu because Chinese 'volunteers' came to the rescue, pushing them back. Sinuiju was now an armed Chinese camp, with both bridges heavily bombed. Although bombed by the US Air Force until 1953, Sinuiju, protected by MIG-15s located safely in Dandong, created 'MIG-alley' along the Yalu. This became the major centre of Chinese supply and for parts of the diplomatic corps. North Korea began renewing its industry in the North-West – although munitions were mainly relegated to inland mountain regions nearby, where they still remain.

After the Korean War, North Korea quickly recovered. Rebuilding existing factories, using power from the joint Chinese-DPRK Sup'ong hydro dam, restored Sinuiju as a major industrial city, with chemical fibre, textiles and pulp factories prioritised alongside footwear, cosmetics and other light manufactures. Meanwhile, Dandong became once again Beijing's distant outpost, although important enough for North Korea to immediately open a consulate there. Sinuiju's urban population approximately doubled from 1953 to 1956, supported by Chinese reconstruction work which supposedly built 100,000 city-centre apartments, while Bulgaria built a hospital. A contemporary described the aim of this policy:

> Eliminating the colonial disproportions in industry, leftovers from Japanese colonial policy toward Korea during the occupation … is the rebuilding plan's goal for the national economy. The plan will anticipate broadening heavy industry, with the simultaneous reconstruction and expansion of those branches of industry that will contribute directly to increasing the working masses' material prosperity. With this goal in mind, the Korean nation should first … direct its efforts to rebuild and develop the metal, machine, mining, electro-technical, railway transportation, war and textile industries. Next … the reconstruction and expansion of agriculture is an equally important issue.
>
> (quoted in Jang 2006; History and Public Policy Program
> Digital Archive 1953)

Sinuiju was seen as a 'safe' location, and various new social institutions emerged, including a national disabled-veterans' college in 1953. It became a training school for disabled soldiers for administrative posts, probably growing faster than Dandong. By 1960, the national urban population had tripled since 1945, and Sinuiju was probably around 150,000, growing under the 7th Plan to 200,000 or more.

Immediately after the war, extensive Chinese aid helped rebuild and extend Sinuiju's factories, but Dandong apparently received little attention. Even by 1979, 71% of the local Chinese population were agricultural – compared with the DPRK's 40%.

This means that Andong (officially Dandong from 1965 – shedding its colonially shadowed old name) remained smaller than in 1945, with around 100,000 inhabitants. It reverted to being a service, rather than manufacturing, centre. Without the old system of Manchukuo, merchants carrying goods down through

Korea to the southern ports, or to Japan by sea from Dandong port, business decayed: Dandong, like many Chinese mid-sized cities, became small and sleepy. Japan had planned Manchukuo for the Japanese, and they were gone. Timber and soy-cakes were no longer exported to the rest of the empire. Non-residents could not legally move into cities at this time; instead, the Manchu population moved into the surrounding countryside where the PRC later created 11 autonomous Manchu zones. Equally, Chinese Koreans had little reason to move to Dandong especially as the relative linguistic freedom of the Yanbian Korean autonomous zone further north created a location where Korean was the official language neighbouring the DPRK's eastern border. In 1961, the Chinese expressed alarm that many Korean ethnic families were crossing into North Korea to reside. A border report of 10 May states: 'according to incomplete statistics, between January and April we have found 240 households and 1,471 people attempting to cross the border; of these, 715 people made it across ... to [North] Korea, but the rest were stopped.' The same report states,

> recently, we have also found people from Anshan, Shenyang, and surrounding areas...actively preparing to go to [North] Korea. There are also 1,843 ethnic Korean households in Andong ... preparing to move ... of these, 153 households have already sold all of their property and attempted to illegally cross.

(History and Public Policy Program Digital Archive 1961)

It was also reported that the North had established reception centres in Sinuiju and was giving a settlement allowance of 40 yuan, thereby attracting attention from ethnic Koreans on the Chinese side.

Things were different in Sinuiju. Sinuiju had one of the three things in short supply in the 1950s, 1960s, 1970s and 1980s in the DPRK: reliable electricity from the Sup'ung dam. The DPRK also has a planning system that tried to work. The five-year plan ending in 1960 was completed in 3.5 years, helped by Chinese, Russian and other communist-nation assistance. The 1961 seven-year plan, wherein Sinuiju was to figure prominently, set the goal of 'GDP' growing 2.5 times, and manufacturing production increasing so that industrial output in 1967 equalled the 1954–1959 total. While not completed successfully, Sinuiju's industrial strength was confirmed. Until the 1980s, the Chinese viewed Korea as the prosperous twin (Lankov 2007). Korea's urban population more than doubled between 1960 and 1980. Sinuiju's population in 1980 was only 271,000, and 289,000 in 1987 (+7%) while nationally the growth was 17% (Eberstadt and Banister 1992).

Zhou Enlai and Kim Il Sung signed a Friendship, Cooperation and Mutual Assistance treaty at Beijing on 11 July 1961. In the same year, the two countries signed a border treaty, based on the ethnicity of people living there. This had some weird results: Korean settlements were found on the river's Chinese side, creating the enclave of Hwanggumpyong well south of Dandong, and the island of Wihwa just north of the bridge. The Chinese worked hard to accommodate North Korea's requests about boundary negotiations, diplomatic cooperation, and economic assistance programs. China's own economic position was not strong, but it agreed a long-term loan in Russian currency (Cheng 2010).

Sinuiju maintained its population, and each factory tried hard to modernise its equipment and increase its equipment quota. Chinese students, especially ethnic Koreans, came to study at universities, although Pyongyang took most. Ships sailed from the Korean side of the river to a variety of destinations, and small ships were now built in the small shipyard, while Koreans proved much more diligent fishermen than Dandong's citizens.

Sinuiju remained the official crossing point for trade with China. But trade then was a matter first for the Trade Ministry in Pyongyang to agree barter deals with its Beijing equivalent. Meanwhile Sinuiju's market area – Chaeha market – was supposedly famous for the range of products on sale from the 1980s through to 2012, when it was moved to a new area to the south (Dormels 2014, 132). These would be long-term agreements wherein bureaucrats ticked off relative product totals assigned by documents. The agreements were handled by State-Trading Companies and the goods flowed through Dandong in railcars up the line to Shenyang and then Beijing, or from Nampho, Pyongyang's port.

A second set of barter agreements were signed between the two adjacent provinces, Liaoning and Pyongan Buk-do, whose goods would all be exchanged through the twin cities. These involved food, machinery, textiles, chemicals and seafood from Sinuiju, and minerals, agricultural produce and machinery from Liaoning.

Zhou Enlai made his famous speech in April 1969 while visiting Pyongyang about the blood-cemented militant friendship, 'China and Korea are neighbours as closely related as lips and teeth', promising the equivalent of USD 562,000,000 in bilateral trade (Cheng 2010). Subsequently, the two countries agreed a Chinese economic aid package, including technical cooperation and long-term commercial transactions for a North Korean Six-Year Plan (1971–1976) and protocols about the border railway and mutual supply chain of goods (Hong 2014). Thereby China began supplying North Korea with heavily subsidised fuel. Two parallel pipelines were laid between Dandong and Sinuiju in 1975–1976, and they remain North Korea's major source of crude and refined fuel (Lankov 2015).

From early 1978 to 2016

But change was about to happen in China, changing the balance of prosperity between the two cities with Deng Xiaoping's promise (18 December 1978) to open the economy through economic reform. First, agriculture was decollectivised and Han Chinese encouraged to move to replace pastoral Manchu farming with arable farming in Dandong's hinterland. Then private entrepreneurship and, finally, removing price controls produced rapid industrial growth. Thereby, Dandong rather than Beijing became the supply base for most of the DPRK, pulling Pyongyang closer to Dandong in terms of a more localised supply chain, especially as North Korea struggled after the 1989 global communist collapse. With collapsing border barter trade and demand for hard currency and cash transactions, this meant that goods, once flowing easily to Sinuiju's Chahae market, now tended to stay in Dandong until purchased by Korean merchants coming directly from Pyongyang. Consequently, Sinuiju lost part of its middle-men trade. In 1990 bilateral trade fell by 20%, reinforcing a reversal of fortunes as Chinese-style capitalism stimulated Dandong, because of its virtual monopoly over entrance and exit to the DPRK.

In 1992 diplomatic mutual recognition between South Korea and the PRC brought South Korean businessmen to Dandong to invest, and South Korean tourists to view North Korea across the Yalu. The growth process, slow initially, rapidly accelerated in the 2000s. The old harbour and dock areas and old Chinese town was swept away and a new city arose. Dandong's municipal GDP grew 15–17% annually, and population rose from 200,000–300,000 in the 1990s to 752,227 in 2003 and an estimated 1,200,000 in 2016 (Dandong Municipal Peoples Government n.d.). Official figures, counting only legal residence, give 854,000 for the urban area in 2018, excluding Dandong Port, when migrant numbers were reduced.

Market reform came late to Sinuiju in 2002–2003; by this time, the lead built by Dandong pulled Sinuiju closer into Dandong's orbit, with the Chinese developing cross-river industries and workforces exploiting outsourced cheaper labour in the DPRK, while cheap Chinese products flowed across the bridge and throughout the DPRK's new market system. Vendors were largely Chinese merchants penetrating through to Pyongyang and converting demand for hard currency from dollars to yuan. Only around 2012–2013 did North Korea begin producing its own basic consumer products competitive with Chinese products.[1]

The Dandong-Sinuiju border-scape 2016–2020

The two bridges, one broken (Figure 13.1), one heavily used, accommodating trains and an alternating one-way column of trucks with occasional workers' buses, are the current foci of inter-city links. One city, a modern one of Chinese-style high-rise, busy day and night; the other, also high-rise, with a new thirty-storey monster building built in 2018–2019 directly opposite the broken bridge, unpassable since the Korean War, lying in apparent DPRK darkness with limited activity after about 8 pm, except at the railway station and customs area which never seem to sleep. Despite famous satellite photos of the DPRK in darkness and China and South Korea brightly lit, Sinuiju is not in darkness, just very dimly lit, especially with limited streetlights which cannot catch the satellite cameras. Dandong feels like a modern city, Sinuiju much more than Pyongyang presents little street life. The taxis plying the streets are from a Chinese Joint Venture. Most industry stretches along the riverbanks from the huge textile factory site upstream from the bridge, with its own berths, and then downstream, several kilometres of old warehouses and industrial buildings with major factories just behind. Between Sinuiju and South Sinuiju is a green belt due for filling under plans announced in late 2018. South Sinuiju is low-rise, with more educational facilities and space. The original walled city of Uiju to the North has been rebuilt as a tourist centre.

Figure 13.1 The broken bridge from the Sinuiju bank.

Source: Rowan Beard Young Pioneers Tours, by permission.

While Sinuiju has retained its 1945 format, Dandong has been transformed by pushing its maritime business southwards down the river with little industry in the centre and emphasis on the residential district, partly eying up property speculators from Beijing and Seoul. Comprehensive development has occurred on an island almost opposite the end of Sinuiju's shoreline industries – known as Moon Island development (Trip.com n.d.). However, the development proved overambitious given the economic slowdown following UN and US sanctions.

While the railway remains in its present position, the opening of the high-speed train to Shenyang, completed in 2017–2018, necessitated the station's massive development as a passenger terminal, pushing freight yards to the side, both towards the sea and inland. Industry had also left the main centre, moving to cheaper land near the intersection of the expressways G1113 running through Benxi to Shenyang in one direction and G11 to the coast and beyond in the other.

Further downstream is the 'bridge to nowhere', currently unfinished, a sign of Chinese confidence and DPRK worries about border fragility. On the DPRK side, the new bridge is far from Sinuiju and in direct line for Ryongchon on the Pyongyang highway. On Dandong's side, a new industrial and commercial estate has been developed including a free-trade area or export zone.

Finally, further south on the Chinese side, bypassing the barbed-wire fences of Hwangyoumpyong, awaiting its second foreign investor, comes the development of the port of Dandong, Donggang City, a sprawling area of grain elevators, fishery factories and small- and medium-enterprise industries plus extensive container yards and railyards. Donggang is a separate legal entity with 640,000 de jure inhabitants in 2018. Dandong airport is nearby. The port was billed to become a big North-East China hub, but this depended on free access to the DPRK and a comprehensive network allowing goods to go by river and coastal craft from the port to Sinuiju's warehouses and industries. It also depended on container lines being established to Shanghai, Hong Kong, Nampo, Incheon, Busan and Japanese ports. This remained far from completion when sanctions hit Dandong and Sinuiju in 2017. As it is, Dandong is outclassed and out-li-nered by Dalian, so that even cargo ships sail from Dalian to Nampo bypassing Dandong: Dalian being the container terminal for the Manchurian provinces' connection with the world. Dandong Port Company defaulted on bonds in 2017 becoming further indebted in 2018 (Leng 2018).

Trade and businesses linking Dandong and Sinuiju are based around the Square near the old bridges in what was the original town (Zhenxing District), now completely rebuilt. Here is the DPRK consulate, plus many business offices – North Korean, joint-venture and Chinese – aiming to deal with trade and investment on both sides of the border. Off-centre is the un-rebuilt Korean town. In these confined areas is a major centre of the DPRK's legal/illegal trade con-ducted under cover of Chinese relative freedom, described in a 2020 report as a 7 billion dollar border (Pardo 2013; Byrne et al., 2020). Here the authors claim about a quarter of the DPRKs external economic trade is handled.

While Dandong has abandoned river traffic, Sinuiju's riverside port remains little altered with rusting ships, some moving only at high tide. But rapid riv-er-mouth growth has given Dandong a new seaport free of tidal constraints. In 2012, Dandong Port achieved a cargo throughput of 96,060,000 tons and a con-tainer volume of 1,250,000 containers. The 2020 cargo throughput was expected pre-COVID to reach 120,000,000 tons, with container throughput sprinting to 1,500,000 TEUs (Twenty Foot Equivalent Units) about the same volume as Southampton. It was claimed as becoming an international commercial port in Northeast Asia (Xu 2013). Zhang Hongjiang, Dandong Port Group's vice-presi-dent, said in 2013: 'Throughput is not the only measure of an international port. Dandong Port aims to build third- and fourth-generation facilities characterised by modern logistics centres, information centres, and resource allocation cen-tres'. He expected goods from Northeast China's 16 cities to flow to Dandong adding 100,000,000 tons of new traffic. (Xu 2013). The year 2013 marked the peak of Dandong's Gross Regional Domestic Product with a slight contraction in 2014 and 2015 as the DPRK began manufacturing its own consumer products (CEIC n.d.).

Sanctions on the DPRK

The foregoing ambitious plan ran into trouble on two counts. First, Dalian's competitive power proved much stronger than Dandong Port Company esti-mated, and the logistics savings minimal. Second were the sanctions imposed on North Korea in steadily increasing severity from the 2006 Banco di Delta case onwards, limiting international finance and scaring off most investors. On 24

May 2010 South Korea prohibited South Korean companies from trading with North Korea, except for processing in the Kaesong Industrial Estate. From 2015 to 2017, a series of United Nations Security Council sanctions were added, making nearly all DPRK exports illegal, backed by secondary US Treasury sanctions virtually freezing all FX exchange between China and the DPRK through the banking system and sanctioned Dandong enterprises. So seafood, agricultural produce, coal, iron and steel, machinery and textiles could now be handled only by smugglers. UNSC bans on exports to North Korea were supposed to prevent almost everything except agricultural products from leaving China for Korea.

Dandong official sources initially estimated its GRDP fell by one-third in 2016, illustrating how much the city's prosperity depended on its twin (or at least relations with the DPRK generally), later adjusted to about 25% (CEIC n.d.). By 2017 the Port of Dandong was bankrupt (Jiang 2020; Bloomberg 2020). Workers formerly subcontracted by Dandong in the clothing business returned from Sinuiju to the countryside in 2018 and 2019. The DPRK closing its borders in February 2020, to prevent COVID-19 spreading from its neighbour, enhanced the problems.

Future relations between the two cities and future plans

If China continues enforcing UNSC sanctions and cannot, alongside Russia, persuade the UNSC to moderate them, Dandong must simply become a Chinese regional city with less additional benefit from its border location and reduced activity with its twin across the river. Essentially, this is already happening. Sinuiju depends on people moving to work in Dandong and beyond, and on Chinese tourists making short trips to the city and neighbourhood.

If sanctions are lifted, then Dandong rebounds, as does the DPRK and all the promising trends of the years 2001–2016. Government plans differ over time. From 2005 onwards, Chinese provincial planning in North-East China emphasised making Dandong a logistic hub for the whole region upgrading expressways and railways, infrastructure completed just as the severe sanctions began. By contrast, until the attention paid by Kim Jong Un to Sinuiju city planning in 2018–2019, little attention had been given to Sinuiju after the collapse of the ambitious plans promoted by making the Chinese billionaire Yang Bin head of a greater Sinuiju administration region in 2002. Greater Sinuiju had included the DPRK SEZ of Hwangyoumpyong, the unexploited stretch of land on the Yalu's Chinese side. The Chinese arrested Yang for tax evasion in 2003. This was followed by Jang Song Thaek's plans – cut short by his execution in December 2013 (Bodeen 2013; O'Neil 2003). In theory, the DPRK could have completed the bridge to nowhere by finishing it any time after 2014. In practice, the bridge would require entirely restructuring Sinuiju's infrastructure, alongside Nam Sinuiju's and Rongchon's, because the direct route to Pyongyang would bypass the first two.

Fragility is a persistent issue, with changes on one side of the border, whether in terms of economic policy or even logistic infrastructure, creating changes in the balance with the other twin. Today the border is again a fragile region with the sanctions burden hanging almost equally on both cities. This is compounded by the DPRK sealing itself off in February 2020, because of fears of COVID-19 crossing the border via Chinese or returning Korean citizens. Each twin has been pushed back on its own resources for the first time since 1911.

Notes

1 Here and in other parts of the text I refer to unpublished interviews I conducted in 1998–2019.

References

Arnold, J. (1918–19), *Commercial Handbook of China* Vol. 1, Washington DC, Government Printing Office, 378-401.

Bank of Chosun, (1948), *Annual Economic Review of Korea Statistical Tables*, Seoul.

Bloomberg (2020), China Port Defaulter's Bankruptcy Ruling Stirs up a Storm. Available: https://www.bloomberg.com/news/articles/2020-01-13/china-port-defaulter-s-bankruptcy-ruling-stirs-up-a-storm (accessed 03.03.2021).

Bodeen, C. (2013). *Jang Song Thaek's Execution leaves China in a Very Difficult Position* Bloomberg. Available: https://www.businessinsider.com/jang-song-thaek-execution-and-china-2013-12 (accessed 03.03.2021).

Byrne, J., Byrne, J., Sommerville, G. and Macdonald, H., (2020), *The Billion-Dollar Border Town Project-Sandstone Report No 7*, London, Royal United Services Institute. Available: https://rusi.org/sites/default/files/billion-dollar_border_town_final_web_version.pdf (accessed 03.03.2021).

CEIC (n.d.) *China GDP per Capita*, Liaoning: Dandong. Available: https://www.ceicdata.com/en/china/gross-domestic-product-per-capita-prefecture-level-city/cn-gdp-per-capita-liaoning-dandong (accessed 03.03.2021).

Cheng, X., (2010), The Evolution of Sino-North Korean Relations in the 1960s, *Asian Perspective* Vol. 34, No. 2 (2010), 173–199.

Dandong Municipal Peoples Government (n.d.). City's official website.

Dormels, R., (2014), *North Korea's Cities: Industrial Facilities, Internal Structures and Typification*, Jimimoondang.

Eberstadt, N. and Banister, J. (1992), *The Population of North Korea*, Berkeley, Institute of East Asian Studies.

Guo, R. and Freeman, C. (2010), *Managing Fragile Regions Methods and Applications*, New York, Springer-Verlag, 1–46.

Han, S.-J., (2005), From Pusan to Fengtian: The Borderline between Korea and Manchukuo in the 1930s, *East Asian History*, No 30, 91–106.

History and Public Policy Program Digital Archive (1953). Report No. 4 of the Embassy of the People's Republic of Poland in the Democratic People's Republic of Korea for the Period of 26 June 1953 to 31 July 1953. Wilson Center. Available: https://digitalarchive.wilsoncenter.org/document/114955 (accessed 03.03.2021)

History and Public Policy Program Digital Archive, (1961), Cable, Ministry of Public Security Party Committee to Zhou Enlai, Deng Xiaoping and Pengg Zhen, Wilson Centre. Available: https://digitalarchive.wilsoncenter.org/document/115323 (accessed 03.03.2021)

Hong, S., (2014), What Does North Korea Want from China? Understanding Pyongyang's Policy Priorities towards Beijing, *The Korean Journal of International Studies* Vol. 12-1, 277–303, doi.org/10.14731/kjis.2014.06.12.1.277.

Jang, S.H. (2006) *Urbanization Process of Big Cities in North Korea: Focusing on the Changes in the Spatial Structure of Chongjin, Sinuiju and Hyesan* in The Society of North Korea, Kyongin-Munhwasa, 455–509.

Jiang, J., (2020), China Merchants Takes Over Dandong Port in Controversial Court Ruling, Available: https://splash247.com/china-merchants-takes-over-dandong-port-in-controversial-court-ruling/ (accessed 03.03.2021)

Kim Il Sung (2011a), Meeting with Self-repatriating Japanese, in *Collected Works of Kim Il Sung 2011*, Vol 2, Dec 21, Pyongyang, 410–415.

Kim Il Sung (2011b), On Organizing the Border Guards, in *Collected Works of Kim Il Sung, 2011*, Vol 2, Nov 27, Pyongyang, 326–327.

Kim Il Sung (2011c), Talk to Traders, Entrepreneurs, Doctors and Christians in Sinuiju, in *Collected Works of Kim Il Sung*, Vol 2, Nov 26, Pyongyang, 311–318.

Lankov, A. (2007), Tale of Two Cities: Dandong and Sinuiju, Available: http://www.koreatimes.co.kr/www/nation/2020/02/165_10839.html (accessed 03.03.2021)

Lankov, A. (2015), Over the Border: What Dandong Means to N. Korea. Available: https://www.nknews.org/2015/08/over-the-border-what-dandong-means-to-n-korea/ (accessed 03.03.2021)

Lautensach, H. (1945), *Korea, eine Landeskunde auf Grund eigener Reisen und Literatur.* Leipzig, Koehler-Verlag.

Leng, S., (2018). China's Biggest Port Trading with North Korea Defaults on US$150 Million Bond. Available: https://www.scmp.com/news/china/article/2117825/chinas-biggest-port-trading-north-korea-defaults-us150-million-bond (accessed 03.03.2021)

Maruyama, P.K., (2016). *Escape from Manchuria.* Litfire Publishing.

Moore, A.S., (2018). From "Constructing" to "Developing" Asia – Japanese Engineers and the Formation of the Postcolonial, Cold War Discourse of Development in Asia, in Mizuno, H., Moore, A.S., and J. DiMoia (eds.), *Engineering Asia: Technology, Colonial Development, and the Cold War Order*, London, Bloomsbury, 85–100.

O'Neill, M., (2003), Tycoon Yang Bin Gets 18 Years Jail, Available: https://www.scmp.com/article/421723/tycoon-yang-bin-gets-18-years-jail (accessed 03.03.2021)

Pardo, R.P., (2013). Dandong and Sinuiju: The Sino-North Korean Border Shadow Economy. In *Dirty Cities*, Palgrave Macmillan, London, 89–109.

Trip.com (n.d.) Moon Island. Available: https://www.trip.com/travel-guide/dandong/moon-island-22867160/ (accessed 03.03.2021)

Xu, Y. (2013), The Throughput of Dandong Port Exceeded 100 Million Tons and Officially Joined the Dagang Club, Available: http://www.chinanews.com/gn/2013/12-23/5652663.shtml (accessed 03.03.2021)

14 Zabaikalsk and Manzhouli
Dynamic asymmetry

Vladimir Kolosov

The twin cities of Zabaikalsk (Trans-Baikal Krai, Russia) and Manzhouli (Inner Mongolia, China) are examples of sharp asymmetry in twin-city development on a contrasting ethno-cultural boundary. They were founded almost simultaneously thanks to the Chinese-Eastern Railway's (CER) construction connecting the then-Russian naval base at the Pacific Ocean's Port Arthur (now part of China's Dalian city) via the Trans-Siberian Railway with Moscow through Manchuria. This was the shortest route. Despite their relatively short history of c100 years, they have undergone tremendous geopolitical changes. Manzhouli and Zabaikalsk emerged as outposts of Russia's colonization of Manchuria. Initially, both were modest villages near railway stations. After the Russian Revolution during China's occupation by Japan, the twin settlements became Soviet-Japanese border posts. With World War II's termination, they became strategically important transit-points for transferring massive Soviet economic aid to communist China. Then, for many years, they were doomed to being closed garrisons facing each other. After the USSR collapsed, they became again transhipment bases for powerful trade flows, and intensive, but peculiar, interactions were restored between them. However, Zabaikalsk has changed little in past decades, while by the early 21st century, Manzhouli had become a substantial city thanks to the skilful use of border rent and post-Soviet Russia's proximity.

This chapter analyses the historical dynamics, and reasons for Zabaikalsk and Manzhouli's asymmetry as a characteristic twin-city feature (Garrard and Mikhailova 2019) and their functions, some derived from Russo-Chinese interactions, others focused on the domestic market.

Vicissitudes of history

Russian settlers built Manzhouli station (Manchuria in Russian) and its nearby village in 1903. Two decades later, just across the international border, a siding No. 86 appeared – the future Zabaikalsk. Throughout the 19th century's second half and most of the 20th century, Russia dominated Russian-Chinese relations, rich in dramatic and abrupt changes. The border between the two countries was formed mainly on Russian terms. From around the 1850s, the Russian Empire actively settled the Far East, developing its resources.

DOI: 10.4324/9781003102526-18

Even after the Aigun and Beijing treaties were signed, wherein the Russo-Chinese border was delimited, until the late 1920s it remained 'transparent', allowing Chinese citizens to freely settle and conduct economic activities on Russian territory. Russian and Chinese settlers (and other nationalities) actively explored the Far East's and Manchuria's sparsely populated territories, facilitated by the CER's construction and the Trans-Siberian's Amur branch to Vladivostok. In 1896, Russia achieved extraterritorial status for the entire right-of-way along the CER, including all settlements where subjects of the empire obeyed only Russian legislation and were protected by Russian troops. The vast region between the two main railways began being called Chinese- or Yellow-Russia, displaying a mixed population and active intercultural contacts (Blakher 2014). On Chinese territory, in cities and towns along the CER, Russians comprised sometimes (at least in Harbin) a third of the population. Nowadays, Chinese cities along the former CER still keep the Russian architectural heritage, including Manzhouli. As Billé (2012) shrewdly notes, this situation corresponded to traditional Chinese understanding of the interstate border not as a rigidly fixed line, but a constantly moving frontier, an interactive zone with neighbouring peoples and states, thereby a space of opportunities.

In Russia, on the contrary, in the 19th and early 20th centuries the Chinese border was viewed as a permanently threatening front line. Russian and other settlers extensively used cheap Chinese labour. However, they constantly felt the 'yellow peril' from Chinese criminal groups, Chinese attitudes towards Russians as occupiers, the closed nature of Chinese communities, and their disobedience to Russian laws and cultural differences. These often aroused sharp rejection from Russian settlers (of unsanitary conditions, widespread opium addiction and gambling, etc.). Russian settlers exhibited hauteur in relations with the Chinese. Unsurprisingly, in Manchuria, anti-colonial actions became anti-Russian (Vendina 2018).

Fear of the 'yellow peril' forced Russian authorities to constantly try tightening the wide frontier. Thus, in 1886, foreigners were theoretically prohibited from settling in the border zone, they were no longer employable for public works, and in 1913 a free-trade ban followed in a 50-*versts* (about 53 km) zone along the border (Kireev 2009). Following Bolshevik victory, most Chinese did not accept Soviet citizenship, and Russians in China were regarded as former colonizers and occupiers.

Armed conflict around the CER (1929), after China's attempt to seize it, induced complete border closure. Manzhouli's vicinity became a main direction of the Soviet offensive, with siding No. 86 being named *Otpor* (Repulse). Japan's occupation of north-eastern China and the creation of a dependent state of Manchukuo (1932), supposedly to combat the communist threat and restore Manchu statehood, further complicated the regional situation. Military conflicts constantly arose on the border. Waves of anti-Chinese repression swept across the Far East. During the great terror, as elsewhere in the USSR, a total cleansing of Chinese, Koreans and other 'unreliable' elements along the mostly 100 km border strip was enacted and a strict border regime introduced. Across the border, Russians and other immigrants from Russia were similarly repressed.

Following World War II, Manchuria again became part of China, and the Soviet-Chinese border was restored to its late 1920s location. Following communist victory in the Chinese civil war and creation of the People's Republic of China, the USSR provided massive and diversified economic assistance.

Trade between the two countries grew rapidly, strengthening the strategic importance of the railway border crossing between Zabaikalsk and Manzhouli. Moreover, after the CER's transfer to China, it became a junction between the Soviet (Russian) 1520 mm gauge railway network and the Chinese (1435 mm) one. The Korean War rendered this especially significant, with much military cargo travelling along the CER. In 1951–1955, freight traffic increased over six times, requiring increased numbers of railway workers, border and customs officers, etc.

Thanks to developing infrastructure, the establishment of border control and related functions, both village populations, then approximately comparable, began growing rapidly. At that time, the main features of present-day Zabaikalsk's social structure were formed, the bulk of its population working on the railway, in border services and administration.

In 1954, Zabaikalsk became an 'urban-type settlement' (an administrative category between town and village; hereafter, it is treated as a *town*). In 1959, Zabaikalsk counted about 6900 inhabitants. In 1966 it became a district center. On the request of the Chinese neighbours, in 1958, during the period of 'indestructible' Soviet-Chinese friendship, the town got its neutral name of Zabaikalsk.

From August 1945, the CER became jointly controlled by the USSR and Republican China. On 31 December 1952 the Soviet government announced its full transfer to the PRC, though a number of Soviet specialists worked on the Chinese side of the border for some years more. As Urbansky (2020) showed in his book, the asymmetry in economic potential and population between Zabaikalsk and Manzhouli emerged just in this period: until 1949, border formalities and cargo handling occurred almost exclusively at Manzhouli station. As a result, in 1949 there were 9000 inhabitants in Manzhouli, and in 1959, 35,000.

The early-1960s deterioration of Soviet-Chinese relations returned the border to impenetrability for many years; flows of goods through Zabaikalsk-Manzhouli decreased by several multiples. After armed clashes with China in 1969, new military units were deployed in the border area. Modest cross-border trade resumed only in 1985. This more-than-20-year crisis in Soviet-Chinese relations led to the decline of Zabaikalsk's population, especially in the 1960s and the early 1970s: in 1979, it was slightly less than it was in 1959.

However, since 1991, after the USSR's collapse, interactions between the neighbours at different levels have intensified sharply. Russia's abrupt transition to market prices made most enterprises producing consumer goods – fabrics, clothes, footwear, household appliances, alongside many other industries – uncompetitive. The economic crisis meant increasing unemployment and impoverishment. It very strongly affected the Far East, whose economy, alongside raw materials, largely rested on military-related products and maintaining numerous garrisons near the border during the period of hostile relations with China. Virtually unregulated, cross-border trade with its southern neighbour was the main source of cheap consumer goods; their delivery and distribution being a survival option for most people. Individuals transported goods across the border supposedly for personal consumption – the so-called *shuttles*. Over time, informal networks for transporting and selling goods arose; among network members, a division of functions and corresponding unequal income distribution developed. Russian economists and sociologists have described this phenomenon well (e.g. Ryzhova 2008). In the 2000s, this informal business began declining for many reasons: the state's constantly effective restrictive measures, Russian citizens' growing incomes and decreasing demand for low-quality goods.

Manzhouli: a leap into the 21st century

In the early 1990s Manzhouli's wholesale markets were flooded with a stormy stream of traders from Eastern Siberia and the Far East. China's government quickly reacted to changed market conditions: in 1992, the PRC State Council granted Manzhouli, alongside three other border cities, 'open border city' status, simplifying border procedures. Naturally, Russian citizens became engaged not only in buying and selling in Manzhouli, but also increasingly using the cheaper services provided by hotels, restaurants, spas, nightclubs and Chinese dentists and doctors. The Chinese border city's attractiveness for Russian citizens was enhanced by its accessibility, cultural difference and the impossibility of visiting it in Soviet years.

In 2010, Manzhouli urban district received the status of a State Experimental Area of Priority Development and Openness. The year 2016 saw a duty-free zone being created. In the same year, about 15,000 Russian private traders used this zone's services. Manzhouli has a 'Trade Center for Russians', where Russian citizens can sell goods and buy Chinese ones.

China could use Manzhouli's unique proximity to Russia and Mongolia after the USSR's collapse in extremely rational and profitable ways. China is by far Russia's largest trading partner, accounting for 16.6% of its foreign-trade turnover in 2019. Sixty-five percent of Russo-Chinese trade passes through the Zabaikalsk-Manzhouli railway crossing, making the border city not only an all-Chinese transhipment base for Russian raw materials and other goods, but also creating conditions for developing their processing, enhancing added value (Wang Qian 2019). Through Manzhouli-Zabaikalsk and further across Russia, Chinese container cargo can transit to European countries.

Manzhouli has become China's only 24-hour dry port, and a major transportation hub with modern airport and reliable hinterland road and rail links. In 2016, regular flights connected Manzhouli to 20 Chinese, and 8 foreign, cities. Dozens of train and intercity bus routes have emerged. An international freight railway park of 15 sq km was created, doubling checkpoint capacity (ibid.).

On Russia's side, the reconstruction of the 365-km-long section of the former CER between the Trans-Siberian Karymskaya station and Zabaikalsk has been long underway. It became double-track, and was due to complete electrification in 2020, at least doubling capacity. In 2015–2019 Russian Railways' capital investment in the Trans-Baikal Krai was 176 billion rubles (over 2.8 billion

USD) (Shevchenko 2019). Russian Railways, alongside carriers, logistics companies and border services, wants to multiply container routes and reduce train downtime at Zabaikalsk station. Average time from receiving a container train on China's 1435 mm gauge to departure on Russian Railways' 1520 mm track is down to seven hours.

Based on imports from Russia to Manzhouli, the PRC's largest transhipment base for timber, oil, liquid natural gas, chemical products and containers was created. Developing Manzhouli's transport, logistics and industrial functions sharply increased its population. The Manzhouli metro area (696 sq km) reached 337,000 as of 1 January 2020, growing about 3% annually, mainly due to migrants from China's interior regions (Macrotrends 2021). Its boundaries embrace several nearby cities and other settlements. Thereby, both inside city boundaries, and even more within the Manzhouli metro area, it has far overtaken Zabaikalsk in demographic potential.

Sharply increasing twin-city asymmetry was facilitated not only by Manzhouli's diversifying functions, but also by its authorities' extremely skilful exploitation of the new, more general asymmetry developing in the 1990s at national and especially at regional level. Almost a decade into Russia's economic crisis and decline following the USSR's collapse (1991–1998), Zabaikalsk's relatively long period of developmental recovery and instability coincided with China's rapid growth, making China a leading world power, claiming new global hegemony. Thus, roles have radically changed: now impulses for economic development and modernization come not from Russia, but China, where hopes are pinned on inflowing investments and participation in the rise of Siberia and the Far East (Kolosov and Zotova 2019).

Zabaikalsk is a double periphery – on both national and regional levels. It is part of Chita region (called the Trans-Baikal Krai since 2008 after merging with Aginsky Buryat Autonomous District), one of the most depressed in Russia. Chita's low living standards are strikingly indicated by its constant population outflow. Economic development is constrained by sparse settlement and huge distances plus a harsh, continental climate.

Transbaikalia cannot host many Chinese tourists, having little modern infrastructure or well-equipped sites of interest, though Russia generally retains a positive image for Chinese people. Yet, in 2018, only 40,000 Chinese tourists visited all of Transbaikalia.

To attract domestic tourists interested in visiting Russia, a mythological, often kitsch, image of Russia was rapidly created in Manzhouli. Numerous buildings have been constructed, evoking Russia's main historical monuments for inexperienced tourists: the Moscow Kremlin, the Red Square Intercession Cathedral, better known as St. Basil Cathedral, the Bolshoi Theatre, St. Petersburg's Smolny Cathedral, etc. True, Russian tourists note that many were built from cheap and short-lived materials imitating the originals. Many buildings look like a kind of Disneyland. Manzhouli even acquired a postmodern mix comprising copies of famous Russian memorials and monuments: 'Motherland' in Volgograd, monuments to Pushkin and Prince Yuri Dolgoruky, founder of Moscow, and Peter the Great ('Bronze Horseman') in St. Petersburg. Visitors can even see a 'typically Russian hut'. The 'Matryoshka Square' usually makes the greatest impression on Russian visitors. Built in 2005–2006 and consisting of giant nesting dolls – outwardly similar to traditional Russian dolls – they also sport distinct Chinese features.

Alongside historical and architectural visual symbols, 'Manchurian Russia' includes a museum of Russia's major historical milestones with interactive attractions, numerous Russian restaurants adapted to Chinese tastes. Most important are abundant Russian foodstuffs, especially flour, honey, chocolate and vegetable oil, all highly valued in China. Some food products, like sausage, seem exotic to people from China's interior. Russian products are available both in Chinese stores and a special market where Zabaikalsk-region residents sell wine, confectionary, black bread and bottles of kefir, all bought the day before across the border.

The city has created all the conditions for 'average' Chinese tourists to feel they are in Russia while remaining in a comfortable environment for them. There is no need to seek a foreign passport, take risks or travel to an unknown and culturally distant neighbouring country, if everything you need is in Manzhouli. If necessary, the real Russia can be viewed from the 40-meter-high National Gate, the glittering marble border-crossing point, China's largest, where border formalities are combined with numerous shops. And the income remains in China.

Chinese associate Russian culture with European culture: in Manzhouli's centre, you can find squares reminiscent of West European cities. Advertising brochures claim the city has almost all architectural styles – Gothic, Renaissance, Baroque.

Despite Manzhouli 'selling' itself to Chinese tourists based on its Russian proximity, Mongolian culture is also commercialized. Zabaikalsk and Manzhouli can be partly considered a tripoint – where three different countries' borders meet. Many Manzhouli residents are of Mongolian background. A tripoint offers more opportunities than a 'ordinary' border because of greater differences in culture, products, prices and services. Besides, the border between Russia on the one hand, China and Mongolia, on the other, represent sharp contrasts in language, religion, cultural traditions, and social mores which favours tourist development (Więckowski 2021). In Manzhouli, you can visit the 'Mongol nomad camp' and taste traditional Mongolian food.

Finds of mammoth remains, well-preserved in a dry steppe climate, served as rationale for creating the Mammoth Park, displaying over 100 mammoth statues in different life-sized poses. Close by Manzhouli is the unique Dalainor Lake, rich in fish and stretching almost 100 km.

This strategy of using the border location to develop mass tourism has been extremely successful. In 2016, a total of 6,820,000 tourists visited Manzhouli, and tourist income was 11.6 billion yuan (1.74 billion USD). Tourism accounted for 15.6% of gross urban product (Xinhua 2017).

Naturally, Manzhouli's exotic image is designed for *both* internal *and* external consumption, portraying China as a powerful and friendly global power. Russian citizens comprise around 10% of visitors (ibid.). Thus, Manzhouli as a whole can be called the 'National Gate', not just the majestic border checkpoint.

Advertisements, names of streets, shops and attractions are everywhere dupli-cated in Russian, sometimes incorrectly, even mirth-inducing. Most establish-ments have Russian-speaking staff. Russians arriving for the first time are greeted by 'helpers' conducting them to shops, nightclubs, etc. for an owner's fee.

All this brightly lit Russian Disneyland occupies only part of Manzhouli city, which is freely spread over the Inner Mongolian steppe. There is a much more modest 'inner' city, where most residents live in high-rise buildings, but the tour-ist part is what many Zabaikalsk citizens see from their homes.

Zabaikalsk: a station village at the border

In contrast, Zabaikalsk's main, even sole, function remains border-crossing maintenance. The main employers are Russian Railways and various federal government sub-divisions: border guards, customs officers, other border-control workers, territorial law-enforcement officers of many kinds, state-bank workers, tax inspectors, etc. To economize in the context of depopulating rural areas and small towns, federal authorities seek to merge state bodies with serve several districts. However, Zabaikalsk town and district administrations strongly want their preservation. The tax inspectorate had rented Russian Railways premises, but higher authorities wanted to close the Zabaikalsky district branch, transfer-ring its functions to another district. Saving about 100 jobs, the city authorities allowed the tax office to move into a building belonging to them for free.[1]

Russian Railways and federal structures mainly employ staff from other regions or cities. So, the border service is now staffed under contract. Many town resi-dents, in particular federal-service employees and rail workers living in the few modern and more comfortable multi-storey buildings, have no great interest in its improvement. They have much higher incomes than 'indigenous' permanent residents. This population structure creates some disunity.

Zabaikalsk district has no industry, excepting a bakery and a small producer of semi-finished food. There are no local construction organizations: companies from Chita and other cities built the accommodation for federal employees. Their staff paid taxes in their home cities.

A big event for Zabaikalsk was the opening in October 2008 of a terminal for container transhipment from Russian broad gauge to standard gauge wagons, designed to process 600 wagons daily – six times more than before. The termi-nal's throughput capacity is over 470,000 containers, or over 1,000.000 tons per year (FederalPress 2008).

An important income source is retail trade and services run by local entrepre-neurs. Some serve individuals crossing the border, others mainly local residents. Amongst small businesses, a significant place belongs to Armenian entrepre-neurs – immigrants from Javakhetia (a Georgian region near Armenia with a predominantly Armenian population), moving to Zabaikalsk after the USSR collapsed. Small businesses are declining, due to large regional retail chains opening four supermarkets in Zabaikalsk. None sell local products. There is little demand for hotels, with two closing during 2018–2019. One reason is that Chi-nese tourists cannot stay in Zabaikalsk due to border-zone restrictions.

Nevertheless, expanding foreign-trade relations with China and consequent railway and federal-service employment caused Zabaikalsk's population to grow from 8600 in 1989, to 11,800 in 2010, to 13,300 in early 2020 (Rosstat 2020). Such increases are unusual for small Russian towns, which are mostly decreasing.

However, Zabaikalsk, despite its border location, remains an unattractive place to live. Its social infrastructure is weak, lacking vocational education institutions. Most local school graduates leave to study and do not return. Almost all teachers and doctors are overworked since the education and health-care sectors need skilled personnel. The deficit is partially mitigated by federal-service employees' wives with pedagogical or medical education. The local municipality's big problem is the decline of engineering infrastructure, built mainly in the 1950s–1960s. Zabaikalsk gets drinking water from the Argun River via a 50 km water conduit, which needs expensive repairs. There is insufficient water. A new boiler house is needed. Municipal funds are insufficient to solve these problems, especially since many workers in small-scale trade and services are uncontracted and pay no taxes. The lack of attractive jobs forces some residents to earn money through small-scale trade in Manzhouli and transporting Chinese goods across the border.

Developments in Zabaikalsk-Manzhouli are very similar to the situation in other border towns in the Far East: booming urbanization on the Chinese side, growing asymmetry and inequality in economic exchanges, poor use of Chinese investment potential, etc. (Mikhailova et al. 2019). Manzhouli entrepreneurs opened sewing and assembly plants and invested in the local market. A Russian-Chinese family has founded a hotel, Chinese restaurant and greenhouse, and runs other small businesses. However, Chinese entrepreneurs want Chinese workers, thereby requiring restricted work permits for foreigners, which have not been issued. Russian control agencies have found financial violations by Chinese entrepreneurs, and their enterprises (hotel and restaurant excepted) have been closed. Nevertheless, the Chinese own some small enterprises through Russian intermediaries.

Zabaikalsk's district administration, alongside specialists from Chita, framed a development strategy. This included constructing social and transport facilities and reconstructing engineering networks. It relied on regional and federal programs. As regional authorities wished, the Trans-Baikal krai was recently included in the Far Eastern Federal District, which commands special federal-government attention. During the post-Soviet period, several Far Eastern development programs have emerged, but without cardinal changes, as evidenced by the constant population outflow.

Several interesting projects aimed at taking advantage of the border location and partly replicating Chinese experience were to be implemented in Zabaikalsk. A grain terminal was planned, based on the possibility of selling Russian grain on the Chinese market. Investors were allocated land plots, the terminal supposedly opening in 2016. No construction has begun.

Another large project – the 'Eastern Gate' complex – was designed for 2015–2018, supposedly located on the border and including an information center, restaurants/cafes, a museum and children's cultural and entertainment complex, an equestrian arena, botanical garden and ice stadium. It aimed to attract at least 500,000 Chinese and other foreign tourists annually, creating about 1000 new jobs. Investments were estimated at 3 billion rubles, c25% federally financed and 5% regionally (Russian Federal Agency for Tourism 2015). Zabaikalsk's location certainly has tourist potential. Chinese tourists could be interested in expositions about Russia's way of life, Soviet-era military fortifications or the Buddhist monastery. However, thus far, everything remains on paper.

A similar, but less ambitious, 'Russian Village' project has been postponed indefinitely. Creating consortiums for implementing such projects is difficult, since regional or local authorities have to co-finance, from non-existent funds. Another problem is coordinating construction with border law-enforcement agencies, whose regime is necessarily restrictive.

However, this regime itself does not prevent development. The problem is rooted in another dramatic asymmetry, an internal one inherited by Russia from the Soviet period: between, on the one hand, powerful 'extraterritorial' federal ministries, now-private companies, with their own infrastructure and little concern for the surrounding space; and, on the other, municipalities with hyper-limited budgets. In Zabaikalsk, the border regime and harsh natural conditions enhance asymmetry. The municipality indirectly receives only an insignificant part of the border rent and is actually 'cut off' from the border. Many objects on its territory do not belong to it.

Another fundamental problem is the conflicting interests of many actors directly or indirectly participating in cross-border interactions – the federal center and its institutions, the regional government, district and town municipalities, state monopolies (Russian Railways), large and small businesses and, finally, local residents. Problems of coordinating their activities, intractable in the borderlands, are also Soviet-inherited.

Local-level cross-border cooperation is limited to traditional official-delegation exchanges, and sporting-competition participation, mainly in Manzhouli.

Lessons from recent history

The historical trajectories of Manzhouli and Zabaikalsk, emerging almost simultaneously and initially performing similar functions, show how strongly global and national factors acting on different sides of the border determine asymmetry in twin-city development.

It clearly reflected internal Russian problems and disproportions. This intercity contrast is symptomatic of depopulation and the Far East's high dependence on government subsidies, raw-materials orientation and the real danger of becoming appendages of the great neighbour's economy. Manzhouli's National Gate and monuments, brightly lit at night, look in Zabaikalsk like evidence of state abandonment. Manzhouli fully uses its border position's obvious possibilities, in a few years becoming a large multifunctional city, incomparably larger than Zabaikalsk, a forgotten station settlement with poor living standards. For Zabaikalsk, the bitter implications of contacts with Manzhouli are exacerbated by widespread Russian perceptions that the neighbouring city rose by distributing and selling Russian raw materials and using money spent in Chinese bazaars by Russian *shuttles*. Furthermore, few Russian visitors travel further than Manzhouli's fairgrounds, especially since a 'border' tourist visa grants no right to do so, nor offers much idea of Chinese life. As Fedorova (2017, 108) notes, Manzhouli's centre is an export space. Such spaces, wherein special practices, norms and rules are common, are evident not only in border cities, but popular resorts in Egypt, Turkey and elsewhere. But unlike them, the culture and traditions displayed and commercialized in Manzhouli's 'export space' are not the host's, but mainly those of the neighbouring countries.

The screaming contrast between the small overcrowded Russian checkpoint building and the National Gate, personifying the mighty Celestial Empire, also evident along other sections of this border, has great demonstrative effect for hundreds of thousands of Russian citizens crossing every year. It seemingly confirms that Russian school graduates in the town have no alternative but to move to Russia's European part or, at least, the regional center. Manzhouli's gigantic monuments remind that China, a backward country in the memory of generations living just a few decades ago, has become a world leader. This impression is reinforced by the neighbouring city's 'European' architecture: for Russians, it was Europe, the West that embodied modernity (Billé 2012, 2014).

Creating a 'symbolic Russia' in Manzhouli was intended, alongside other motivations, to secure it from excessively depending on Russian interaction, to orient its economy not only to external relations, but also internal needs. However, dependence remains strong. Manzhouli's well-being is hardly guaranteed, and determined by external factors – Russia's economic condition, the ruble-exchange rate and the purchasing power of Siberian and Far Eastern region residents. Like Heihe and several other Chinese cities bordering Russia, Manzhouli experienced significant difficulties due to the ruble's sharp fall against the yuan, reducing the flow of Russian visitors. The reorientation of transit flows from the Trans-Siberian Railway and its branch through Zabaikalsk to the reconstructed shorter railway leading through China's Xinjiang Uygur Autonomous Region to the Alashankou-Dostyk crossing (Kazakhstan) also unfavourably affected Manzhouli. Competition between overland routes connecting China's most developed coastal regions with European countries will definitely grow. It is unclear how far e-commerce, boosted by the COVID-19 pandemic, will impact Manzhouli's trading business for Russian consumers.

The twin cities belong to different economic and cultural worlds, almost never mixing: in Zabaikalsk, it is difficult to find traces of China's proximity; in Manzhouli, Russia's presence is limited to markets and hotels in the city centre (Billé 2014). Mixed marriages or joint businesses remain exceptional. Informal interactive inter-city networks between residents – guaranteeing sustainable and developable local cooperation – remain unformed. The roots of its weakness lie in mutual mistrust: the Chinese side accuses Russia's bureaucracy of indifference, sluggishness and restrictiveness hindering interaction (Mikhailova et al. 2019); Russia's side believes the Chinese are concerned only with their own benefit and take little account of their neighbours' needs. Nonetheless, the turnover of the Russian-Chinese border-trade zone is growing.

Notes

1 Information on Zabaikalsk's situation comes from interviews with the Deputy Head of Zabaikalsk administration for economy and the district newspaper's editor-in-chief (25.06.2018).

References

Billé F. (2012) 'On ideas of the border in the Russian and Chinese social imaginaries' in Billé, F., Delaplace, G. and Humphrey, C. (eds.) *Frontier Encounters: Knowledge and Practice at the Russian, Chinese and Mongolian Border*. Cambridge, UK: Open Book Publishers, 19–32.

Billé, F. (2014) 'Surface modernities: Open-air markets, containment and verticality in two border towns of Russia and China,' *Economic Sociology*, 15(2), 154–171.

Blakher, L.E. (2014) 'Big history and slow existence on the frontiers of civilization. Casus of the Amur Region', *Politeya*, 3(74), 92–109 [in Russ.].

FederalPress (2008), 'A large container terminal was put into operation at Zabaikalsk station', 07.10.2008. Available: https://fedpress.ru/federal/econom/train/id_116499.html (Accessed: 26.02.2021).[in Russ.].

Fedorova, K. (2017). Manzhouli or Manchzhuriya? Linguistic and cultural hybridization in the border city. In *Intercultural Communication with China*, Singapore: Springer, 91–110.

Garrard, J. and Mikhailova, E. (eds.) (2019) *Twin Cities. Urban Communities, Borders and Relationships over Time* London-New York: Routledge.

Kireev, A.A. (2009) 'Specificity of the Far Eastern border of Russia: Theory and history', *Oecumena*, 2, 70–81 [in Russ.].

Kolosov, V. and Zotova, M. (2019) 'China and Russian-Chinese relations in the mirror of Russian discourse' in Kotlyakov, V.M. and Shuper, V.A. (eds.) *Questions of Geography. N 148. Russia in the Emerging Greater Eurasia*. Moscow: Kodeks, 281–309 [in Russ.].

Macrotrends (2021) 'Manzhouli, China Metro area population 1950–2020'. Available: https://www.macrotrends.net/cities/20616/manzhouli/population (Accessed: 26.02.2021).

Mikhailova, E., Chung-Tong Wu and Chubarov, I. (2019) 'Blagoveshchensk and Heihe: (Un)contested twin cities on the Sino-Russian border?' in Garrard J. and Mikhailova, E. (eds.) *Twin Cities. Urban Communities, Borders and Relationships over Time*, London and New York: Routledge, 288–300.

Wang Qian (2019) 'Stronghold of Sino-Russian trade', *Kitai*, 19.12.2019. Available at: http://www.kitaichina.com/rjingji/201912/t20191212_800187478.html (Accessed: 26.02.2021).

Rosstat (2020), 'Russian Federation's resident population by municipalities January 1, 2020'. Available: https://rosstat.gov.ru/compendium/document/13282 (Accessed: 26.02.2021).

Russian Federal Agency for Tourism (2015), 'Eastern gates of Russia Zabaikalsk – Manchuria'. Available: https://bit.ly/3ebBSIq (Accessed: 26.02.2021).[in Russ.].

Ryzhova, N. (2008) 'Informal economy of translocations. The case of the twin city of Blagoveshensk-Heihe' *Inner Asia*, 10(2), 323–351.

Shevchenko, A. (2019) 'In 2020, the electrification of the border railway in Transbaikalia will be completed', *Nefetgaz.ru*, 16.12.2019. Available: https://neftegaz.ru/news/Oborudovanie/513753-v-2020-g-zavershat-elektrifikatsiyu-prigranichnoy-zheleznoy-dorogi-v-zabaykale (Accessed: 26.02.2021).[In Russ.].

Urbansky, S. (2020). *Beyond the Steppe Frontier: A History of the Sino-Russian Border*. Princeton: Princeton University Press.

Vendina, O.I. (2018) 'Historical memory and the symbolic landscape of the borderland' in Kolosov, V. (ed.) *Russian Borders: Neighborhood Challenges*. Moscow: IP Matushkina, 200–232 [in Russ].

Więckowski, M. (2021) 'How border tripoints offer opportunities for transboundary tourism development', *Tourism Geographies*. https://doi.org/10.1080/14616688.2021.1878268

Xinhua (2017), 'Tourism thrives on China-Russia border', 19.08.2017. Available: https://www.chinadaily.com.cn/business/2017-08/19/content_30818796.htm (Accessed: 26.02.2021).

15 Khorgos

The making of an equal twin on the Sino-Kazakh border

Verena La Mela

ICBC Khorgos, international twin city

Kazakhstan is characterized by hectares of barren steppe. Approaching the Sino-Kazakh border in its south-eastern corner, the unbuilt landscape suddenly changes from sandy steppe and occasional corn field to a labyrinth of high-rise buildings. The urban center, constituting the International Center of Boundary Cooperation (ICBC Khorgos[1]), sits right on the border, advertised as a joint economic and cultural project by the Kazakh and Chinese governments. Its physical appearance, however, suggests the contrary, indicating the coalescence of two sides with a barbed wire border fence in between, demarcating the different national territory. Passing from one side to the other is possible only through a large gate spanning the open main square. The ICBC Khorgos is a gigantic trading estate with long-term aspirations to become more than that (Figure 15.1). There are facilities for people to stay, eat, trade and buy. Efforts to add features of genuine urbanity, like a permanent population, are planned for the future.

In 2005 the then Chinese president, Hu Jintao, and Kazakhstan's first president Nursultan Nazarbaev signed an agreement to establish ICBC Khorgos, thereby fostering mutual economic cooperation. ICBC Khorgos straddles the border and is administered as a free-trade zone (FTZ). Visitors from any country as of 2019 may stay up to 30 days without a Chinese or Kazakh visa and purchase tax-free goods. Until 2013, when ICBC Khorgos began, there was just a non-descript border village named Khorgos on the Kazakh side with an estimated several hundred inhabitants. The adjacent settlement on the Chinese side, Huo'erguosi, was granted city status in 2014 when ICBC Khorgos emerged.[2] It now has a population of some 85,000.

ICBC Khorgos qualifies as an international twin city as its two parts physically melt into each other. There is, however, a remarkable feature to it which leaves most visitors puzzled upon arrival: the economy and infrastructure on the Chinese side far exceed that on the Kazakh side in terms of size, number and capacity. Thus, ICBC Khorgos is not typical of internal twin cities 'often merging and becoming indistinguishable' (Garrard and Mikhailova 2019: 3) but does exemplify both internal and cross-border pairs in its dominant-subordinate relationship (Garrard and Mikhailova 2019: 16). ICBC Khorgos's construction

DOI: 10.4324/9781003102526-19

Figure 15.1 Khorgos poster in a suburb of Zharkent with the slogan 'Two countries – one goal' (economic cooperation).

Source: Photo by Verena La Mela, 2017.

formally aimed to establish two equal parts; eight years after its foundation, the two separate entities remain visible and the gap seemingly widens. With 217 hectares,[3] the Kazakh side is notably smaller than the Chinese (343 hectares). This inequality is also reflected in the distribution of shopping centres and the speed of construction: much faster on the Chinese side (cf. Chien and Woodworth 2018). During the first years of my research, the Kazakh side featured just one half-finished building, an ever-locked welcome center plus a couple of yurts. This humble scenery stood against the impressive backdrop of six large multi-storey shopping malls on the Chinese side.

China sports several twin cities along its bordered lands, like Heihe/Blagoveshchensk on the Sino-Russian border (cf. Mikhailova et al. 2019; Saxer and Zhang 2017) or Zhuhai/Macao on China's east coast. However, none match Khorgos in scope and connectivity. Khorgos, globally speaking, connects China with Central Asia and is the main logistical transit land corridor between Asia and Europe. The difference between its two sides, however, is striking: compared to the commercially busy Chinese side with its high-rise buildings, the sleepy and flat Kazakh side looks underdeveloped. As of 2019, there were no visible efforts to emulate Chinese-side development levels. In fact, I observed a puzzling absence of what Taussig called mimesis or imitation (Taussig 1993), a concept Billé (2017) appropriated to describe the competing development of Blagoveshchensk and Heihe. That is to say, underdevelopment is an idea only emerging alongside the term development (Esteva 2010: 2). In this chapter, I explore the case of ICBC Khorgos to show how development and interdependence produced a dominant-subordinate twin.

The origin of the name Khorgos is unclear. The PR version of ICBC Khorgos suggests that 'Khorgas' means 'camel station' in Mongolian, an idea fitting well into the 'New Silk Road' narrative. A native Mongolian speaker I consulted confirmed the existence of a Mongolian word 'khorgo', meaning '(animal) shelter,' A native Kazakh-speaking colleague, however, suggested it is more likely a Russified version of the Kazakh word 'qorgaw' meaning 'to defend' (correspondingly, Khorgos is referred to as Qorgas or Khorgas in Kazakh and other Turkic languages). This could also make sense given that Khorgos was a military border post during Soviet times. This conjecture, though, has been vehemently denied by ICBC staff I approached with this idea. In fact, all we know for certain is that the border river and the adjacent village are also named Khorgos.

ICBC Khorgos should not be confused with what people in Kazakhstan and neighbouring countries generally refer to as 'Khorgos,' a rather ambiguous designation indicating any of the following places: (1) the railway station Altynköl which opened in 2011, functioning as the main drop-off point for shoppers visiting ICBC Khorgos; (2) a dry port logistical hub attached to the railway station where shipping containers are forwarded between East Asia and Europe and also reloaded on to wagons operating on a different railway gauge; (3) another Special Economic Zone called 'Khorgos Eastern Gate'; (4) a new settlement, Nurkent, constructed particularly for migrant workers involved in constructions in and around Khorgos. All these places lie within a radius of around 20 km of ICBC Khorgos and can be considered part of its larger infrastructure.

ICBC Khorgos is a cross-border twin, viewed with love by its initiators and ambivalence by local Kazakhstanis; still a baby with a short biography. Can it live up to its promise and expectations? Its champions promise 'a huge new city of the future' and 'a city of dreams' [4] with, amongst other things, an international university, a cultural theme park, entertainment and sports complexes. Infrastructure promises evoke expectations (Anand et al. 2018). Local populations, however, are often disappointed by the gap between planners' promises and what seems likely in a foreseeable future. Media reports extensively covered the glamorous aspects of ICBC, while pointing out the disparity.

The material here is based on observing Khorgos's development during ethnographic field research during 2016–2019 for my doctorate. This was mainly conducted in south-eastern Kazakhstan with occasional trips into China. My main field site was the small border town of Zharkent, a mainly Uyghur settlement, 30 km from ICBC Khorgos. Living in Zharkent allowed me to observe and document social and economic change among the local Uyghur population induced by the large-scale infrastructural development around them. Two research assistants worked in ICBC Khorgos: Rollan in the administration, and Sasha as a customs broker (all names anonymized). Both worked in Khorgos for many years and shared their knowledge of internal structures and border procedures. Data also emerged from informal conversations and unstructured interviews with people who had visited Khorgos.

A brief history of Khorgos and its setting

From any angle, Khorgos is an outpost in the vast steppe-land around the Ili River, far from any political centre. An account of its Soviet military history mentions a fortress in the village of Khorgos along the trade route Kuldhza-Zharkent which turned into two border watchtowers in 1924 (*Administration of the Army of the Krasnoznamennogo Eastern Border Division of the KGB of the USSR 1984, 3*). Under the USSR, Khorgos was an important military border post. In the 21st century, it was crafted into one of several pioneering joint economic hubs of two historical opponents. From Kazakhstan's former capital, Almaty, a stretch of the new Western Europe–Western China highway leads 300 km through steppe and desert to Khorgos. Driving 600 km further on leads to Urumqi, capital of the Chinese province Xinjiang. The flat transboundary corridor is framed by the Dzhungarian Mountains in the north and the Tianshan range further south; in between flows the mighty Ili River originating in the Chinese Tianshan Mountains and winding across the border into Kazakhstan's Lake Balkhash. Social and economic exchange occurred here long before the Belt and Road Initiative (BRI), a China-induced mega-construction and investment project reviving and enlarging old networks of the overland and maritime Silk Roads. Khorgos's strategic geopolitical and economic position at the crossroads of the BRI turned it into its flagship in Central Asia. China's government advertises Khorgos as a 'benchmark and model example project'.[5]

In the last century, border demarcations and people – Uyghurs and other Muslims inhabiting the Sino-Soviet borderlands – were moved back and forth, wedged between the Russian and Qing Empires and later the Republic of China. Cross-border flows of people, goods and ideas ceased (at least officially) from the mid-1960s until the mid-1980s, a period when China-USSR relations were at an historical low. For people on both sides of the border, the Sino-Soviet split was traumatizing, particularly for those with border-divided families. Only in the late 1980s and through the 1990s when the border opened, my interlocutors reminisced, were Uyghur businessmen able to revitalize their contacts in China.

For many Chinese and Kazakhstani Uyghurs, the 1990s were the heyday of the open Khorgos border. According to Alim, a 50-something Almaty lawyer, some businessmen profited from the early border trade, facilitated by relatives in China:

> Chinese goods were cheap and we had problems obtaining goods. That's how the first people got rich in Panfilov [Zharkent's former name]. Their relatives in China would send goods to them. They had well-equipped warehouses in Panfilov and they would sell the goods in the bazaar. At that time Panfilov was a closed town because it was a border region.

Alim described the beginnings of trade in Khorgos, recalling an early 1990s scene when Uyghur traders met on the border bridge spanning the Khorgos River, exchanging goods from their trucks. The image of trucks parked on the bridge is important because it later developed into the idea of founding a FTZ in

Khorgos. Political chaos characterized the early years after the USSR collapsed. Uyghurs got the chance to profit from the open border through cooperation with cross-border kin. However, they lost their advantageous position with the border's increasing formalization in the 2000s. Border infrastructure emerged, and subsequent procedural formalization made informal cross-border interaction difficult, instead regulating it through more controlled border-passing procedures.

As mentioned, what is currently designated Khorgos is a conglomeration of geographical names and administrative entities, economic institutions and infrastructure projects. Basically, Khorgos is a tiny border village on Kazakh territory. It is also a border city in China where it is given its Chinese name *Huo'erguosi* (Chinese phonetic notation for Khorgos). According to Rollan, the Chinese part of ICBC Khorgos is administered by the Huo'erguosi city council exercising considerable autonomy from Beijing. On the Kazakh side, the main shareholder is the railway company *Kazakhstan Temir Zholy*. Khorgos is also the name of the river, nowadays dammed north of Khorgos, demarcating the Sino-Kazakh border.

The larger setting also includes the brand new four-lane highway connecting Almaty and Urumqi. The highway's quality matches Chinese roads and is remarkable compared to other, poorer-quality Kazakhstan highways. The international border crossing on China's side was originally inside the city of Huo'erguosi, visually almost glued to ICBC Khorgos. On the Kazakh side, the old border had been inside Khorgos village, its lack of grandeur starkly contrasting with that on China's side and accessible only to village inhabitants and business permit holders.[6] In 2018, an impressive new border post was erected on the recently built highway circumventing the older Khorgos institutions, though connecting them through a distributor road.

Alongside what is already constructed, Khorgos also constitutes a vision, hope and promise (see Anand et al. 2018). Politicians, construction engineers and architects deploy the narrative of a 'second Dubai' to illustrate their ambitions for Khorgos, involving its development into one large settlement reaching as far as Zharkent. Kairat, a planning engineer responsible for developing Khorgos, told me Khorgos would eventually get an international airport. As of 2020, neither an airport nor a 'second Dubai' has materialized, but construction proceeds little by little.

Visual appearance of Khorgos today

A taxi driver brings me and three other passengers from Zharkent to the border. The handles of a foldable pushcart stretching across our laps nearly hit my face whenever we confront a pothole or the driver swerves to avoid a crash. The pushcart is Mahira's, a shuttle-trader (cf. Mukhina 2009) in ICBC Khorgos. She goes there regularly for cheap goods, reselling them at a profit on the local bazaar.

Arriving at the parking lot in front of the shiny new customs building painted in Kazakh blue, we pay the driver 500 Tenge (c1.20 USD) without any receipt and proceed through the crowd to passport control. Shuttle-traders queue with occasional tourists in front of one open booth (from an estimated ten) where my passport gets stamped and the shuttle-traders show their identity cards. I then must buy an expensive ticket (c3.60 USD) for an obligatory shuttle bus driving me another 5 km to the shopping complexes, the FTZ's core. The whole procedure of buying and scanning the shuttle-bus ticket is new. Rollan described the procedure as being introduced to limit the shuttle-trade the Kazakh state considers illegal.

On the shuttle bus we drive through a long stretch of no man's land, passing the bridge Alim mentioned earlier spanning the wide Khorgos River bed. My mobile-phone network provider sends me a 'Welcome to China' message. On the horizon the vista of ICBC Khorgos emerges like a mirage, looking otherworldly, contrasting with the surrounding unbuilt area. In both countries, space is aplenty; but since China started its 'Open Up the West' campaign around 2000, the Chinese state began overtly showing its physical presence in western-border areas. This also gets manifested in the exclusively Chinese mobile-phone network coverage inside the whole ICBC Khorgos. The sheer number and size of buildings on the Chinese side is overwhelming. Meanwhile, on Kazakhstan's side of the zone, there is one flat shopping mall framed by four trudging clay camels. Behind them, the tips of three blue yurts. That is about it.

The camels serve as 'metaphor for connection and trade' (Winter 2020, 2), part of a Silk Road narrative China seeks to establish along the BRI. A 'geo-cultural Silk Road' (ibid.) displays Chinese power along the BRI under the pretext of 'Chinese civilization' (ibid.). We pass a construction site (a new shopping center on the Kazakh side) and reach ICBC Khorgos's core, a central square, where I and the shuttle-traders leave the bus. The crowd of shuttle-traders immediately rushes to the FTZ's Chinese side. Signs point out the zone's inner boundary. A large open arch marks the different national territories extended by a militarized barbed-wire fence demarcating the border. The arch actually comprises two towers shaped like the ancient Chinese ritual cauldron, the Ding, simultaneously symbolizing the letter 'H' indicating 'harmony' (see Figure 15.2).[7] This symbolism, referring to ancient Chinese history, was appropriated by the Chinese side with no equivalent for Kazakhstan's past. Once through the arch, I am clearly in an entirely different economic, cultural and political space. A threateningly red sign warns against taking pictures. People arriving from Khorgos's Kazakh side need a security check and x-ray of their belongings. Not the other way round, though. People in Zharkent said this was to single out long-bearded men and veiled women suspected of being extremists in China.

Physically entering Chinese territory also entails a different sensorial landscape. Visual differences are the most striking: six large shopping complexes, several warehouses, offices, banks, restaurants, hotels and several buildings under construction. A two-lane highway has taxis and buses operating. However, sound contrasts are also important. China greets me with pervading Chinese pop music blasting from loudspeakers. Only a few metres further on, a chorus of pitched Chinese and Russian voices adds more: advertisements (always maximally loud) for fur sold in the multi-storey shopping complex in front of me. Uyghur and Dungan noodle stalls temptingly smelling of Chinese spices and smoked meat make me hungry.

Figure 15.2 ICBC Khorgos's central square with the two towers shaped like the ancient Chinese ritual cauldron – the Ding, with the Kazakh shopping mall 'Samruk' in the background.

Source: Photo by Verena La Mela, 2017.

While I am standing forlornly on a large square in the open, I wonder which shopping center to enter first. The Kazakh choice is simple: there is only one shopping mall. I decide to start with the 'Golden Port,' whose name is boldly written in Russian and Chinese. To enter, I must pass yet another security control, typical for most Chinese shopping malls in Xinjiang these days. 'Samruk', the lonely Kazakh shopping mall, can be accessed without trouble or security checks. Inside the Golden Port, the smell of cheap Chinese plastic I know too well from my research in China is all-pervasive. One source is a shoe shop run by a Han-Chinese man from a city in China's northeast. Another is a shop selling plastic bags run by a Han-Chinese woman claiming to be from an eastern Chinese city.

The Han-Chinese shop owners, comprising most sellers beside the occasional Uyghurs, seldom speak Kazakh or Russian and thus hire Chinese Kazakhs (bilingual Kazakhs born and raised in China, and one of China's 55 recognized ethnic minorities) to communicate with Kazakh-speaking customers. Sometimes, these shop owners do speak Russian, particularly if coming from China's north-east, close to the Russian border. Paradoxically, in the one shopping center on the Kazakh side, the situation is similar because most shops in the mall (advertised as Kazakh) are nevertheless run by Han-Chinese. I remember Rollan once saying: 'I think through the economic project [ICBC Khorgos] people can actually know about China because now a lot of people [Kazakhstanis] think of China like a country who wants to "eat" Kazakhstan'. Chinese intentions are uncertain because Chinese investments on the Kazakh side rather seem to reinforce the contrary view and deepen the friction.

Inside the Golden Port, I hear the steady sound of a humming device resembling a vacuum cleaner, placed in front of every shop. It sucks the air out of packages, thus reducing their size ready for export. The wrapping of bundles is a deafening whirling sound. As of 2019, the rule was that individual traders, like the shuttle-traders, were allowed to import only a maximum of 50 kg per person, and packages could not exceed 60 x 40 x 20 cm. At Kazakh customs, packages are weighed, then pulled through an opening in a fence: if it fits into the hole, it passes; if not, the trader is in trouble. So the 'vacuumers' are crucial equipment for everyone.

The Kazakh side is silent. The absence of ear-piercing noise in the (basically) Chinese shopping complex on the Kazakh side can be explained because different products are sold. Textiles, like linen, clothes and fur, are particularly popular with Kazakhstani customers visiting the malls on the Chinese side. They are easy to vacuumize, making them exportable in large quantities. Other products flowing through the border into Kazakhstan include Korean cosmetics, car parts, camping equipment, electrical appliances, electronics, crockery, jewellery and accessories, toys, household goods, tobacco and Chinese medicine. In the Kazakh-Chinese direction flow products associated with natural and chemical-free production, like honey and dairy products. Also on sale are European cosmetics and chocolate, plus alcohol (like Georgian wine or Russian vodka).

Apart from shuttle-traders, there are few other customers in ICBC Khorgos, and entire floors of the shopping malls are empty. The territory seems oversized for the amount of people visiting daily during opening hours: 7 a.m. until 7 p.m. After browsing through nearly empty shopping malls dispersed across the Chinese side for the whole day, I journey back to Kazakh customs control. The first challenge is to enter one of the buses filled close to bursting with shuttle-traders. For double the cost of a bus ticket, I find a taxi which drives me back the 5 km to the customs building. We pass a long queue of nearly 20 shuttle buses with engines turned off. The passengers squat stoically on the roadside smoking and waiting for their turn to move towards the checkpost. My taxi driver, however, waves at an official and is through without waiting or security check.

We are back to the blue Kazakh customs building, where a knot of people with their goods fights for a place in the non-existent queue in front of passport control. I recognize Mahira, the shuttle-trader I met earlier in the shared taxi. She waves me over. I climb over bundles of packages. She smiles at the customs officer, who seems to know her very well, and he opens a door in his fence for me to reach the passport booth. The shoving continues where the traders get their goods weighed and registered on their ID cards. I finally reach the exit, glancing back at the opening in the fence through which the bundles in the last regulatory procedure need to fit. Visitors entering and exiting from the Kazakh side often complained about the 'disorder' caused by shuttle-traders at Kazakh customs, giving an impression of an uncivilized and thus less-developed place.

An unbalanced growth

Much of the infrastructure surrounding our daily lives is invisible to untrained eyes: underground and undersea cables, water pipes, drains, wireless communication and so on. Khorgos, however, is an infrastructure project meant to be seen (cf. Larkin 2013). This reveals its unbalanced growth. Appel et al. write that 'attention to infrastructure makes visible the world as both already structured and always in formation' (Appel et al. 2015). In ICBC Khorgos, pre-existing political, economic and social structures became visible during its formation as a global economic and cultural hub. The story of Khorgos's emergence, however, is a contested narrative. Rollan explains:

> [The] ICBC was a Kazakh idea, it was in 2002 when [Kazakhstan's] president came to China and suggested a free-trade zone in Khorgos. This was not ICBC; he just suggested a *free trade zone*, and then in 2004 [the] governments of China and Kazakhstan signed an agreement to create ICBC. And then in 2005 they signed an agreement for operating [it]. And this was before One Belt One Road ... So ICBC [actually] helped China to create this Silk Road program. ... Because ICBC was very early, in 2002. At that time no one talked about [the] Silk Road or anything.

Even though Khorgos is associated with China's BRI, it is considered a Kazakh project in Kazakhstan. Kazakhstan's self-sustaining role in the BRI was also reflected in the answers of my Kazakhstani interlocutors whom I asked: 'Who built the highway [between Almaty and Khorgos]?' Unanimously, people answered that it was Kazakhstan helped by a bunch of foreign companies. They stressed China was not involved.

Upon construction in 2012, ICBC Khorgos was divided into Chinese and Kazakh parts separated by a large square with an open arch. This internal division was visible from the beginning, even though official rhetoric spoke of a joint project, as this observation from my research assistant reveals:

> Our two parts always say ICBC doesn't need to [be] separate[d] into [a] Kazakh and Chinese side. It's one single project. ... Last week the mayor of Khorgos city said that ICBC is neither a Chinese nor a Kazakh project, it's *our* project. So we are like one family.

Nevertheless, the internal division became even more obvious, when Xinjiang's political situation deteriorated in 2017 and the province faced serious tightening of security controls, especially targeting Muslim ethnic minorities. Back then, China erected an internal security checkpost to control visitors coming from the Kazakh side. Establishing this checkpost came with the emergence of re-education camps in Xinjiang (which Chinese state rhetoric calls 'vocational education and training centers') into which Uyghurs and other Muslim ethnic minorities in China were put under the pretext of susceptibility to extremism.

The Chinese and Kazakh sides of ICBC Khorgos were, at the point of research and writing, an unequal twin. Whereas one showed itself in stunning outfits, the other seemed to lag behind the fashion trends its sibling set. Most international media writing about Khorgos noticed this inequality. Why are the two parts perceived as unequal? Only in the shadow of Chinese magnificence, the Kazakh side appears much less glamorous, showing comparisons are always relative. It reveals that the all-pervasive nature of a Chinese 'propaganda of success' added by Western reports downgrades the Kazakh sibling to seeming 'less developed'. In fact, only through this powerful contrast does the Kazakh side look dull. The construction of this unequal relationship forces Kazakhstanis to seek reasons and explanations. ICBC Khorgos Vice-President Sakengali Nurtazin in a news article from 2014 legitimized the situation: 'The Chinese are ahead of us because their government has closer ties to their businesses' (Trilling 2014). In other words, China is displayed as more experienced in implementing business-related procedures. The quote shows that inherent parts of the political system are blamed for being unable to keep up to speed.

The unequal administrative relations and cooperation with the respective governments are also expressed in terms of autonomy. The Chinese side's streamlined administrative structure allows them to make decisions and implement new projects quickly. The Kazakh side seemingly depends on longer communication paths with Nur-Sultan (until 2019, Astana). In the beginning, and indeed now, incentives to invest in the Kazakh side were low. Thus, the Kazakhstani administration decided to turn it into a separate Special Economic Zone, independent from SEZ 'Khorgos Eastern Gate,' Rollan illustrates the slow procedure: 'We worked very hard to change Kazakh legislation because ICBC is very special. They went to our government and said that ICBC needs specific laws for the management. And it took a lot of time [three years],'

The unequal rise of Khorgos also shapes local Kazakhstani peoples' perceptions about China. The exponential growth of infrastructure on the Chinese side and encroachment into Kazakh territory are viewed with suspicion by local people who fear an acquisition of Kazakh property by Han-Chinese. Already, the threat of losing territory may create feelings of loss (Billé 2014). Fears of surrendering territory to the Chinese were often expressed by my Kazakhstani interlocutors. For most, however, the relationship with China was ambivalent because many traders depended on trade with China.

Conclusion

ICBC Khorgos is branded as a symbol of 'cross-border cooperation'. Spatial, administrative and economic differences, nevertheless, are manifold. Inequality, however, characterizes most twin cities and, as such, deserves being looked upon as less idiosyncratic. Remarkable, however, are the two distinct ways of representing ICBC Khorgos in the two countries. China makes it part of its New Silk Road narrative, while Kazakhstan stresses that it had the idea first to create a free trade zone and, in this way, initiated its representation as a node of the New Silk Road. Khorgos is celebrated by international media as the wunderkind of the BRI, thus nurturing China's BRI propaganda of success while Kazakhstan is represented as the underdeveloped part. The visual disparities do not go either unnoticed or uncommented. Such painful comparisons provoke Kazakh justifications blaming political and economic differences. Meanwhile, high expectations among local people were induced by Kazakhstani media and state propaganda, which then produced disappointment because hardly any of the promises materialized up to the point of writing in 2021. The 'second Dubai' remains a dream, and perhaps will remain an infrastructure promise ever awaiting realization (Anand et al. 2018), even though some are optimistic, pointing out the birth of modern Nur-Sultan, Kazakhstan's capital.

Billé impressively demonstrated, for the twin cities of Blagoveshchensk and Heihe, the mutual imitation between the two sides and identified this practice as a process producing commensurability (Billé 2017, 35). Interestingly, such a process never got going between the Kazakh and the Chinese sides of ICBC Khorgos. The Kazakh side could not attract investments, and thus the Chinese government recently took over the investment role. The result is that the two sides look entirely different, with no visible Kazakh aspiration to outperform the Chinese side. The dominance of China's idea of a New Silk Road based on its long peaceful history – as interpreted by China – is visible in the arch and in the remarkable lack of Kazakh effort to demonstrate a counter-narrative. Rather, we see a fragmental compliance as represented in the clay camels on the Kazakh side. The size and number of shopping malls and the aspirations to construct vertical buildings forcefully show China's economic pre-eminence. The security checkpost is a political statement of China's ideological supremacy. All these visible markers make Kazakhstan look dull. Underdevelopment, however, is always relative to what is perceived/claimed to be development and, since the Kazakh side made no efforts to compete with the Chinese side, the question arises over whether the two sides are comparable at all. I cannot answer this question but conclude the chapter with the insight that as long as China is considered a reference point, Kazakhstan will have a difficult time keeping up.

Notes

1 In Kazakhstan, people refer to it by the abbreviation of its Russian name 'm-tse-pe-s'.
2 The State Council ratifies the establishment of Huo'erguosi City in Xinjiang, http://www.gov.cn/xinwen/2014-07/11/content_2715998.htm (accessed 08.03.2021)[In Chinese].
3 ICBC Khorgos website, http://www.mcps-khorgos.kz/en/project/geographical (accessed 06.06.2020).
4 Video presentation of JSC 'ICBC' Khorgos, https://www.youtube.com/watch?v=vEeZrdQ6p6I (accessed 08.03.2021).
5 A summary of Huo'erguosi, http://www.xjhegs.gov.cn/zjka/hegsgk/sqjs.htm (accessed 08.03.2021) [In Chinese].
6 The permits resembled ID cards containing a note saying one was entitled to pass for business reasons only or as a village resident. Permits were received at local authorities. The dirt road to the village was locked by a barrier and military-controlled access.
7 This information was retrieved from the website of Khorgos Municipal People's Government (http://www.xjhegs.gov.cn/info/1166/17377.htm, accessed 16.06.2020) that has been under maintenance during the final stages of the chapter writing.

References

Anand, Nikhil, Akhil Gupta, and Hannah Appel (eds.). 2018. *The Promise of Infrastructure*. Durham: Duke University Press.

Appel, Hannah, Nikhil Anand, and Akhil Gupta. 2015. 'Introduction: The Infrastructure Toolbox', *Theorizing the Contemporary*, *Fieldsights* (blog). 2015. https://culanth.org/fieldsights/introduction-the-infrastructure-toolbox.

Billé, Franck. 2014. 'Territorial Phantom Pains (and Other Cartographic Anxieties)', *Environment and Planning D: Society and Space* 32(1): 163–78. https://doi.org/10.1068/d20112.

Billé, Franck. 2017. 'Bright Lights across the River. Competing Modernities at China's Edge', In *The Art of Neighbouring: Making Relations across China's Borders*, (eds.) Martin Saxer and Juan Zhang, 33–56. Asian Borderlands 2. Amsterdam: Amsterdam University Press.

Chien, Shiuh-shen, and Max D. Woodworth. 2018. 'China's Urban Speed Machine: The Politics of Speed and Time in a Period of Rapid Urban Growth: China's Urban Speed Machine,' *International Journal of Urban and Regional Research* 42(4): 723–737. https://doi.org/10.1111/1468-2427.12610.

Esteva, Gustavo. 2010. 'Development', In *The Development Dictionary: A Guide to Knowledge as Power*, (ed.) Wolfgang Sachs, 2nd ed, 1–23. London; New York: Zed Books.

Garrard, John, and Ekaterina Mikhailova, eds. 2019. *Twin Cities: Urban Communities, Borders and Relationships over Time*. Global Urban Studies. Abingdon, New York: Routledge.

Larkin, Brian. 2013. 'The Politics and Poetics of Infrastructure', *Annual Review of Anthropology* 42: 327–343.

Mikhailova, Ekaterina, Chung-Tong Wu, and Ilya Chubarov. 2019. 'Blagoveshchensk and Heihe. (Un)Contested Twin Cities on the Sino-Russian Border?' In *Twin Cities: Urban Communities, Borders and Relationships over Time*, (eds.) John Garrard and Ekaterina Mikhailova, 288–300. Global Urban Studies. Abingdon, New York: Routledge.

Mukhina, Irina. 2009. 'New Losses, New Opportunities: (Soviet) Women in the Shuttle Trade, 1987–1998', *Journal of Social History* 43(2): 341–359.

Saxer, Martin, and Juan Zhang, eds. 2017. *The Art of Neighbouring: Making Relations across China's Borders*. Asian Borderlands 2. Amsterdam: Amsterdam University Press.

Taussig, Michael T. 1993. *Mimesis and Alterity: A Particular History of the Senses*. New York: Routledge.

Trilling, David. 2014. 'On China-Kazakhstan Border Lies a Lopsided Free-Trade Zone,' Eurasianet.

Winter, Tim. 2020. 'Geocultural Power: China's Belt and Road Initiative', *Geopolitics*, January, 1–24. https://doi.org/10.1080/14650045.2020.1718656.

Part II

International Twin Cities

D. In South America

16 Transborder dwelling in Albina (Suriname) and Saint-Laurent (French Guiana) on the Lower Maroni

Clémence Léobal

Until very recently, the Maroni River was a porous border between French Guiana and Suriname, a now-independent former Dutch colony. Like many post-imperial Caribbean borderlands, this porosity existed despite the authorities' best efforts at immigration and commercial-exchange control. However, in March 2020, the global COVID-19 pandemic succeeded where French authorities had failed, closing the border for the first time in its history. In March 2020, Surinamese authorities imposed a near-total ban on Albina-to-Saint-Laurent pirogue (wooden boat) crossings. As of June 2020, this has caused major food shortages on the French Guianese riverside and mass unemployment on the Surinamese side.

Before the closure, local populations both crossed the river and also inhabited its banks and islands. As such, according to the *habiter* concept, meaning the activity of building a territory in the world (Hoyaux 2002), they were creating an intimate relationship with opposite places.

Situated on the Maroni estuary, the twin towns of Saint-Laurent-du-Maroni and Albina are inhabited by people who continually crossed this five-kilometres-wide river. The Ndyuka are one of the Maroon groups that fled (and fought) Suriname plantation owners. Nowadays, they say they 'wear' French or Surinamese nationality, meanwhile identifying as Ndyuka, an independently conceptualized identity connected to territories upriver where the Dutch Republic through the 1762 peace treaty recognized their sovereignty. The Maroni Basin is populated by Amerindians and Maroons, both of whom resisted colonialism and lived largely autonomously from their respective state governments (Piantoni 2009). In recent decades, these ethnic groups have undergone profound social transformations due to mass migration to towns along the river. Maroon ethnic groups have created a transborder territory along this political border: characterized by the omnipresence of the two 'sides', sharing a common language (Léglise and Migge 2019). The Ndyuka concept of *liba* means both 'river' and 'inhabited territory'. The river is thus considered less a division than a link between two banks. While the Ndyuka (among Maroon and Amerindian

DOI: 10.4324/9781003102526-21

groups) inhabit both banks, the administrative border linking them separates a European country (French Guiana is part of France) from Suriname, an independent Third World republic.

The Maroni border is 525 kilometres long, and contested by Suriname in the upriver zones. Immigration rules applying here are based on French law but are much stricter than those in metropolitan France: there are many derogations reinforcing immigration control (Benoît 2015). However, the rules have very low effectivity due to high cross-border mobility. The *Police aux Frontières* certainly do not check all boats crossing the river. Since 1985, the official border post has been located in Iracoubo, 110 kilometres up the Saint-Laurent-to-Cayenne road.

Transborder ways of dwelling are broadly shared among all those on the estuary, including the working classes, Maroons, Kali'na Amerindians (Collomb and Renault-Lescure 2014), Haitians (Laëthier 2014), Brazilians (Heemskerk and de Theije 2009) and the local transnational elite (Chinese traders and Surinamese entrepreneurs). I focus on these towns' day-to-day life, where common living space became administratively separated upon Surinamese independence in 1975, examining the concrete, day-to-day implications of living there for Ndyuka working-class inhabitants along with their perceptions and experiences. At the local level, transnationalism (Glick Schiller et al. 1992) involves borderlanders' pride in being able to navigate between Suriname and French Guiana, but also ethnic pride in being Maroon. It involves kinship relations as well. Inhabiting the Maroni thus entails ways of dwelling within larger kinship groups, involving solidarity but also tensions, power relations, conflict and dysfunction. In her study of the Mexican-US border, Whiteford invented the concept of 'extended community' to analyse how kinship networks transcended barbed wire (Whiteford 1979). Here, I draw on this but also build on Marcelin's concept (Marcelin 1996), where people do not live just in single houses but multiple houses. Connecting kinship anthropology and residential geography, it helped analyse local ways of border negotiation. The concept was associated with mobility, circulations and tensions (Motta 2014) and included governing modes that were both private (governing the house) and public (border policies). These configurations were shaped by familial conflicts involving class and national divisions intertwined with border policies.

This analysis is based on a case-study of one large family I met in Saint-Laurent-du-Maroni, the largest town on the French Guianese side. This family group had also built a small village on the Surinamese side, close to the twin town of Albina. I will show how they dealt with the political border and how the border affected them. Their residential choices were underpinned by administrative opportunities and constraints. I conducted an ethnographic study in 2013–2014, following one family member, Lena, through her various displacements and pursuit of official documents.

First, I describe this group's conceptualization of living *on* the border in a configuration of houses straddling both sides. I then explore the circumstances, both private and administrative, forcing Lena to return to her maternal family on the Surinamese side. Finally, I examine the experiences of Lena and other border inhabitants facing the complex entanglement of state administration policies on houses, healthcare, social benefits and land access.

'Eating from both lands': Transnational lives

'If the State demolishes my house, I'll build a stilt house *on* the river'. These were the words of Lili, a female elder living in Saint-Laurent-du-Maroni. Her house was threatened with demolition because the French Guianese authorities considered it illegal. Lili had no building permit, no title deed or land tenure, no personal authorization to live on French territory. Beyond her proposal's humorous overtones (stilt dwellings are no longer constructed in the town), she underscored her conceptualization of inhabiting the border with her projection 'on the river'.

I first went to *Lebi Doti*, a peripheral neighbourhood, to visit Lili's niece-in-law, Lena, a 25-year-old Ndyuka woman. I had made contact through Diane Vernon, an anthropologist working in the local hospital culturally mediating between patients and medical staff. She introduced me to Lena, who lived in a sector set for demolition. Lena regularly visited the hospital because of her disabled child. Lena was being accommodated by Lili, her husband's aunt, and they lived with many relatives in two houses. Over frequent visits, I came to understand that Lena occupied a marginalized position among them. A year later, she moved back to the Surinamese bank due to familial pressure alongside the administrative deadlock she faced as an undocumented person.

Although telling her many times I was not working for either local administration but for a Paris university, Lena expected me to help her secure social housing and residence documents, seeing me as a White (*bakaa*) woman somehow connected with the state. I felt guilty being unable to help more. The little I could provide for these inhabitants (e.g., occasionally purchasing food and medicines) was incommensurate with the invaluable thoughts they shared on border life during our afternoon discussions.

Lili had spent her whole life on one or the other riverside. She never called this space a border, referring instead to its 'two sides'. One afternoon, as we sat on her terrace with her sister, she said, laughing: 'I'm eating from both lands'. This expression normally justifies male polygamy: 'eating from' different women. Lili, a widow, had taken it from its gender-relations context: inversing the implied power relation, rendering herself superior in this two-sided territory, enjoying both lands simultaneously like polygamous Maroon men. This reveals her sense of belonging and mastery within the Maroni River territory, much greater than the simple boundary depicted in state cartography (see Figure 16.1).

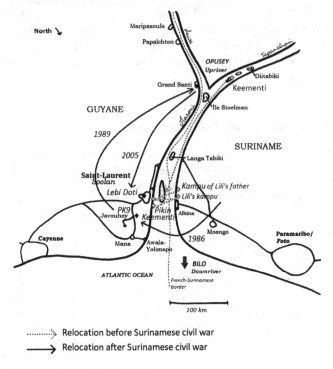

Figure 16.1 Lili's trajectory through civil war.

Source: Map by Clémence Léobal.

Mapping the border

Based on cognitive mapping (Aberley 1993), these drawings spatially represent elements from my ethnographic research. The handmade representation underlines the constructional nature of any cartography. My drawings, do not attempt precise topography, though respectful of approximate kilometric scales. I forego the conventional northerly orientation of maps, instead orienting upriver, respecting inhabitants' perceptions conceiving places according to their river location. In the Ndyuka language, people go either upriver (*opusey*) or downriver (*bilo*).

Living together in several houses

Lebi Doti was reached by a lateritic path called *Piste des Vampires*. Winding through dense vegetation, this path was not exactly an urban landscape. Lena and Lili lived in two houses perched on a hilltop. This 'hill' was actually an earth mound bequeathed by a development company's bulldozers, which were already levelling the ground for three-storey buildings. The two wooden houses were divided into four homes. Franz lived in one house with his wife and ten children. The other was partitioned into three homes for Lili, Lena (plus husband and three children) and one of Lili's nephews' wives.

According to Marcelin, within such a configuration, the houses' inhabitants engage in relations comprising exchanges of goods and services. They can eat, sleep, circulate and leave children to be looked after within each other's houses. These configurations nevertheless can be changed by life events and conflicts within the group. Lena was dwelling together with other people. There were many exchanges between the four homes. Children circulated from one to another hoping to be fed whenever dishes were prepared. All the women would periodically care for each other's children, especially Lena's third child, Gabriella, whose illness rendered her insensitive to pain. Lena was often without resources because neither she nor husband Bobby had residence documents or received state benefits. In using the terms 'husband', 'wife' or 'marriage' here, I refer to a non-administrative Ndyuka union, contracted in a ceremony between the couple's respective matrilineages. Bobby sometimes worked a 'job' driving someone's boat or building a house. Lili sometimes gave Lena a little of what she needed to feed her children. Franz and his wife also gave her products from their fields. However, this mutual assistance had limits. Lena had to go to community services that distributed food, mostly macaroni (which she hated). She could not count solely on neighbours' help.

Configurations of houses across urban and national borders

Lena actually lived within two distinct configurations: her husband's in *Lebi Doti* and those of maternal relatives. Both extended across urban boundaries into other neighbourhoods and over the river.

Lebi Doti's inhabitants also lived closely connected with other houses in the town. They often visited Ine, Lili's sister and Lena's mother-in-law. She lived on the *Piste des Vampires*'s opposite side. One of Lena's daughters took me there on my first visit, leading me along a winding path. Just 5 years old, she was used to going to see her grandmother by herself without asking permission, showing how often Ine looked after her. The extensive exchanges between Ine and Lili also resulted from past cohabitation. A few years before, after separating from her husband, Ine had lived at her sister's until one of Lili's friends 'gave' Ine some land so she could build a house – even though, as with *Lebi Doti* and most of the rest of French Guiana, the land was legally French state property. When I returned to the field a year later, Lili was unwell and being accommodated and looked after by Ine. One of Ine's daughters had replaced Lili in her former house in *Lebi Doti*. The houses' occupants thus frequently moved in and out depending on life events (see Figure 16.2). This changing configuration resulted from changing relationships. Lena and Bobby had come to *Lebi Doti* in 2011, replacing one of Lili's other nieces, Kiki. Bobby had lived there with Lili as a young boy. The present configuration thus resulted partly from past arrangements.

Lena also visited the configuration inhabited by her foster sisters in Baka Lycée, a social-housing estate 15 minutes' walk from *Lebi Doti*. Lena had lived there in

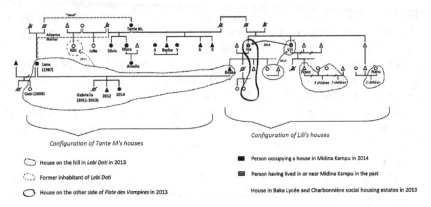

Figure 16.2 Lena's kinship tree.

Source: Drawing by Clémence Léobal.

the past with her *sissa* (sister or cousin) Silvia. Silvia was Lena's foster-mother's daughter, who had raised Lena in Suriname and whom she called *Tante* ('aunt' in Dutch). Fostering was common in Maroon societies resulting in children being redistributed within or outside the matrilineage (Vernon 1992). Silvia lived in an individual state-owned house, which she had extended to accommodate several homes. Lena and Bobby visited several times weekly, for example to charge their phones, *Lebi Doti* being without electricity. The relatives had also gathered at Silvia's house when Lena's disabled child died in July 2013. During family crises, Lena also visited another estate further away called Charbonnière to see Silvia's sister, Mora. When Mora's daughter Amalia was hospitalized following a miscarriage, Lena visited her several times. Lena thus certainly lived in a configuration of houses: indeed, as I later discovered, moving back and forth between two configurations, that of Bobby's relatives, including Lili, Franz and Ine, and that of her foster mother's relatives.

These configurations extended not only over urban boundaries but also over the national border. Lili, Lena and her relatives had houses in a village across the river called *Midina kampu*. When I visited, both women spoke fondly of this village, describing it as a peaceful place with a beautiful creek to bathe in. Both had lived there, Lili in her own *kampu* very close to the current *Midina kampu*, until the Surinamese 1986 civil war destroyed it. She said 'the war dispersed the house', meaning everyone had gone their own ways. She subsequently moved to the French side, and her sister Ine moved to Paramaribo.

Lena and Lili lived closely connected with the surviving *Midina kampu*, which was included in their configuration. Lena visited the *kampu* very regularly. Her *Tante* had a house there as well as in Paramaribo. Many of the *Tante's* children had houses in the *kampu*, including Silvia and Mora, who also had social-housing accommodation in Saint-Laurent. The *kampu* inhabitants regularly came to Saint-Laurent, only 30 minutes away by boat. The *kampu* was also linked to Albina by a muddy track. The 15-minute taxi ride from the *kampu* to Albina cost 20 euros, and the five-minute boat trip from Albina to Saint-Laurent three

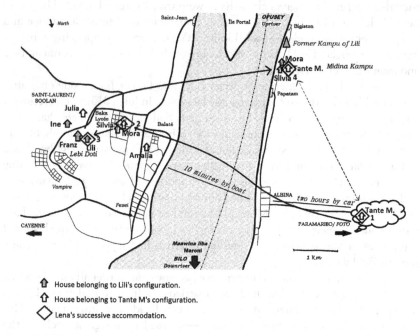

Figure 16.3 Lena's transborder configuration of houses and trajectory.
Source: Map by Clémence Léobal.

euros. A man once came from the *kampu* to *Lebi Doti* in his boat to deliver his sisters to the hospital to look after Amalia, who had miscarried. Their modes of dwelling thus involved solidary exchanges between houses from both sides of the river.

This family group was appropriating this transborder space. Some like Lena's *Tante*, had multiple residences (see Figure 16.3, which reflects the situation during my research and especially from Lena's viewpoint). Mora and Silvia had houses in the *kampu* and on the French side, where their children attended school. They came to the *kampu* for holidays and weekends to relax or work in the fields. Both Surinamese and French mobile-phone networks were available in the lower Maroni region facilitating cross-river integration. Most people had two phones enabling calls in either French Guiana or Suriname.

Border policies and familial conflicts intertwined

Why did Lena move between banks in 2014? The configurations were shaped by conflict and combined different logics, including immigration and border-control policies making it impossible for Lena to obtain residence documents. Furthermore, internal pressures within the configuration placed many demands on Lena regarding her financial autonomy, motherhood skills and general respectability. These family-group tensions were unspoken in my presence. Only during the crisis triggered by Lena's child dying were they expressed aloud. They were nevertheless constitutive of the housing configurations. Mutual assistance and solidarity entail individual costs (Gollac 2003), even accompanying forms of 'tyranny' (Douglas 1991). Lena was doubly excluded, from Lili's configuration and from the French riverbank.

I understood the distinction between Lili's and Lena's *Tante's* side only after Lena's child died at the end of my first fieldwork stay in July 2013: when I returned after the ensuing crisis in 2014, I noticed Lena had moved from *Lebi Doti* to her *Tante's* in *Midina kampu*. The configuration of houses, including Mora's, Silvia's and Lena's, was reproduced on a different scale here. One entire neighbourhood of the *kampu* constituted what they called Lena's *Tante's Ma bee: bee* meaning 'belly' also refers to the 'lineage' or the neighbourhood formed by a *Ma* (woman) and her descendants. Her house stood amidst an array of houses belonging to her children and other relatives. Mora and Silvia thus supported Lena in Saint-Laurent because of their common relationship to *Tante's* as attested by the layout of the houses in *Midina kampu*. Lili and her relatives had supported Lena because she was Bobby's wife.

This distinction was difficult to perceive because of an earlier alliance between these two groups. Lena's *Tante* had been married to Lili's maternal uncle. Hence, all members of both configurations where Lena lived were inter-related. This alliance explains how Bobby, Ine's son, was raised by his great-uncle in the *kampu*, where he met Lena. Lili had not returned to that side of the river since the war, but she still had relatives there. Lena was raised by her *Tante*, to-ing and fro-ing between her houses in the *kampu* and Paramaribo. This was how Lena met her two husbands. She had first come to Saint-Laurent in 2003 to have her eldest daughter treated at the hospital (she died a few years before I met Lena). When her third daughter, Gabriella, began suffering the same illness, Lena's relatives held her responsible, believing her a bad mother, too lazy to work or get her residence documents. Diane Vernon told me later Lena was held responsible for her child's illness because her first husband was a maternal cousin, strongly prohibited by Ndyuka custom. The child's death was interpreted as ancestral punishment.

Lena's childcare practices shocked me. I thought I was being normative but later realized her family also believed her a poor mother. Thus, I felt disgusted by seeing how Gabriella generally spent the whole day dirty because of dust from their house's dirt floor in *Lebi Doti*. A nurse visiting them regularly had recommended Lena putting socks on Gabriella's hands to prevent her biting them and consequent infection, but the socks were dirty. The way Lena treated her five-year-old eldest daughter, Gabi, also horrified me. Lena constantly and peremptorily asked her to do domestic chores, like caring for the younger children or buying provisions from neighbours. Gabi had not yet attended school, and I felt she should do so because teachers told me how crucial nursery school was for French-Guianese children to learn French.

Over the months, I discovered that many relatives also criticized Lena and felt she was over-dependent. These criticisms surfaced only after Gabriella's death. I went to Silvia's house, where everyone gathered in mourning. Lena sat in the living room, wearing black-and-white mourning clothes, surrounded by many relatives, including Silvia and Mora. Lena told me her daughter had undergone surgery when her hand became infected, and never awoke. I immediately thought there had been a medical error. However, Lili and Franz's wife provided a different version when I visited them later. They claimed it was Lena's fault because she should have entrusted Gabriella to the state (meaning the children's hospital).

Upon returning to the field a year later, I visited Lena at Midina *kampu*. She had returned to her *Tante*'s configuration after exclusion from Lili's. Lena rejected Lili's and Franz's accusations: 'They say I killed her, but it was the doctor's fault'. She suggested Franz refused to drive her to the hospital that night, saying she was always bothering him for something. She found a taxi instead. The lack of support from members of her previous configuration was why she had returned to 'her' family in the *kampu*. She added she had not entrusted Gabriella to the doctors because they took her first child to Martinique without her permission, where he died, and she could not bury him. She had not wanted that happening again.

However, Lena's support within her *Tante*'s configuration was also limited. Neither Mora nor Silvia had offered accommodation in Saint-Laurent. Her *Tante* provisionally lent her half a house to stay in with her husband and their children. However, she had to prove willingness to become financially independent. Her husband found a security-guard job in one of Albina's Chinese shops, and she started planting a field near the *kampu*. She also planned to build a house. Silvia would give her some land she had bought nearby, between the *kampu* and Albina. Lena was, however, Silvia's second choice; Silvia had wanted to give it to her own daughter, but she already had a house in Saint-Laurent. This shows Lena's marginality within her *Tante*'s configuration. The two riverbanks did not share the same status in this transborder configuration. Although living on the French side meant enhanced social-benefit access, Lena had to return to the Surinamese side because of her marginality.

The border territory had two unequal sides. This was clear from people's everyday journeys to different institutions. Lena never considered her move beyond the border permanent. After my *kampu* visit she crossed to the French side several times, visiting relatives or government offices. Her disabled child's death had reduced her chances of getting her residence documents because the hospital's social workers had been supporting her regularization application. However, she still hoped to return to French Guiana. Her eldest daughter, Gabi, was still being schooled there even though she had moved to the *kampu*. At first, she had entrusted Gabi to a *sissa* in Saint-Laurent who eventually threw the little girl onto the street, where the police found her. During my visit, Gabi was staying at someone else's house in Albina so she could cross the river every morning for school in French Guiana. A schooling certificate can serve as proof in regularization applications that someone has stayed five years or more on French territory. Lena had not totally relinquished the idea of living on the French-Guianese side.

National lines and urban planning: Facing eviction

The border also became concrete in *Lebi Doti*'s inhabitants' everyday lives through the denial of land rights. In 2011, French authorities had demolished their fields, asking them to leave. For a long time, urban planners had failed to take into account the presence of many existing residents in designing a huge project for 4000 social-housing dwellings. Also in 2011, an NGO commissioned a study of 'spontaneous dwellings' in the zone. Helped by aerial images, the authors estimated over 6000 people lived in the zone and surrounding areas (Colombier, Deluc, and Rachmuhl, 2012). The project's implementation implied their eviction. Demolition had occurred at the site where the first buildings were being constructed. During initial groundworks, the project's building company – the Senog, *Société d'Economie Mixte de l'Ouest Guyane*, which is 80% municipally owned – had instructed bulldozer drivers to dig very close to existing dwellings aiming to intimidate residents into leaving. Those with French citizenship or French-residence documents were sometimes offered social housing. 'Illegal' immigrants, however, had no right to social housing. Intimidation mostly worked. In 2013, only a few houses remained amidst a flattened space, including Lili's.

Lili described the bulldozer arriving, their crops being destroyed and how she looked on helplessly. She denounced the injustice of demolition, comparing the country with Suriname, where this would never happen: 'they give you the land, and they cannot demolish your work'. Her son Franz was fighting the clearance, using French law to demonstrate his goodwill and good faith. They showed me photos Franz had taken upon learning his field would be destroyed. Franz recalled the same thing happened when the municipality evicted them from a neighbourhood called Vietnam, transferring them to Djakata. The mayor had not come himself but sent his son, as municipal representative, along with his colleagues. This time, however, Franz believed he was well within his rights. In Vietnam, many people were occupying the land, but here he had gone through the official procedure to apply to the French Guiana public development agency (EPAG) for the title deed. This procedure was not completed, but he had retained the application receipt, which he showed me proving his goodwill even though he knew it did not constitute a title deed:

> When I saw the place, before I cleared it, I went to the EPAG. The agents looked at the map and said, this land belongs to state property, so if I want it, what do I want to do with it? I said I wanted to cultivate the land and build a house, I wanted to live there. They said OK. They gave me a form, I filled it out and sent it. Then I received this receipt. This meant I had to wait a little while.

Franz was illiterate, but he knew exactly what the procedure meant. He had recently obtained French nationality having been stateless his whole life – he was born on his mother's *kampu*, never being registered by any state. In 2011, he had tried to go to the tax office to pay the land tax, like his neighbours to prove occupation, but was told he should have done so long before because it was now impossible with the planned urban-development project.

Franz claimed he knew nothing about the project before 2011, protesting the injustice he was subject to. He argued the EPAG was willing to give him the land, but the commune's mayor blocked it: certainly all EPAG land applications were first unofficially municipally reviewed. He was angry about other Ndyuka people coming to settle on 'his' land without his authorization. Upon seeing the workers demolishing his field, he visited the mayor's son, whom he knew personally. The latter said Franz could receive compensation for his fruit trees and EPAG officials had been sent to count them. Franz saw them but did not know they would destroy everything. He had to accept this meagre consolation and return home. Nevertheless, he continued defending his rights using the register of French administrative law. He was thus arguing (on behalf of his group) their defence in French legal terms even though judicial success appeared slim. French-Guianese power structures made this procedure nearly impossible: working-class Maroons were marginalized by state officials applying French law in ways providing no chance of winning their right to their land. Franz's action left *Lebi Doti* people in a wait-and-see situation, withstanding threat of demolition thanks to their own resources and agriculture. Six years later, Franz was still there, surrounded by new buildings and awaiting a deal to legalize his situation.

***

In these twin cross-river, cross-border towns, Ndyuka inhabitants negotiated a combination of state opportunities and constraints. They had created a transborder space, connecting Saint-Laurent and Albina, where national hierarchies were in play. The two sides did not have the same value. Compared with Suriname, French Guiana offered the Ndyuka more healthcare, benefits and housing-prospects, but also greater risks around residence documents and denial of land rights. National borders and social boundaries intertwined with kinship groups. Ndyuka people did not just inhabit single houses but entire configurations, where residents actively engaged with each other based on matriliny. While exchanges and solidarity were evidenced between the evolving housing networks, these ways of dwelling were also structured by tensions. Lena's story shows that belonging to a configuration entailed individual costs. People had to follow certain rules, like proving they would try to become financially independent or be good parents. If someone transgressed, they could be excluded or marginalized. However, in the case of evictions, for example, kinship ties allowed mobility to the Surinamese side of the river and the capacity to return depending on administrative constraints. These configurations thus provided solidarity resources, helping Ndyuka people cope with state constraints.

References

Aberley, D (ed.) 1993, *Boundaries of Home: Mapping for Local Empowerment*, New Society Publication, Philadelphia.

Benoît, C 2015, 'Sans-papiers amérindiens et noirs marrons en Guyane: La fabrication de l'étranger sur le fleuve Maroni', in Moomou, J (ed.), *Sociétés marronnes des Amériques:*

Mémoires, patrimoines, identités et histoire du XVIIe au XXe siècles, Ibis Rouge, Matoury (French Guiana).

Collomb, G & Renault-Lescure, O 2014, 'Setting up frontiers, crossing the border: The making of Kari'na Tyrewuju', in Carlin, EB, Léglise, I., Migge, B., Tjon Sie Fat, B (eds.), *In and Out of Suriname: Language, Mobility and Identity*, Brill, Amsterdam.

Colombier, R, Deluc, B, Rachmuhl, V 2012, *L'urbanisation spontanée en Guyane: Appui à la mise en œuvre de modes d'aménagement alternatifs*, Groupe de Recherches et d'Echanges Technologiques, DEAL Guyane, 3 reports.

Douglas, M 1991, 'The idea of home: A kind of space', *Social Research*, vol.58(1), 287–307.

Glick Schiller, N, Basch, L & Blanc Szanton, CB 1992, 'Transnationalism: A new analytic framework for understanding migration', *Annals of the New York Academy of Sciences*, vol. 645(1); 1–24.

Gollac, S 2003, 'Maisonnée et cause commune: Une prise en charge familiale', in Gramain, A & Weber, F, *Charges de famille*, La Découverte, Paris.

Heemskerk, M & de Theije, M 2009, 'Moving frontiers in the Amazon: Brazilian small-scale gold miners in Suriname', *European Review of Latin American and Caribbean Studies*, No.87, 5–25.

Hoyaux, AF 2002, 'Entre construction territoriale et constitution ontologique de l'habitant: Introduction épistémologique aux apports de la phénoménologie au concept d'habiter', *Cybergeo: European Journal of Geography*, 29 May 2002 (accessed 03.03.2021), https://doi.org/10.4000/cybergeo.1824 .

Laëthier, M 2014, 'The role of Suriname in Haitian migration to French Guiana: Identities on the move and border crossing', in Carlin, E.B., Léglise, I., Migge, B., Tjon Sie Fat, B. (eds.), *In and Out of Suriname: Language, Mobility and Identity*, Brill, Amsterdam.

Léglise, I, Migge, B. 2019, 'Language and identity construction on the French Guiana-Suriname border', *International Journal of Multilingualism*, 1–15, doi.org/10.1080/14790718.2019.1633332

Marcelin, LH 1996, *L'invention de la famille afro-américaine: Famille, parenté et domesticité parmi les Noirs du Recôncavo da Bahia, Brésil*, PhD Thesis, Universidade Federal do Rio de Janeiro, Museu Nacional, PPGAS, Rio de Janeiro.

Motta, E 2014, 'Houses and economy in the favela', *Vibrant – Virtual Brazilian Anthropology*, vol. 11(1), 118–158.

Piantoni, F 2009, *L'enjeu migratoire en Guyane française: Une géographie politique*, Ibis Rouge, Matoury (French Guiana).

Vernon, D 1992, *Les représentations du corps chez les Noirs marrons Ndyuka du Surinam et de la Guyane française*, Paris: ORSTOM, 1992.

Whiteford, L 1979, 'The borderland as an extended community', in Kemper, R (ed.), *Migration across Frontiers: Mexico and the United States*, State University of New York at Albany, Institute for Mesoamerican Studies, Albany, 127–137.

17 The Oyapock River Bridge as a one-way street

(Un)bridgeable inequalities in Saint-Georges (French Guiana) and Oiapoque (Brazil)

Fabio Santos

Saint-Georges and Oiapoque in the Amazon rainforest's north-eastern corner stand out for their peculiar connection between South America and Europe, Brazil and France.[1] Amongst all outermost regions, as the EU calls its forgotten colonies (Boatcă 2019), French Guiana is the only continental one, bordering Brazil eastwards and southwards and Suriname to the west. Considering that Guyane (shorthand for French Guiana) is a full-fledged part of France and the EU, the *New York Times* depicted the borderland in 1992 as the 'latest contact point between two tectonic plates: the economically developed north and the economically deprived south'. Twenty-five years later, these 'tectonic plates', divided by the Oyapock River, were literally bridged: on 18 March 2017, the monumental Oyapock River Bridge was ceremonially inaugurated after a five-year delay. With a length of 378 metres, the cable-stayed bridge connects the twin cities of Saint-Georges and Oiapoque approximately 40 kilometres upstream the river mouth in the Atlantic Ocean.

This chapter traces the conflicting steps producing this watershed in the French-Brazilian borderland: long-awaited by some, mistrusted by others. Setting the stage, I first introduce Saint-Georges and Oiapoque, their diverse populations, entangled histories and inequalities. Second, I explore how the Bridge was politically forged in the 1990s and 2000s after political neglect by the two respective 'centres' in Paris and Brasília. Third, I examine its inauguration, offering explanations for the striking delay by drawing connections to mobility inequalities experienced and voiced by borderland inhabitants and politicians with differently situated positionings. Finally, I reinforce the argument that visa non-reciprocity not only underpinned the delay: after inauguration, it became a one-way street, perpetuating inequalities on the basis of citizenship and its intersections with other criteria. Though now constructed and opened (with a pandemic-related closure at the time of writing), it disrupts a historically entangled region and contradicts prevailing political framings of a connecting bridge. Having conducted six months of ethnographic fieldwork alongside document analysis, I argue that, on the French-Brazilian border, consultation with marginalised populations should have outweighed geopolitical interests. Without it, the Bridge continues acting as a one-way street, unilaterally favouring a privileged group on the northern 'tectonic plate'.

***

DOI: 10.4324/9781003102526-22

The borderland shared by Guyane and Brazil's state of Amapá is a space of inextricable entanglements overlapping throughout history, with different groups crossing and settling along the Oyapock River: indigenous communities, fugitive enslaved persons whose descendants identify as Maroons, convicts from the French metropole, French Caribbean islands and south-eastern Brazil, small-scale gold miners from different Brazilian regions and of different ethnicities have turned the borderland into a highly diverse microcosm with fluid layers of belonging deriving from past decades and centuries (Espelt-Bombin 2018; Gomes 2003; Santos 2020). Likewise, the border was hazy and shifting, with its current river-based location resting on a Swiss arbitration court decision in 1900 favouring Brazil's territorial claims (Granger 2011; Romani 2013). Even though ethnic markers remain important, many people's definitions today derive from a mixed cultural heritage (*créoles*) or cross-ethnic and cross-border histories of living and practising rural production along the river's effectively shared space (*ribeirinhos*). Regarding the characteristics of twin cities described in the introduction of this book, Saint-Georges and Oiapoque are best characterised by the notion of interdependence, which should not be mistaken as harmonious togetherness, but rather seen as historically evolved conviviality crucially shaped by everyday negotiations, including conflictual ones (Costa 2019). This interdependence also becomes manifest in the language use that goes beyond a simple binary model of French and Portuguese, since Créole and indigenous languages are also spoken by several communities.

How do Saint-Georges and Oiapoque appear nowadays? Despite commonalities like population diversity and the precarious wooden houses prevailing in both cities' impoverished areas, there are clear differences, partly resting on different population levels: 4,220 registered in Saint-Georges (2019), 27,906 in Oiapoque (2020). The latter's bustling tropical urbanity contrasts with Saint-Georges's calm, especially nocturnally. Oiapoque's streets teem with mopeds and vans hosting oversized loudspeakers announcing parties and evangelical meetings alongside beggars, street vendors, and food stands. Also evident are brothels and the clandestine small-scale gold-mining industry secluded in the rainforest, with cheap hotels and shops selling mining equipment. In sleepy Saint-Georges, by contrast, public life revolves around the town hall and main square with a restaurant, bar, hotel, tiny market for local produce, two Chinese-owned minimarkets and a few snack stalls within walking distance. Unlike Oiapoque, Saint-Georges has some well-preserved, impressive Créole buildings as well as some neighbourhoods with modern townhouses built for the predominantly white officials from the metropole stationed in Saint-Georges, especially police and customs officers, health personnel, and teachers.

While economically dependent and disadvantaged relative to the French metropole, Guyane has a high standard of living relative to its South American and Caribbean neighbourhood. Because of intersecting motives involving better financial prospects, security and political stability, Guyane has become a privileged destination for many migrants, with ever-increasing immigration in recent decades (e.g. Mam Lam Fouck 2015). Today, over one in three French Guianese residents does not hold French citizenship, and these migrants come mostly from Haiti, Suriname and Brazil. Richard Price rightfully argues this diversity distinguishes Guyane from other French Caribbean departments, giving 'its society an edgy feel that is absent in the more French-style (bourgeois) islands of Martinique and Guadeloupe. Much greater inequality of wealth, many more non-French speakers, more crime and tremendous unemployment make Guyane feel Third World' (2018: 19). Its unique status as continental EU's outermost region in South America complicates integration into regional networks like the Caribbean Community, Mercosul or UNASUL (Hoefte et al. 2015). Saint-Georges, Oiapoque and the Oyapock River Bridge thus need examining in light of these regional integration efforts and Guyane-bound migration patterns, with Oiapoque and Saint-Georges being not only important stopovers, but also places of arrival and return due to border controls and deportations (Benoît 2020; Martins 2015). Additionally, however, any truly cross-border cooperation must focus on tackling inequalities. It is uncertain whether top-down initiatives like the Bridge are able to achieve this and whether there is still room for policy adjustment.

Although individuals on both sides of the river face striking inequalities, Brazilians and other non-EU citizens living in Oiapoque (oftentimes clandestinely in Saint-Georges) are particularly excluded, lacking access to fundamental resources like quality healthcare and education, in turn triggering attempts at cross-border movements and activities. Yet, physical mobility across the French-Brazilian border is legally restricted, at least for most wishing to enter Guyane. While French citizens can usually easily enter Brazil, Brazilians and many other non-EU citizens must undergo complicated, costly and oftentimes fruitless visa application procedures. Since 2015, registered Saint-Georges and Oiapoque residents can request so-called transborder cards, enabling 72-hour access to the opposite twin city without needing a valid visa. Nevertheless, mobility inequalities persist: all French citizens have visa-free 90-day access to all of Brazil, thus not necessarily needing transborder cards. Meanwhile, only Brazilian citizens officially registered in Oiapoque can legally enter a small fraction of Guyane (Saint-Georges). These regulations thus display constant reminders of how global inequalities are perpetuated on the basis of citizenship, complexly interacting with other ascriptive criteria like gender, race and class (Boatcă and Roth 2016; Korzeniewicz and Moran 2009; Shachar 2009).

***

To understand how the Oyapock River Bridge evolved, we must start with regional integration efforts advocated by two rising left-wing politicians from Guyane and Amapá in the early 1990s: Antoine Karam and João 'Capi' Capiberibe, arguably the Bridge's 'founding fathers'. Belonging to the *Parti Socialiste Guyanais* and former advocate for Guyane's independence from France, Karam was President of the Regional Council of Guyane (1992–2010) and one of two French Guianese members of the French senate from 2014 until 2020. Capi, from the *Partido Socialista Brasileiro*, spent several years in exile during the military dictatorship before becoming a notable Brazilian politician, first as mayor of Amapá's capital Macapá (1989–1992), then as Amapá's governor (1995–2002) and one of three senators from Amapá in the Brazilian senate (2011–2018).

In Guyane, Amapá and the shared borderland, Capi and Karam identified similar problems of neglect and distance from their respective nation-states' political and economic centres. Karam, seeking increased Amazonian integration, highlighted the shared problem of political neglect by drawing an analogy and describing Amapá as Brazil's 'outermost region' in our 2016 interview. Likewise, Capi emphasised the need for cross-border cooperation in a senate speech: 'by perceiving our isolation (Amapá distant from Brasília and Guiana very distant from Paris), Antoine Karam and I decided to call the attention of our central governments' (Capiberibe 2017). Indeed, the resultant framework agreement of cooperation constitutes the first bilateral agreement allowing decentralised cross-border cooperation between Amapá and Guyane. Signed by then-Presidents Chirac and Cardoso, the document aimed at promoting opportunities for economic cooperation and 'development' in the two border towns (Nonato Júnior 2015; Silva 2017). Moreover, Capi and Karam convinced the Presidents to celebrate revived Brazilian-French friendship in Saint-Georges. This ceremony occurred on 25 November 1997, producing an unprecedented media coverage of the borderland, putting it briefly in the spotlight after centuries of neglect. But what, besides affirming strong bilateral relations between the two countries, was the meeting's agenda?

The original plan was to announce paving roads connecting Oiapoque to Macapá and Saint-Georges to Cayenne. However, having repeatedly reaffirmed their will to strengthen ties across all of the Guianas, Capi's and Karam's ambitions went significantly beyond constructing connections to the respective regional capitals: they envisioned a 'Trans-Guianese Highway', that is, a road interconnecting Amapá, the three Guianas, Venezuela and Colombia (Capiberibe 2003). This aspiration influenced decisions during the 1997 ceremony. In our interview, Karam stated that the Bridge's announcement was relatively spontaneous, after he approached Chirac requesting a bridge. Accordingly, the two Presidents endorsed the idea, making it public while in Saint-Georges. Given this spontaneous endorsement (the Bridge was first agreed on in one of the proposals of the 1997 Mixed Transborder Commission, a bilateral commission promoting cross-border cooperation), it is not surprising that it took eight more years for the official treaty to be signed in 2005. While Chirac remained French President, Brazilian politics had changed considerably, with Lula becoming Brazilian President in 2003. The 2005 treaty thus occurred under their presidencies, additionally coinciding with the 'Year of Brazil' in France. This provided for further actions like consulting construction companies. Such steps, in turn, would take more years to be completed. In the meantime, archaeological excavations unearthed artefacts found in burial grounds dating back to pre-Colombian at the site of the Bridge (CNRS Images 2010).

This did not play a role in a second bilateral meeting which occurred in Saint-Georges on 12 February 2008, when Presidents Lula and Sarkozy (Chirac's successor from 2007) reaffirmed their partnership in various fields by adopting a strategic partnership agreement. Strikingly, the Bridge and other regional issues played secondary roles, emphasis instead resting on how Brazil and France could support each other in world politics: France reaffirmed its support for Brazil becoming a permanent member of the UN Security Council; Brazil recognised France's interest in joining the Amazon Cooperation Treaty Organization (ACTO), an international organisation seeking to promote sustainable development of the Amazon Basin. This illustrates how local issues became pretexts for discussing international topics more relevant to political actors in Paris and Brasília (Hoefte et al. 2015: 98). Still, more than a decade after declaring their mutual support, neither objective (Brazil's Security Council seat, France joining ACTO) has materialised. However, another of Lula's and Sarkozy's projects in Saint-Georges has been reached, even if behind schedule: in 2008, they decided to 'begin immediately' constructing the Bridge with inauguration scheduled for completion in 2010, thereby 'symbolising the proximity between France and Brazil', permitting 'the junction of Macapá and Cayenne by road' and bearing 'multiple benefits for the economic and social development of the region' (Déclaration conjointe 2008).

This developmental emphasis from its advocates notwithstanding, what is striking is that local people, many of them marginalised and impoverished, have at no stage of the planning process been involved. No impact study about possible societal consequences was conducted, and Saint-Georges's and Oiapoque's inhabitants, already on the societal fringes, had no say in the rare moments when political attention shifted towards the borderland. What the border means to them, how they use the river and what knowledge they have of it – none of this was included in the decision-making processes leading to the Oyapock River Bridge (Figure 17.1). Similarly, the importance of entangled histories in and of the region, as represented forcefully by the excavation of indigenous tombs and artefacts precisely in the area of the bridge, was at no point adequately considered or in any way stressed by the relevant decision-making actors. At least on a discursive level, this changed at the inauguration ceremony.

***

The partial inauguration of the Bridge did not occur until 18 March 2017, in the early stages of what became a general strike and the largest demonstrations ever held in Guyane. It was overshadowed by a scandal on the previous day: at the meeting of several Caribbean contracting parties to the Cartagena Convention, hosted by then-Environment Minister Ségolène Royal in Cayenne, a group of men known as *500 frères* (500 brothers) entered the negotiation room, demanding equal opportunities vis-à-vis the French metropole. This incident, broadcast on television, made visible and strengthened a mass social movement that had emerged shortly before from a homicide in Cayenne. This social discontent resulted in a general strike lasting over a month in all of Guyane. Such discontent shows that notwithstanding Guyane's relatively high standard of living compared to its geographical neighbours, the territory suffers several problems compared to other members of the northern 'tectonic plate': economic dependency, high unemployment and security issues alongside the lack of comprehensive education and healthcare were all made explicit during the protests (Mam Lam Fouck and Moomou 2017). Though supposedly opening the Bridge the following day, Royal left Guyane prematurely. Likewise, no Brazilian minister attended the opening ceremony. This lack of ministers from either side strikingly

Figure 17.1 The Oyapock River Bridge as viewed from a *catraia*, connecting/dividing Saint-Georges (French Guiana/France) and Oiapoque (Amapá/Brazil).

Source: Photo by Fabio Santos.

reflects the general sensation in Guyane and Amapá of being irrelevant to their respective national governments. Yet, numerous Amapá and Guyane politicians did participate, offering important hints about why this opening was arguably *incomplete*, with only some borderlanders benefitting unless reciprocity measures were extended.

Despite portraying the Oyapock River Bridge as a symbol of friendship allowing further economic integration, several actors from both sides openly stressed ongoing disagreements, especially concerning unequal mobilities. This departs from earlier explanations which saw Brazil's widespread corruption and alleged lack of reliability in international politics as main reasons behind the Bridge's delay. In my earlier conversations with political actors and during participant observation at diplomatic meetings of the Mixed Transborder Commission, disputes about differential visa policies were treated with reserve. However, whisperings on hallways and behind closed doors made clear that some Brazilian delegates were dissatisfied with the French policy of closed borders in its overseas department. This was not the only reason behind Brazil's delay in finalising construction on its side and opening the Bridge; however, unresolved conflicts around visa non-reciprocity clearly contributed to Brazil's dilatoriness about opening a bridge effectively blocked to most Brazilians.

Asked about her impressions and future expectations by a TV reporter, Brazil's consul in Cayenne, Vera Campetti, said negotiations on the circulation of goods and people were unfinished, but the Bridge was a first step. When further asked if free access across the Bridge was the next step, she replied: 'It depends on the negotiations'. Asked if negotiations about free movement were 'tough',

Campetti abruptly ended the exchange by replying 'yes'. The reporter took these remarks as starting points when talking to other political actors at the ceremony, including the two left-wing Guyanese politicians Georges Patient, a French senator, and Gabriel Serville, a French National Assembly member.[2] Both admitted reciprocity was a problem, saying they would try to increase it; Serville added that he would attempt to allay fears of mass Brazilian migration into Guyane. Shortly afterwards, both extensively reinforced the point when delivering speeches to the audience, with Serville saying they would lobby to ensure that 'visa problems, insurance problems, and traffic problems can be resolved as soon as possible'. In pushing for mutual access across the Bridge, both acted in an unprecedented way. Although they, alongside other politicians like Karam, had demonstrated their sympathy for Brazilian claims on earlier occasions, their ceremony pronouncements bespoke a critique novel in its clarity.[3] They unambiguously displayed the deep differences between many French Guianese politicians and most of their French metropolitan counterparts.

Amongst other speeches that day, Roger Labonté, chief of an indigenous Saint-Georges neighbourhood, brought a historical perspective into the discussion:

> we fought a lot and so, for us, the indigenous, the border doesn't matter for us, because men have created this border. But also, for us, the river is connected with the mountain. And the mountain is connected with the sea.

Drawing connections between different natural formations, Labonté evoked a 'fluid' conception of borders where different elements are intrinsically connected: one cannot exist without the others. The fact that today one element, the river, constitutes a national border reflects human action ('men have created this border'). Labonté and indigenous communities have long criticised strict border controls conflicting their daily practices.

Finally, two left-wing Brazilian politicians echoed this historical grounding: Randolfe Rodrigues called the border an 'invention of France and Portugal'. Before colonisation, he continued, 'the peoples Karipuna, Galibi, and Galibi Marworno were in dialogue, talking without the existence of any artificial border between them'. Like many others, Randolfe framed the Bridge as a symbol of friendship, overcoming borders, spurring development, ending isolation and conflict. By also pointing out the borderland's entangled histories not only across the river but also far beyond the Guianas, he framed it in a critical way which had previously been marginal(ised). This emphasis was reiterated in Capi's speech. One of the Bridge's founding fathers, he described borders as 'fictitious':

> Before the Europeans came, there was no border ... But I want to announce the end of these borders. It doesn't make sense that we are divided by the river, by the road, by any landmark. There is only one humanity, there is no need for borders, and these borders will cease one day.

He also gave practical examples illustrating why he started fostering cross-border cooperation with Karam in the 1990s: 'It did not make sense not to fight together when the malaria mosquito bites those who live in Oiapoque and those in Saint-Georges'.[4]

Most guests supported the Oyapock River Bridge: the linkage with French-Brazilian friendship was continuously referenced, alongside the wish for rising cross-border cooperation, particularly in economic terms. Yet, celebrating the inauguration of the Bridge still allowed several speakers to emphasise inequalities and the borderland's entangled histories. Criticism of unequal mobilities strengthened a view most politicians had previously avoided. Highlighting these inequalities and calling for them to end was an unexpected novelty amidst Guyane's incipient social unrest.

***

Four years after inauguration, what remains of the promises about mutually beneficial cross-border cooperation, including visa reciprocity? Who has used the Bridge, and what for? Before inauguration, it was conceived as a 'geopolitical bridge' (Théry 2011) bringing 'all sorts of further changes' (Price 2018: 30). Local opinions were divided, and it is difficult to generalise. Most Brazilians (across ethnic, gender and class lines) held negative views ranging from sceptical indifference to vehement rejection. The fiercest critics were *catraieiros* (boat drivers) who organised protests and strikes fearing loss of income (Martins 2015: 188ff.). All of the more than 100 *catraieiros* were Brazilians, even though some few held French residence permits.[5] Moreover, since many Brazilians on both sides have no car, they questioned the usefulness of a bridge constructed a few kilometres outside the two city centres. Walking from one town centre to the other across the bridge takes approximately 90 minutes. In contrast, crossing the border by car or *catraia* (a small, motorised boat) takes only 10 minutes. The French population had no uniform opinion, though many showed solidarity with their Brazilian neighbours who are oftentimes colleagues, friends or family members. Established borderlanders simply saw no added value. Taking a *catraia* across the river was a daily routine for many, one usually described as sufficient, sometimes pleasant. However, the Bridge had proponents, not least those with jobs directly connected to its opening: police officers and customs personnel who would conduct controls on the checkpoints located on each side of the Bridge. Importantly, the vast majority of these public servants are white, without a long history of borderland habitation. Usually, they are stationed in Saint-Georges for a few years before returning to the metropole. With their presence considerably increased over the past years, the borderland nowadays has a 'climate of permanent tension' (Martins 2015: 186), and they are among the high-income earners in the heterogeneous and inequality-ridden borderland.

It is precisely this group and their families that benefit most from the inaugurated Oyapock River Bridge: they cross for tourist trips in northern Brazil, eat in Oiapoque's restaurants, and buy lower-priced food and beverages. Locally, it is an open secret that Frenchmen also take advantage of their resources and mobility opportunities to frequent Oiapoque's many brothels. On their return to Saint-Georges, I noticed that the cars of this 'inner circle' are rarely inspected by colleagues who are oftentimes also friends. While this group (mainly white metropolitans, a few

Créoles and established Brazilians living legally in Guyane) can take advantage of the opportunities the Bridge provides, others rarely or never use it.

Many from both sides, but especially Brazilians and non-EU citizens, continue taking a *catraia*. One reason they still can is because the Bridge initially opened only from 8 am to 12 pm and from 2 pm to 6 pm as well as on Saturdays from 8 am to 12 pm. Since August 2019, it opens daily from 8 am to 6 pm. At the time of finalising this chapter (January 2021), the Bridge was closed, and strict border controls were enacted by French and Brazilian police along the river due to the COVID-19 pandemic, by which Amapá and Guyane have been heavily affected. Although it is expected that *catraieiros* will be able to resume work as soon as the crisis is over, their fears have become partially realised, independent of the pandemic-induced unemployment. Although the Bridge has not become the only legal means of border-crossing, the river traffic has diminished to the extent that competition has become sharp, with several *catraieiros* seeking new employment (pre-COVID). Many Amerindians, Maroons and *ribeirinhos* never use the Bridge and do not intend to. Apart from their intimate knowledge of the river and its resources, most would not be entitled to cross, in case they hold Brazilian instead of French citizenship: visa policies have not changed and remain unequal. To the detriment of the non-EU majority of borderlanders, legal access to Guyane remains restricted, while French and other EU citizens can easily enter Brazil via Oiapoque. Moreover, Brazilian car holders entering Guyane must pay up to 175 euros insurance, a fortune in Brazil's impoverished North. The consequence is clear: in 2019, of an average 2000 cars crossing the Bridge per month, only 5 had Brazilian registration numbers (Seles Nafes 2019).

A one-way street favouring French and EU citizens, the Bridge is marked by its rupture with traditional forms of border crossings, deepening inequalities especially in terms of unequal mobilities, in turn based on one's citizenship. 'Randomly' ascribed mostly by bloodline (*jus sanguinis*) or territory (*jus soli*) in the majority of cases, citizenship has proven most relevant in reproducing global inequalities, impacting access to fields like quality education, dignified work, healthcare, safe water and nutrition (Korzeniewicz and Moran 2009; Shachar 2009). This becomes still more relevant when set in historical perspective: citizenship granted rights only to some, while denying them to racialised and gendered others (Boatcă and Roth 2016). To this day, Brazilian and French Guianese 'minorities' suffer from the unequal distribution of civil, political and social rights linked to citizenship. In the French-Brazilian borderland, this has produced a complex nesting: Amapá is one of Brazil's poorest and remotest states, home to a majority of persons of colour often suffering unemployment, inadequate healthcare and running domestic water, amongst others. While the non-white majority in Guyane faces similar challenges, as the protesting thousands in 2017 demonstrated, French citizens know that when worse comes to the worst, their chances of upward mobility, even survival in extreme cases (like the COVID-19 pandemic), are undoubtedly higher in this South American EU exclave. Citizenship, intersecting with other axes of stratification like gender, race and class, thus centrally shapes social life in Saint-Georges and Oiapoque as well as between these unequal twins. That it plays such a prominent role results from colonially

Figure 17.2 The French and EU sign after crossing the Oyapock River Bridge from the Brazilian side, with border checkpoints in the background.

Source: Photo by Fabio Santos.

entrenched power struggles between France and Brazil, with France transferring and even reinforcing its 'deportation regime' (Benoît 2020) in Amazonia, and Brazil reluctantly accepting the rules or simply not caring about its impoverished periphery (Figure 17.2).

***

In a senate speech long before inauguration, Capi envisioned 'a bridge that will unite the people' (Capiberibe 2015). Likewise, his counterpart Karam insisted in our interview on its transformative power between structurally disadvantaged territories and populations. Four years after inauguration, the contrast could hardly be sharper. Notwithstanding their optimism, the border along the Oyapock River appears evermore demarcated, unveiling tensions between different notions of space: borderless in imagination, but increasingly confined in practice; rhetorically connecting, yet effectively separating. The citizenship and visa regime limits access across the river, and its underlying notion of clear-cut, absolutist territorial space contradicts prevailing local understandings of shared space, fluid borders and belongings.

Today, the unresolved issue of visa non-reciprocity producing highly unequal access across the Bridge verifies earlier predictions. The 'perverse effects of exclusion and disempowerment' (Kramsch 2016: 209) have become manifest, with a regional elite being the sole beneficiary. This elite not only has the right 'papers' but also the relevant social and economic capital to cross unproblematically. Though widely lamented for years by various inhabitants as well as politicians at the inauguration ceremony, the lack of visa reciprocity remains: the Bridge acts as a one-way street perpetuating mobility inequalities primarily based on citizenship. Though in place, the Bridge is therefore out of place, historically disrupting a deeply entangled region. In contrast to prevailing political framings of a connecting bridge, the bitter reality is that the opening has so far enhanced closure, except for a relatively small group, mostly those with French citizenship or residence permits. Most other(ed) inhabitants and migrants, by contrast, have encountered more difficulties in continuing with their historically routinised practices of exchange and mobility across the river. Consultation with these marginalised groups in the region – Amerindians, Maroons, *ribeirinhos* and others – is long overdue. Neglected in the planning process alongside most political processes, these groups and their knowledge, practices and needs should guide any further steps towards new types of mutually beneficial cross-border cooperation and exchange. Otherwise, the Bridge will remain a one-way street in a region traditionally characterised by rhizomatic lines of movement, wherein an 'intersection' with roots and routes running in multiple directions would be a more apt symbol. Rather than being bridged, the two 'tectonic plates' risk collision, unveiling widespread societal disenchantment with politics cementing inequalities instead of undoing them.

Notes

1 A longer version of this chapter is included in my monograph (Santos 2022). This research received funding from the German Research Foundation (IRTG 'Between Spaces') and the OHM Oyapock/LABEX DRIIHM.
2 Most elected politicians in and from Guyane and Amapá belong to the centre-left political spectrum. Their push for visa-free access is therefore informed by not only the regional particularities outlined in this chapter, but also their political affiliations.
3 Karam, a 'founding father', could not attend the inauguration.
4 This plea for sanitary cooperation has gained renewed relevance in light of the pandemic currently ravaging the borderland.
5 The number of *catraieiros* varied seasonally; some had multiple jobs at different times.

References

Benoît, C. 2020. Fortress Europe's Far-Flung Borderlands: 'Illegality' and the 'Deportation Regime' in France's Caribbean and Indian Ocean Territories, *Mobilities* 15(2), 220–240.

Boatcă, M. 2019. Forgotten Europes. Rethinking Regional Entanglements from the Caribbean, in: Bringel, B., and Cairo, H. (Eds.). *Critical Geopolitics and Regional (Re)Configurations: Interregionalism and Transnationalism between Latin America and Europe*, London, Routledge, 96–116.

Boatcă, M., and Roth, J. 2016. Unequal and Gendered: Notes on the Coloniality of Citizenship, *Current Sociology* 64(2), 191–212.

Capiberibe, J. 2003. Pronunciamento de João Capiberibe em 21/03/2003. https://www25.senado.leg.br/web/atividade/pronunciamentos/-/p/texto/331197. [Accessed 20/03/2020].

Capiberibe, J. 2015. Pronunciamento de João Capiberibe em 02/03/2015. https://www25.senado.leg.br/web/atividade/pronunciamentos/-/p/texto/411189. [Accessed 20/03/2020].

Capiberibe, J. 2017. Pronunciamento de João Capiberibe em 21/03/2017. https://www25.senado.leg.br/web/atividade/pronunciamentos/-/p/texto/429639. [Accessed 20/03/2020].

CNRS Images. 2010. Les archéologues de l'Oyapock. https://images.cnrs.fr/video/2230. [Accessed 20/03/2020].

Costa, S. 2019. The Neglected Nexus between Conviviality and Inequality, *Novos Estud. CEBRAP* 38(1), 15–32.

Espelt-Bombin, S. 2018. Frontier Politics: French, Portuguese and Indigenous Interactions between Cayenne and the Amazon, 1680–1697, in: Wood, S. and MacLeod, C. (Eds.). *Locating Guyane*, Liverpool, Liverpool University Press, 69–90.

Gomes, F. 2003. Other Black Atlantic Borders: Escape Routes, 'Mocambos', and Fears of Sedition in Brazil and French Guiana (Eighteenth to Nineteenth Centuries), *New West Indian Guide/Nieuwe West-Indische Gids* 77(3–4), 253–287.

Granger, S. 2011. Le contesté franco-brésilien : enjeux et conséquences d'un conflit oublié entre la France et le Brésil, *Outre-mers* 98(372), 157–177.

Hoefte, R., Bishop, M. L., and Clegg, P. 2015. Still Lonely After All These Years? Contemporary Development in the 'Three Guianas', *Caribbean Studies* 43(2), 83–113.

Korzeniewicz, R. P. and Moran, T. P. 2009. *Unveiling Inequality: A World-Historical Perspective*. London, Russell Sage Foundation.

Kramsch, O. T. 2016. 'Spatial Play' at the Ends of Europe: Oyapock Bridge, Amazonia, *Tijdschrift voor Economische en Sociale Geografie* 107(2): 209–213.

Mam Lam Fouck, S. 2015. *La société guyanaise à l'épreuve des migrations du dernier demi-siècle, 1965–2015*. Matoury, Ibis Rouge Éditions.

Mam Lam Fouck, S., and Moomou, J. 2017. Les racines de la 'mobilisation' de mars/avril 2017 en Guyane. *Amerika* 16.

Martins, C. 2015. Cooperação internacional em território fronteiriço: novas sociabilidades e novos controles. *Textos e Debates* 1(27), 177–196.

New York Times. 1992. Perilous Jungle Passage Leads Poor to 'France'. https://www.nytimes.com/1992/07/04/world/oiapoque-journal-perilous-jungle-passage-leads-poor-to-france.html. [Accessed 20/03/2020].

Nonato Júnior, R. N. 2015. La France et le Brésil de l'Oyapock, quels enjeux bilatéraux entre développement et durabilité ? *Confins* 24.

Price, R. 2018. The Oldest Daughter of Overseas France, in: Wood, S. and MacLeod, C. (Eds.). *Locating Guyane*. Liverpool, Liverpool University Press, 17–32.

Romani, C. 2013. *Aqui começa o Brasil! Histórias das gentes e dos poderes na fronteira do Oiapoque*. Rio de Janeiro, Multifoco.

Santos, F. 2020. Crisscrossing the Oyapock River: Entangled Histories and Fluid Identities in the French-Brazilian Borderland, in: Rein, R., Rinke and S. Sheinin, D. (Eds.). *Migrants, Refugees and Asylum Seekers in Latin America*. Leiden, Brill, 217–241.

Santos, F. 2022. *Bridging Fluid Borders: Entanglements in the French-Brazilian Borderland*. London, Routledge.

Seles Nafes. 2019. Por mês, 2 mil carros cruzam a Ponte Binacional e apenas 5 são Brasileiros. https://selesnafes.com/2019/03/por-mes-2-mil-carros-cruzam-a-ponte-binacional-e-apenas-5-sao-brasileiros/. [Accessed 20/03/2020].

Shachar, A. 2009. *The Birthright Lottery: Citizenship and Global Inequality*. Cambridge, MA, Harvard University Press.

Silva, G. 2017. France-Brazil Cross-Border Cooperation Strategies: Experiences and Perspectives on Migration and Trade. *Journal of Borderlands Studies* 32(3), 325–343.

Théry, H. 2011. "France-Brésil: un pont géopolitique." *Diploweb.com, La Revue Géopolitique*. https://www.diploweb.com/France-Bresil-un-pont-geopolitique.html. [Accessed 20/03/2020].

18 The everyday of the twin cities of Chuí (Brazil) and Chuy (Uruguay)

A semiotic analysis

Gianlluca Simi

Throughout the centuries, Chuí (Brazil) and Chuy (Uruguay) have always formed a continuous space and shared the same name. This area in the extreme South of South America was subject to erratic territorial claims between the then-imperial powers of Portugal and Spain and to subsequent conflicts for regional dominance between their successor states, Brazil and Argentina. Following the short-lived 1777 Treaty of San Ildefonso, the area Chuy now occupies became a buffer zone between Portuguese and Spanish dominions, known as The Neutral Fields – a conceptual void created to limit communication and avoid further conflict.

Only at the beginning of the 20th century was the Brazil-Uruguay border finally established, by which time the local residents, known as Chuienses, had developed strong social, cultural and economic networks mostly disregarding the official, and historically unstable, border. This legacy is still perceptible in the everyday life of *Chuy* – as I will henceforth refer to these twin cities – still marked by uninterrupted flows of people and goods across the nominal border (Figure 18.1).

This chapter discusses the results of empirical research conducted in 2016 in Chuy against a theoretical background drawing on Semiotics and Everyday Life Studies. The first section offers a brief historical overview of the Brazil-Uruguay borderlands, elucidating how relations between the two countries shaped present-day Chuy. The second uses key theoretical ideas to parse out the reasons why exploring how meaning is made and conveyed in everyday life is a prime lens through which to understand how history still influences these cities, even though the border's very meaning is itself susceptible to some elasticity that interacts with individuals' own personal narratives and histories as well as with discursive tendencies emerging in today's Chuy. The chapter's final section starts by noting that, when asked about what the border meant to them, Chuienses tended to describe it positively, thus challenging dominant Latin American perceptions of borderlands as threats to the nation due to illegal and criminal activities. It then presents some patterns in Chuienses' responses when asked to elaborate on values attributed to the border, which offer a more nuanced picture of how people's relationship to the border is just as often about reinstating it as it is about subverting it.

DOI: 10.4324/9781003102526-23

Figure 18.1 The International Avenue where both sides of Chuy meet.
Source: Photo by Gianlluca Simi, 2016.

By the end, I hope to demonstrate that, in Chuy, the border is explicitly seen in positive terms due to the openness its current setting engenders. In everyday life, nevertheless, it is equally a mechanism to appraise individuals, practices and products according to their position in relation to the border when this extreme openness is deemed uncontrollable or menacing for local residents. Ultimately, this experience suggests that understanding the reality of open borderlands might be less about cataloguing perceptions of the border and more about analysing the ways wherein meanings declared by local residents inform and/or differ from real-life instances when what the border enacts is at play.

The Neutral Fields: a brief history of the Brazil-Uruguay borderlands

Brazil is gigantic, and Uruguay, tiny. To understand their relationship, we must peruse where they meet, along 1069 km of a single Brazilian state, Rio Grande do Sul. Both Uruguay and Rio Grande do Sul have historically been somewhat of a buffer zone between Portuguese and Spanish dominions and, later, as Garcia (2011) and Goes Filho (2013) highlight, between their heirs, Brazil and Argentina. This historical entanglement is particularly relevant when studying Chuy, as it lies in The Neutral Fields, a toponym as often used today as during its conception in 1777, as we will soon establish.

Pucci (2010, 75) argues that Rio Grande do Sul and Uruguay have been intertwined at least since Portugal's first 1530–1532 expedition, when the first mention of the homonymous Chuy stream fixed the region's location along the incipient coastal route between Portuguese and Spanish dominions. Every claim

to possession thereafter was inextricably connected to political projects to col-
onise and mobilise the scattered population of the *pampas* – a strategy official-
ised by the 1750 Treaty of Madrid, under the Roman concept of *uti possidetis ita
possideatis*, meaning something like 'as you possess, so you may possess': i.e. the
prerogative to *retain* control at the moment of dispute over land (Amaral 1973;
Vargas 2017).

Thus, any later claim to allegiance in establishing the border must be comprehended in this context of instrumental mobilisation of human presence,
effectively manufacturing the idea of a unified people on whose behalf territory was claimed. What matters most in understanding how the Brazil-Uruguay
border was formed is, therefore, 'the interpretations assumed by statespeople
in each socio-political era to support theses … pertinent to the conjuncture in
which conversations to fix the boundary lines were established' (Golin 2002,
165) within a wider history of conflictual dominion that ultimately failed to
create a sense of cultural belonging extending too far beyond people's immediate
horizons.

Following the failed 1750 Treaty of Madrid and the 1761 Treaty of El Pardo,
Chuy was finally described as The Neutral Fields by the 1777 Treaty of San Ildefonso, whose underlying notion was that, given the severity and frequency of
conflict in this area, the solution was to create a buffer zone between the two
dominions (Cesar 1979, 200). However, far from resolving Spanish-Portuguese
conflicts, emerging from the very volatility of boundaries, it complicated things
further. In Europe, relations looked no better, war being imminent between
Portugal, one of England's remaining allies, and Spain, allied to France under
Napoléon Bonaparte. Animosity in the Old Continent prevented the evolution
of boundary-marking in South America, where bands of people kept criss-crossing the fields, neutral and otherwise, building settlements and creating the case
for future territorial claims.

In 1821, one year before independence, Brazil showed signs of the imperial
lust that would dominate its foreign relations until the early 20th century by
annexing the river Plate's Eastern Bank as its Cisplatina Province. For a few
years thereafter, Uruguay was officially and entirely part of Portuguese, and later
Brazilian, dominion. In 1828, Argentina and Brazil mutually agreed (Teixeira
Soares 1975, 189) to grant the Cisplatina Province independence via the Treaty
of Montevideo, creating the new Oriental Republic of Uruguay. This was the
beginning of the end of nearly 400 years of dispute over these lands, culminating
in the 1851 Treaty of Limits between Brazil and Uruguay, which, aside from a
minor 1909 amendment, still remains in force.

Since then, Chuy has retained an aspect of incomplete modernity and relative
isolation due to the different degrees of state involvement each side has engen-
dered from their respective countries. Thus, whereas Uruguay's recognition of
its portion of Chuy can be traced back to 1888, Brazil recognised Chuí only in
1939 (Azambuja 1978, 149–152), highlighting a pattern that has not changed
massively even now due to an undeniable distinction between these two: while
Chuí Brazil is a tiny community within a gargantuan country, Chuy Uruguay is a
small town in an already-small country.

This is also marked in the different flows merging across Chuy. The completely open border creates an interstitial space where all Chuienses, regardless of citizenship or potential national allegiance, come together to form an everyday landscape characterised precisely by how the border is paradoxically irrelevant because of its very presence. Thousands of seasonal shopping tourists, mostly Brazilians, are attracted by the marvels of luxury goods offered at significantly lower prices at Uruguayan duty-free shops where only non-Uruguay citizens can shop. At weekends, however, it is Uruguayans from elsewhere in the country who visit the Brazilian supermarkets for their cheaper groceries, Brazil's industry being much stronger and more diverse than Uruguay's. Finally, in between these bursts of visitors, Chuienses themselves take advantage of both offerings, without even needing to go through customs, as both Brazil's and Uruguay's offices are located effectively outside the urban area, posing few obstacles to those living *inside* Chuy.

At the local level, in other words, Chuy itself is still marked by 'both urban and rural relations [that] have a long common history ... with strong cultural and labour exchange' (Ministério da Integração Nacional 2005, 147), thus highlighting a defining aspect of present-day Chuy: the predominance of locality amidst frequent displacements between the levels of the local, regional, national and global. If Chuy has been scarred by different borders since the colonial era, it has likewise always been a small town, whose combined population today does not exceed 16,000 according to 2018 data from both countries' statistics bureaux (Uruguay's INE and Brazil's IBGE) and whose everyday, for those living here, can be understood only through this overlapping between the two countries: their formation, economies, political systems and particularly how neither Brazil nor Uruguay have ever been able to claim unmatched dominion over Chuy.

Ultimately, modern-day Chuy's identity has gradually morphed from isolation and even abandonment to the scalar friction permeating all aspects of life here. We cannot state which set of national references, if any, is at work unless we look at particular events: the people therein involved and, especially, the meanings that they might assign to the border depending on what they intend to achieve or what they find most important to enact.

A semiotic approach to the everyday in the borderlands

As this research focused on how the border was perceived and used in Chuy's everyday, surveys emerged as appropriate tools to collect Chuienses' responses in a systematic and structured manner promoting 'a situated activity that locates the observer in the world' (Brinkmann 2012, 20), whilst Chuienses speak of the borderlands as their primary space of socialisation. In other words, surveys allowed the observer to ask Chuienses to ultimately make sense of their own lives: since people can speak only from their own position, even if the question demanded some universality, whatever Chuienses said of their lives *in* the borderlands would be what was said *of* the border.

Their answers highlight speech acts constituting not a pristine copy of reality but, rather, discourses thereof, ways of speaking about life in the borderlands emerging from how people experience and reflect it. The reference to life in the borderlands therefore yields various texts about reality in its inseparability from the border, understood as 'a structure of messages, or message traces, which has a socially ascribed unity' (Hodge and Kress 1988, 6). Texts become discourses as they interconnect: i.e. arrangements allowing language to be produced inside the same reality about which it speaks, evoking a long-established understanding that discourses produce knowledge about reality (Fiorin 2017; Benveniste 1974).

Every statement about borderland life contains, therefore, a proposition about reality. This draws on Saussure's understanding of language as far more than 'a list of terms corresponding to a list of things' (Saussure 1983, 65). Words do not describe reality as make sense of what it can become in the process. Chuienses are not offering mere nomenclature for the physical space; their many names for the border and the borderlands produce the very reality of these borderlands. Rather than an image of reality, the meaning attached to the border is, itself, already an action: that of naming. This use of language to render the border. meaningful is, therefore, the very use of the border (Wittgenstein 1958, sec. 43).

In this sense, a semiotic approach to the border understands simultaneously that no fragment or element of the world is self-explanatory: i.e., to be considered part of reality, it must be rendered meaningful (see Barthes 1968; Randviir 2004). This approach is based not on absolutely relativistic views of reality whereby only what is perceived is deemed to exist; but, rather, on understanding that reality, from a semiotic perspective, is 'concerned not with absolute truth but with truth as speakers and writers and other sign-producers see it, and with the semiotic resources they use to express it' (Van Leeuwen 2005, 160).

This is particularly valid in the border's case, because the boundary line itself has no magnitude (see Cameron 2011). As no dividing line is really fully natural, but rather a naturalisation of motivations to divide (Barthes 1968, 53–54), the border does not even refer to a real object – it must be assigned some meaning to be perceptible: the border exists only as sense is made of it. In this sense, 'the conceptual emptiness of the line', Cameron (2011, 419) writes, 'is both central to its meaning and fundamentally contradictory to its function'. This is to say that the border is always a function, an action comprehensible only as part of a signification process.

Furthermore, in studying not only the meanings of borders but, likewise, how these meanings operate in Chuy's everyday, we must establish that the everyday is not a set of neatly recognisable phenomena that can be scrutinised in isolation. 'The everyday', writes Sheringham (2006, 16), 'is a dimension of human experience rather than an abstract category'. The everyday is, in other words, a perspective from which to access the real, to analyse the meanings Chuienses associate with the border. It therefore conveys 'insider's knowledge' of particular social processes (Gardiner 2000, 4), which are, here, the dynamic socio-semiotic relations established around and across the border.

The founding ambiguity of Chuy's twin cities produces clear tension between two levels of the border. Drawing on Certeau's (1990) definitions of strategy and tactics, on the one hand, the line remains a strategy of division and control whereby, for the state, there is no uncertainty about what portions of Chuy belong to which country. On the other hand, the everyday around the line is permeated with tactics Chuienses have been developing for generations, allowing them to test the line's malleability without necessarily jeopardising their gains. Or, as Colebrook (2002, 699) puts it, these are 'unintended dilations', fairly uncontrolled circumstantial responses to the degree of profitability that individuals calculate within the scope of what is deemed proper without necessarily dismantling power structures. If the border as a strategic mechanism of division goes rather unquestioned in uninhabited areas, in Chuy it is very different, as the border is precisely what enables certain practices that subvert the line's original function.

Here, people play with the border in their everyday in accordance with their calculus of convenience without necessarily hoping to annihilate it. These everyday procedures, writes Certeau (1990, 40), 'imply a logic of the operation of actions relative to types of situations'. These procedures do not forcibly imply rejection of the original strategy; they are part of 'a never complete process of b/ ordering space along state-centred lines, whose success in containing and moulding the spatial properties of events needs to be verified as it unfolds in its actuality' (Novak 2011, 745).

The names of the line: naming and attributing value

The survey used to collect Chuienses' perceptions and uses of the border included four questions, three of which are analysed in this chapter: Q2, Q3, and Q4. The first question, asking people to indicate which nationality they identified themselves with the most, is not included here, as it yielded conflicting results that could not validate any statistically significant claims.

Q2 asked people what came to mind when they thought of the border; Q3 asked what aspects they found positive about living in the borderlands; Q4 asked what they found negative about borderland life. In practice, Q2 provided an inventory of meanings associated with both the geopolitical line and the region's lived reality. It also ignited a semiotic process whereby respondents' attention was directed towards Q3 and Q4, which effectively classified aspects of everyday life in the borderlands as either positive or negative, thus eliciting how the border is used, rather than simply described, by cataloguing possible instances when the line, as a strategy of division, is either respected or resisted.

From a population of approximately 16,000, a total of 426 valid surveys were collected amongst those aged 18 or older, yielding 59 *in-vivo* codes later arranged into 275 categories. The 11 main codes – read 'primary meanings of the border' – represented 59.4% of all code occurrences. The categories formed by the 843 interactions between these main codes represented, in turn, 24% of all category occurrences. The main categories – read 'actionable meanings of the border' – highlights a generalised perception that the border works as a pulling factor for people who relocate to Chuy looking for job opportunities as well as for shopping tourists from both Brazil and Uruguay. Through this flow of people, the very diversity of origins and of cultural heritage become apparent, converging into a mostly positive experience of this cross-border space of socialisation. In other words, the border is not merely a characteristic of Chuy; it is what makes Chuy; it is an inseparable element of the city. The border founds Chuy.

These are, however, descriptive perceptions: generic mentions to 'shopping', 'culture', 'diversity', 'unity', etc. (the main codes). The next step in analysing results is, therefore, to examine responses to Q3 and Q4 – about positive and negative aspects of borderland life. These provide a more nuanced view of the topography of meanings associated with the border for two reasons.

The first is positionality. These latter questions obliged Chuienses to take a stance in relation to the border, to classify their experiences as advantageous or disadvantageous, ultimately preventing them from offering idealistic interpretations. The second reason is the resulting tension between levels of advantages and disadvantages, from which emerge, on the one hand, instances when the border's meanings correspond to a general strategy and, on the other, instances when it is tactically reappropriated.

The most remarkable aspect, amongst the main codes in Q3 (advantages), is the prominence of references to 'cheap', highlighting a common topic in Chuy. Living costs are definitely lower than elsewhere in both Brazil and Uruguay. This is partly because this is a small town, where costs are usually lower anyway. However, another factor bringing most costs down is that the (open) border offers opportunities to take advantage of the gap between fiscal jurisdictions and national economies. This refers both to those people who might relocate to Chuy seeking job opportunities as well as to waves of shopping tourists, middle-class Brazilians seeking duty-free luxury goods and working-class Uruguayans shopping for cheaper groceries. What is noteworthy is the assumption that 'being cheap' is, therefore, not exclusively attractive to those coming to Chuy to shop but, perhaps more significantly, to those who live here and can, day-by-day, profit from this imbalance between the two economies, either by taking advantage of lower living costs or by making some extra money from temporary jobs as waiters, salesclerks or street guards during busy summer months. As Chuienses often said, 'when one side is doing badly, the other one is thriving'.

Intriguingly, seven of the main codes in Q3 mirror the main ones from the free-association lists from Q2. Ideas related to lower prices, diversity, culture, shopping, internationality, work and language all feature both in the main codes in Q2 and Q3, from which there seems to be an intrinsic proclivity to define the border in positive terms at first. Conversely, amongst responses to Q4 (disadvantages), mentions of scarcity-related issues are, by far, the most noticeable. The idea of 'scarcity' is completely understandable here, as the cities' lack of services and poor infrastructure are painfully visible and have been officially recognised by Brazil's government (Brazil 2005, 61). Chuy is surrounded by an aura of 'incomplete development': many buildings look derelict; the streets are pothole-filled and bereft of proper signs; pavements are narrow and overcrowded with hangers and mannequins from shops; fly-tipping is common, as shoppers dispose of packaging haphazardly to make it through customs more easily. Not to mention the most contemporary of horrors: poor internet connection, as companies find little economic incentive to invest in coverage for such a small, isolated market.

Poor infrastructure is, in other words, what Chuienses touch on when mentioning 'dirtiness', 'lack of urban planning', 'insecurity', and 'chaos'. The highly frequent mentions of 'bureaucracy' also highlight a potential connection between these aspects of poor infrastructure and the government, which does not, however, simply imply local government. Chuienses seem deeply aware there is only so much local government can do if there is little financial and political support from regional and national governments. One likely cause of what effectively translates into a feeling of abandonment, heightened by the city's geographical isolation, is the fact that Chuy is a small town, even when both sides are considered.

What is clear from examining how disadvantages are described separately is that, besides helping not to essentialise meanings associated with the border, they are mostly polarised. If ideas mentioned as positive replicate those mentioned on the free-association lists, ideas mentioned as disadvantages do not evoke any declared idiosyncrasies of these borderlands as such. In fact, the negative aspects of living in Chuy could easily be used to describe the everyday of countless other small Brazilian or Uruguayan towns.

There is one exception, nonetheless, that needs noticing: fluidity. On the one hand, responses to Q3 mostly identified the same ideas used to describe the borderlands in Q2 by saying that some of the advantages are 'the possibility to buy imported products' (Respondent 180); 'being close to the culture of both countries and absorbing the best from each of them' (R110); as well as 'the cultural diversity that is typical of Latin America and the respect for diversity' (R203). On the other hand, however, responses to Q4 effectively flipped these same ideas upside down, claiming that some of the disadvantages here are 'insecurity because of the mixing of different cultures' (R189); 'loss of cultural identity' (R019); and 'foreign threats' (R192). The concept of fluidity is, thus, the lowest common denominator between responses to Q3 and Q4.

Overall, mentions of fluidity were measured by how often ideas insinuating movement were stated by using words like 'flow', 'fluid', 'access', 'passing', 'crossing', 'openness', 'free', etc. As a code, 'fluidity' featured only in the bottom half of responses to Q2 only to rise to the main occurrences for Q3 and, especially, for Q4. If we consider that the absolute top codes for each Q3 and Q4 are complete opposites (i.e., ideas most often presented as positive are, in contrast, amongst those at bottom of the list of negative ideas, and vice versa), it is striking that ideas around fluidity are where responses to Q3 and Q4 meet despite a slight tendency towards being described as a negative aspect of life in Chuy.

This is, of course, a tendency identified after consolidating the data obtained from all 426 surveys. However, one respondent managed to illustrate the ambivalence of the fluidity concept perfectly. Asked about the advantages of living in these borderlands, R067 replied that it was 'the knowledge you acquire about different cultures as well as the exchange between them'. Immediately after that, R067 said that the disadvantage of living in Chuy was 'the miscegenation of undesired cultures and the risks that badly behaved, unknown people pose to those who live in these borderlands'. This ambivalence is, as results show, intrinsic to the very idea of fluidity in Chuy; one allowing different people and different cultures to come together, enjoy the same space, live next to one another, but one which, likewise, allows for criminals to enter the city, for undesired mixtures to take place and for the establishment of a widespread feeling of insecurity to take over.

Generally speaking, whereas fluidity was usually combined positively with ideas of diversity, shopping, culture, and the condition of 'being in two countries', it was also negatively associated with ideas of danger and insecurity, chaos and impunity, alongside smuggling. In itself, 'fluidity' is quite broad, certainly, but the ways wherein it combined with other ideas suggests that the general notion of movement and openness is actually the most evident aspect of everyday life in Chuy precisely because it is so ambivalent, echoing Nail's argument that borders are best understood in terms of 'flows' (2016). That is, as a conspicuous aspect of everyday life in these borderlands, 'fluidity' and its related meanings are central to Chuy's identity, replicating Coelho's claim that borderlands 'behave as a space of transitory characteristics' (2014, 220). The border's meaning in Chuy is, therefore, a variable for consideration alongside the individual who utters such meaning and the use or stance this utterance potentially enables.

Data thus show that the permissiveness characterising and enabling certain flows that are then described positively, like cultural diversity and shopping, is immediately seen as a downside when seen to facilitate practices threating local people or apparently allowing others to take advantage of the city's very border condition. In other words, what this does is reinforce the hypothesis that the border has no universal meaning, positive or negative, even in a community apparently marked by the ways it transcends the line.

Instances when the border is actively evoked as a means of separation can thus be as frequent and widespread as those when it is rejected. This ultimately highlights the interdependence of strategy and tactics rather than the predominance of tactics as a mere inventory of meanings could have suggested. Chuienses tend to endorse the idealised view of the borderlands as a space of cultural diversity as much as they reject or suspend it depending on how their stance might impact on their own everyday lives, when it enables illegal or harmful behaviour from outsiders, for example.

Final remarks

The border's meanings in Chuy are therefore fleeting, but still perceptible. The ways wherein they change, acquiring different practical senses according to different individual perspectives, compose an intricate topography of the everyday, doing more than offer an inventory of commonly used expressions to refer to the border. Most importantly, it shows the patterns into which meaning is arranged and activated.

Overall, results indicate the border tends to be described in positive terms, pointing to the construction of a concept of boundary line resisting the divisionary strategy of lines at large. Broader notions of diversity, unity and hybridity were coupled with more site-specific aspects related to job offer, shopping tourism and lower prices. When asked to effectively categorise these notions as either positive or negative aspects of borderland life, Chuienses provided nuanced views of the everyday in which stand out, on the one hand, replications of the main ideas and words used to describe the line in positive terms as well as, on the other, issues of poor infrastructure and poor services generally associated with small peripheral towns.

Fluidity therefore stood out as a categorically ambivalent notion, highlighting ideas of movement and openness as fundamental to analysing the positionality of responses as circumstantial acts of enunciation. This positionality in turn relies on the very understanding individuals might have of the border and the history they might have already established therewith until the point of enunciation, ultimately echoing Mikhailova and Németh's findings on what they called 'personal linkages across the border and the perceptions of the other place' (Mikhailova and Németh 2019, 279). These linkages and perceptions define the everyday in Chuy, simultaneously allowing for the subversion of the border's strategic role as well as for its reinstatement when deemed convenient or necessary.

In other words, the border in Chuy is as meaningful as its uses, making it impossible to say to which extreme Chuy tends because neither subversion nor reclaiming the border is fully operative at all times. If there is one thing Chuy demonstrates very poignantly, it is that, perhaps, the success or failure of borders depends less on economic and military power, and much more on how effectively discourse can be mobilised and polarised towards a particular agenda to the detriment of an everyday that has always been as open and fluid as marked by attempts to shut down communication between different sides.

References

Amaral, A. F. (1973). *Os campos neutrais.* Porto Alegre: Grafisilk.

Azambuja, P. (1978). *História das terras e mares do Chuí.* Caxias do Sul: UCS/EST.

Barthes, R. (1968). *Elements of semiology.* New York: Hill and Wang.

Benveniste, É. (1974). *Problèmes de linguistique générale, 2.* Paris: Éditions Gallimard.

Brinkmann, S. (2012). *Qualitative inquiry in everyday life: working with everyday life materials.* London: Sage.

Cameron, A. (2011). Ground zero – the semiotics of the boundary line. *Social Semiotics* 21(3), 417–434, DOI: 10.1080/10350330.2011.564391

Certeau, M. d. (1990). *L'invention du quotidien.* Paris: Gallimard.

Cesar, G. (1979). *O Contrabando no Sul do Brasil.* Porto Alegre: Editora Globo.

Colebrook, C. (2002). The politics and potential of everyday life. *New Literary Review* 33(4), 687–706.

Fiorin, J. L. (2017). Uma teoria da enunciação: Benveniste e Greimas. *Gragoatá* 22(44), 970–985, DOI: 10.22409/gragoata.v22i44.33544

Garcia, F. C. D. (2011). *Fronteira iluminada: história do povoamento, conquista e limites do Rio Grande do Sul a partir do Tratado de Tordesilhas (1420–1920).* Porto Alegre: Sulina.

Gardiner, M. E. (2000). *Critiques of everyday life.* London: Routledge.

Goes Filho, S. S. (2013). *As fronteiras do Brasil.* Brasília: FUNAG.

Golin, T. (2002). *A Fronteira 1.* Porto Alegre: L&PM.

Hodge, R. and G. Kress (1988). *Social semiotics.* Cambridge: Polity Press.

Mikhailova, E. and Németh, S. (2019). Impacts of town-twinning on the communities of Imatra and Svetogorsk through different fields of cross-border cooperation. In: J.

Garrard and E. Mikhailova (eds.) *Twin cities: urban communities, borders and relationships over time*. London: Routledge, 272–287.

Ministério da Integração Nacional. (2005). *Proposta de Reestruturação do Programa de Desenvolvimento da Faixa de Fronteira*. Brasília: MI.

Novak, P. (2011). The flexible territoriality of borders. *Geopolitics* 16(4), 741–767, DOI: 10.1080/14650045.2010.494190

Pucci, A. S. (2010). *O estatuto da fronteira Brasil-Uruguai*. Brasília: FUNAG.

Randviir, A. (2004). *Mapping the world: towards a sociosemiotic approach to culture*. PhD Thesis. Tartu, University of Tartu.

Saussure, F. D. (1983). *Course in general linguistics*. London: Duckworth.

Sheringham, M. (2006). *Everyday life: theories and practices from Surrealism to the Present*. Oxford: Oxford University Press.

Teixeira Soares, Á. (1975). *História da formação das fronteiras do Brasil*. Rio de Janeiro: Conquista.

Van Leeuwen, T. (2005). *Introducing social semiotics*. Abingdon: Routledge.

Vargas, F. A. (2017). *Formação das fronteiras latino-americanas*. Brasília: FUNAG.

Wittgenstein, L. (1958). *Philosophical investigations*. Oxford: Basil Blackwell.

19 'You can't have one without the other'

Bilateral relations between Paraguay's Ciudad del Este and Brazil's Foz do Iguaçu

Omri Elmaleh

Introduction

The Triple Frontier between Brazil, Argentina and Paraguay is best known for its impressive waterfall cluster, which partly underpins political borders between the three states. Alongside this natural wonder is a multicultural human wonder. Populated by some 800,000 people, originating in dozens of countries and adhering to over 20 religions, the Triple Frontier consists of three border cities, river-divided and bridge-linked.[1]

Paraguay's transition from Argentina's long-lasting sphere of influence to Brazil's economic patronage during the 20th century's second half transformed the border region dramatically. This chapter depicts how foreign relations, through a series of agreements between Brazil and Paraguay, turned Ciudad del Este and Foz do Iguaçu into an economic emporium based on interdependent and inseparable twin cities. Meanwhile, their Argentinian 'stepsister', Puerto Iguazu, missed out on evolving regional dynamics, remaining a small town dependent upon waterfall tourism left out of the demographic growth and economic boom of the Triple Frontier.

When 'East' meets 'West'

Some international twin cities have gained attention due to efforts by national and regional authorities, seeking new centres for economic development (Garrard and Mikhailova 2019, 1). One of these is the conurbation of Foz do Iguacu and Ciudad del Este which, in the late 20th century, became a leading Latin American trading venue. It is almost impossible to understand the relationship between these neighbouring cities without placing them in the broader context of changing Brazilian-Paraguayan relations in the 20th century's second half. Geographically, Paraguay is an inland country without access to the sea. This caused its unilateral dependence on Argentina, which controlled Paraguay's passage to the Atlantic Ocean (Doratioto 2012, 522–523). This situation altered in August 1941 with Brazilian President Getúlio Vargas's historic visit to Paraguay, causing 'Paraguay's geopolitical compass to move slowly from the south [Argentina] to the east [Brazil]' (Ynsfrán 2012, 64).

Vargas's policy, in pursuit of his national vision, 'March to the West',[2] which he believed did not stop at the Brazil-Paraguay border but continued westward,

DOI: 10.4324/9781003102526-24

underpinned more tangible measures occurring during the next decade. Over ten years later, in May 1954, pro-Brazilian general Alfredo Stroessner carried out a military coup in Paraguay which marked a historic turn. Stroessner's foreign policy was characterized by various strategies for overcoming Paraguay's geographic barriers, including an appeal to Brazil, which had recently expressed great interest in becoming an alternative to Argentina (*New York Times*, hereafter *NYT*, 1956, 12). The region most impacted by the three states' inter-relationships was their actual meeting point – the Triple Frontier.

A road, a bridge and a city: building a future

Joint economic and geopolitical interests, alongside the deep friendship developing between Stroessner and Juscelino Kubitschek, the newly elected Brazilian president, underpinned a series of agreements and understandings largely concerned with three main themes: infrastructure, trade and movement of people.[3] The first stage in creating Paraguay's new continental corridor to the Atlantic Ocean was to fill the missing gaps, creating a continuous highway between two capitals, Asunción and Curitiba (Paraná State capital), and Paranaguá Port to the west. The paving of Route 7, running through an inaccessible eastern area of Paraguay, all the way to the Paraná River banks opposite Foz do Iguaçu, ended in 1959 (Silva 2018, 322–329). Meanwhile on the Brazilian side, work began on constructing Road BR-277 cutting east to west through the state of Paraná, from Paranaguá to Foz do Iguaçu (*O Estado de São Paulo* 1969).

Despite good intentions, there were many obstacles, including a natural one: the Paraná River. To overcome this challenge, an idea emerged to build what would be at the time the world's largest arch bridge. On 6 October 1956, a ceremony was held in Foz do Iguaçu to inaugurate work on the future 'Friendship Bridge', with General Stroessner and President Kubitschek participating (*NYT* 1956, 12). Close to where the heads of state met, the Friendship Bridge would join two national roads leading to the Paraná's banks from east and west (Figure 19.1).

In the second half of the 1950s, the small town of Foz do Iguaçu (population c15,000) became the main regional hub and main hope for 'March to the West' across the Paraná River. Meanwhile, in Paraguay, 'March to the East', the Paraguayan response to the Brazilian vision (*La Patria* 1957), began emerging with plans to establish a new city beside the future arch bridge, giving Paraguay access to the Atlantic. At the cornerstone-laying ceremony in February 1956, the future city's name was revealed: 'to be called from now and forever Puerto Presidante Stroessner (later to be renamed Ciudad del Este), in recognition of the leader by the people' (*El País* 1957). Under a year later, vision became reality with a decree for the city's establishment, whose purpose was defined as a settlement point where all the region's opportunities and resources would be concentrated for those coming to live there.[4]

As the joint regional development plan got underway, Brazilians and Paraguayans constructed roads and bridges, and even founded a city; but one component was still missing – people. This induced the need to transform the border

Figure 19.1 The Friendship Bridge, looking westward to Ciudad del Este.
Source: Photo by Luis Ignacio Carrasco Aranda, by permission.

region from remote periphery into an economic hub for diverse populations of migrants. A series of economic agreements regulating the two countries' trade granted Paraguay free access to, and customs-free warehouses (*depositos francos*) in, the port city of Paranaguá, on the road connecting Asunción to the Atlantic. Paraguay's imports and exports in the warehouses, would be exempt from Brazilian taxation. Brazil, meanwhile, saw Paraguay as a significant destination for Brazilian exports (O *Estado de São Paulo* 1956a, 1956b). It was decided to facilitate movement of citizens and tourists at both countries' border crossings.[5]

'When Puerto met Foz'

The small Brazilian border town of Foz was founded in 1889 as a military colony (*Colônia Militar da Foz do Iguaçu* in Portuguese) close to the Iguazu Falls, to establish Brazil's sovereign presence in the region (de Brito 2005 [1938], 11 and chapter 3). Throughout the twentieth century's first half, both Foz do Iguaçu and the Argentine Puerto Iguazu were the departure points for the famous Iguazu Falls. From the late 1950s, the former became the departure-point for another attraction – neighbouring Puerto Presidente Stroessner. Thanks to the economic opportunities created, many traders settled in Foz do Iguaçu intending to be at the forefront of Brazilian industrial exports to Paraguayan markets. Evidently, in 1958 Paraguay was Brazil's largest export destination in the textile industry (Birch 1992, 225). Their gamble paid off, following the Friendship Bridge's inauguration on 27 March 1965, which, according to Paraguayan and Brazilian media, would mark one of the most important events ever recorded in the two nations' histories (*La Tribuna* 1965, 3; *O Globo* 1965). Consequently, the bridge, the new road system, and developing regional trade transformed Foz do Iguaçu and Puerto Presidente Stroessner into a 'revolving door' for people and goods.

To understand the commercial dynamics formed between the two cities, the two countries' broader economic context in the mid-1960s needs considering. Thus, while Paraguay's economy was based on an outward-oriented strategy, Brazil continued its conservative model of import-substitution industrialization (Baer and Birch 1987, 601–602; Weisskoff 1980, 647–675). These economic perceptions were both competing and complementary: although designed in the two countries' distant capitals, they intersected on the Friendship Bridge, heralding the beginnings of a commercial emporium to challenge the world's largest trading centres.

Exports to Paraguay increased significantly following Brazil's 1964 military takeover. Government-initiated incentives and benefits for the export industry (Baer 1983, 157) slowly transformed Foz do Iguaçu into one of Brazil's major export localities. The Friendship Bridge and its eastbound road were hundreds of meters north of Foz historic city centre and, over time, the neighbourhoods of Jardim Jupira and Vila Portes alongside the river mouth became a wholesale distribution centre for tax-free Brazilian commodities, mainly food and textiles (*A Gazeta do Iguaçu* 2006).

The late 1960s saw the inauguration of Interstate BR-27; an important event in the neighbouring border cities' history, since it would facilitate movements of merchandise and people to the Triple Frontier, boosting regional trade (*A Gazeta do Iguaçu* 1994). Meanwhile, the small city of Puerto Presidente Stroessner, defined as a free-trade zone, developed into a cheap and accessible commercial hub for imported, tax-free international brands of consumer goods. This 'shopping tourism' shaped the new city's character as a showcase for products unpurchaseable or too expensive in neighbouring countries.

Brands and dams

In the 1970s, Brazil exported not only domestically produced industrial goods to Paraguay but also citizens. As part of Paraguay's agricultural development program, large waves of Brazilian immigrants, later becoming known as *Brasiguiaos*, settled Paraguay's uninhabited eastern regions. The urban and commercial centre for most Brasiguiaos was the Foz do Iguaçu–Puerto Presidente Stroessner axis.

Above all, however, the 1970s are memorable for another colossal project. In April 1973, Brazil and Paraguay signed the Itaipú Agreement to build a hydroelectric dam on the Paraná River (at the time, the world's largest) several kilometres north of the Triple Frontier. Many saw it as an important historical milestone for the Rio de la Plata upper basin and, possibly, the most significant project in Latin American history (*O Globo* 1973; *NYT* 1973, 20). The consequences of April 1973 were almost immediate and touched every aspect of life.

Demographically, the 10-year construction project of Itaipú created human turbulence around it with massive migration. According to some estimates, within about a decade, Foz do Iguaçu's population grew more quickly than any other Brazilian city, from 30,000 to some 200,000 (*Veja* 1987, 65), while Puerto Presidante Stroessner had a population of 49,423 in 1982 compared to only 7,069 in 1972 (DGEEC 2002, 146). Thus, Itaipú generated a huge market for workers, migrants and tourists, rapidly expanding regional commercial activity (Baer and Birch 1987, 607). This was expressed most saliently in the two main trading centres on either side of the Friendship Bridge.

On the Paraguayan side, burgeoning 'shopping tourism' was transformed from boutique to mass tourism. The city's commercial hub near the bridge's termination point became a tourist paradise (mainly Brazilian), offering Chinese toys, expensive English and Italian clothing brands, American jeans and whiskey, Japanese and German electronic devices, Swiss and Dutch chocolates and much more. Based on the 1975 annual average, around 50,000 Brazilians crossed the bridge monthly, leaving some $5,000,000 per month in Paraguay (Da Mota Menezes 1990, 27, 31).

Other 1970s measures also boosted regional trade. These included Paraguay's exclusion from the monetary deposit restriction applying to Brazilians leaving the country for tourist purposes, and extending Brazilian permits to stay in Paraguay (*O Globo* 1975). These produced enormous flows of buyers and migrants to the Triple Frontier, changing the scope and form of trading in the Paraguayan city due to the establishment of luxurious new shopping centres.

On the bridge's other side, as wholesale exports to Paraguay grew during the 1970s, the commercial infrastructure in Foz do Iguaçu developed further. The Jardim Jupira and Vila Portes neighbourhoods expanded greatly due to Paraguay buying from Brazil 'everything from toilet paper to a plane' (Da Mota Menezes 1990, 19–20, and 43, note 42). Meanwhile, Paraguayan shoppers, mostly the poor or those with limited means, traversed the bridge to the Foz do Iguaçu trading houses to buy basic goods and clothing (Da Mota Menezes 1990, 32).

The golden age of the Triple Frontier

Thanks to its ongoing economic relations with Foz do Iguaçu, Puerto Presidente Stroessner became Paraguay's second largest city during the 1970s and, in the early 1980s, its commercial sector earned more national income from trade than agriculture or industry (Baer and Birch 1987, 604). During this decade, the volume of shopping tourism increased dramatically: housewives, senior citizens, students, street artists and thousands more came from across Brazil to shop in Puerto Presidente Stroessner (*Nosso Tempo* 1986).

Shopping tourism drove both border cities' economies, via people purchasing goods on the Paraguayan side and enjoying the tourist and civilian infrastructure on the Brazilian one. As shopping in Paraguay grew, so too did the civic, urban and tourist infrastructure of Foz do Iguaçu. In 1986, 1.2 million tourists passed through the city (*Veja* 1987, 66). In 1987, Foz do Iguaçu was Brazil's third largest city in terms of numbers of beds per capita.

The end of the Itaipú project left a large unemployed population which, alongside increasing Brazilian middle-class demand for consumer products, created even more significant purchasing power. The unemployed became *Sacoleiros* (bag-carriers), Brazilian peddlers and petty traders, who purchased merchandise in Puerto Presidente Stroessner shopping centres, carried it in big sacks across the Friendship Bridge, evading the tax authorities, and sold it covertly throughout Brazil. The *Sacoleiros* fall under the definition of Shuttle trade which refers to imports and exports by individuals who travel to neighbouring countries to purchase goods for resale in street markets or small shops (OECD 2002, 89). In fact, if Paraguay's mid-1980s shopping experience was mostly of Brazilian middle-class consumers, the *sacoleiros* dramatically changed its character into industrial-scale spending (Rabossi 2004, 52–55). The sacoleiros trend intensified over the years and, as of 1989, tens of thousands of Brazilians flooded the city's thousands of stores, resulting in Puerto Presidente Stroessner coming second after Asunción in contributing to national income in taxes (Ynsfrán 2012, 166–167).

Meanwhile, in Foz do Iguaçu, wholesale exports to Paraguay grew continuously, becoming so significant that in 1982 Brazil replaced Argentina as Paraguay's largest economic partner (Baer and Birch 1987, 606, 613; *Revista Painel* 1980). Furthermore, many Paraguayans continued crossing the bridge daily to shop in Foz do Iguaçu, where they always found Spanish speakers or people who would accept Paraguayan currency (*Revista Painel* 1981).

However, rapid commercial prosperity and tourist development had a price. Municipal authorities on both sides of the border could not keep pace with dizzying demographic growth and the accompanying challenges, like unemployment, inadequate health and education services and crime (*Nosso Tempo* 1983, 1987).

The beginning of the end, or a new era?

In April 1989, South America's longest dictatorship ended with General Stroessner's overthrow. In the early 1990s, the city founded by him and bearing his name was renamed Ciudad del Este (the city of the east), after its location at Paraguay's eastern end.[6] Regime and name change had little impact on regional development. The two cities' urban populations, numbering just over 200,000 in the early 1980s, now exceeded 320,000.[7]

Economically, shopping continued being central in Ciudad del Este (*Geographical Magazine* 1996), peaking in terms of sales volume and number of buyers in 1994–1995. Over 60,000 Brazilian buyers visited the city every week (Rabossi 2010, 241). A year later, it was noted that Ciudad del Este attracted 12,000,000 visitors per year, more than double Paraguay's population, and some 20,000 Brazilians made their living there (*Veja* 1995). The city's revenues in 1995 were estimated in billions of dollars, making it, seemingly, the world's third largest trading city after Miami and Hong Kong (*La Nación* 1997; *Revista ACIFI* 2015).

However, the year Ciudad del Este was declared one of the world's largest commercial centres also marked the culmination of a series of globalizing, commercializing, neo-liberalising, and international collaboration efforts seeking to turn the region's binational economic model into a multinational integration one. The most important step was the establishment of the Mercosur (Southern Common Market – Brazil, Argentina, Uruguay and Paraguay) regional trade agreement, aimed at unifying the area by removing trade barriers and implementing a joint passport (Folch 2018, 268). This action undermined Ciudad del Este's very existence. From the second half of 1995, a few months after the Common External Tariff (customs tariff) came into effect, the city's economy shrank and commercial traffic declined. Many businesses closed and shoppers' numbers fell significantly (Penner 1998, 45).

If trading on one side of the border was harmed, the other side would probably also suffer. As of 1995, with the Mercosur's introduction alongside other structural changes in Brazil's export model, a direct link was created between Brazilian production centres and the Paraguayan market. Thereby, the Brazilian *casas exportadas* (export houses) in the neighbourhoods by the Friendship Bridge became irrelevant, significantly diminishing their historical trading role with Paraguay (Rabossi 2010, 13). Hence, the border cities' glamorous past was replaced by an uncertain future.

'Out of the Triple Frontier evil shall emerge ...?'[8]

The 1990s witnessed not only important transformations in the commercial arena but, most importantly, a steep decline in the Triple Frontier's reputation. Over the past three decades, the media both in South America and around the world have portrayed the region as an international space defying cultural, social, political, and economic logic – a modern version of the American Wild West, concentrating within it all the sins of modern times, including smuggling, money laundering, human and drug trafficking, forgery, international crime, and Islamic terrorism and its financing (Ferradas 2010, 35).

Its epicentre was Ciudad del Este which, according to one description, was nothing more than a shelter for the entire world, 'from Paraguayan pickpockets to Arab terrorists' (*Veja* 1995, 77). However – and as the chapter's title infers – everything that happens on one side of the border affects the other. Therefore, while Ciudad del Este was portrayed as the 'anus of the Earth' (Robinson 2000, 13) and a safe haven for international immunity and delinquency (*Clarín* 1997), Foz do Iguaçu also shared this dubious honour.

In criminal terms, smuggling came first. An illegal activity that moved from the Triple Frontier's urban and rural fringes to the beating heart of the bustling shopping centres became a foundation of the regional economy. Although goods smuggling is criminal in both countries, many Brazilians and Paraguayans saw it as a legitimate source of livelihood (like the activity of the *sacoleiros*). However, with the transformation of the official regional trading model, from the mid-1990s smugglers increasingly turned to more extreme criminality, like money laundering, document forgery and arms and drug trafficking, with Ciudad de Este increasingly becoming 'the smuggling capital of South America' (*La Nación* 1997). A US official described it well: "Paraguay has always been known to be a smuggling center ... two new problems have emerged in recent years: drug trafficking and the counterfeiting of visas and passports, which serve as an open door for terrorism' (*El País* 2001). And terrorism, in many people's opinion, did enter.

The common assumption that the Triple Frontier is infested with Muslim terrorist cells is often associated, with little proof, with the region's significant Muslim-immigrant population. This began when they were individually and collectively linked to terrorist attacks against the Buenos Aires Israeli embassy (March 1992) and the Jewish Community Center (July 1994). The allegations spread with the 9/11 terrorist attacks, and the Triple Frontier became one of many fronts in the US 'War on Terror' (*NYT* 2002). Some even claimed Usama Bin Laden visited Foz do Iguaçu, even perhaps the famous Iguazu Falls (*Veja* 2003; *Washington Post* 2003).

Now and then

For Buursink, border cities pretty much depend on the border for their existence (2001, 1–2). The present cases demonstrate how they fit this generalization, since throughout their shared history there was mutual interdependence, while each developed its own urban character. In general, while in Ciudad del Este, the emphasis was on creating commercial infrastructure to feed tourism, Foz do Iguaçu developed a tertiary sector, providing services, citizens and tourists. The tourist-oriented *Geographical Magazine* described their situation well in 1996:

> Although one street runs from Brazil across the open border to Paraguay, the two sides of the Bridge are in different worlds. Foz could be a European town, orderly with pavements and tree-lined residential streets. Ciudad del Este is stereotype South American – loud and disorganized, scruffy but cheerful – and poor.
>
> (13)

This portrayal was reasonably accurate. Over the years, Foz do Iguaçu created a relatively developed urban and tourist infrastructure, a good education system and an active community life. Meanwhile, Ciudad del Este's public space, or at least the city's commercial centre, was neglected, with minimal infrastructure, for example unpaved or dirt roads, no law enforcement, poor municipal services and deficient education.

For many inhabitants of the region, it seemed that Ciudad del Este's existence necessarily became evident only during the daytime hours of lively commercial activity. Upon nightfall, its streets and alleys emptied of people, and only the tons of garbage outside the shops testified to the shopping sprees of countless tourists. Merchants and service providers preferred returning to their homes in Foz do Iguaçu. Since the urban character of each city met different needs, a transnational daily life was born: 'living in Foz, working in Ciudad'. Thus, community boundaries and self- and collective identity could be constructed across two sides of an international border.

According to historian Alfredo da Mota Menezes, Brazil and Paraguay's burgeoning relations are sometimes presented as a long-term marriage (Da

Mota Menezes 1990, 225). Continuing this metaphor, one might suggest that, if the capital cities Brasilia and Asunción are in-laws, Foz do Iguaçu and Ciudad del Este are the married couple. Although border-city ties met each country's economic and historic needs and symbolized the fulfilment of a national ideal, the cities also grew increasingly interdependent, much more so than on their respective capitals. The joint road and energy infrastructures were the foundations for their bilateral economic relations around border trade and shopping tourism.

Although the initiative to transform the two cities into a meeting point for national and economic interests was a product of South American cooperation, those executing the plan were mainly immigrant groups, mostly of Middle Eastern and East Asian origins. While a minority of the general population, these migrants recognized the economic potential of changing geopolitics and, by 'representing' each country's economic interests, they became integral to making the Triple Frontier. In a way, immigrant groups have negotiated between two sets of transnational identities: both linking their country of origin and country of settlement (Glick Shiller et al. 1992), and, experiencing how the concept of immigrant citizenship and their national affinities to Brazil and/or Paraguay have been implemented within Ciudad del Este and Foz do Iguaçu.

This chapter has demonstrated the deep interconnectedness and mutual dependence of two Latin American border-crossing cities, which were operating quite autonomously of their national centres throughout the second half of the 20th century. However, it is also evident that the fate of these two interconnected cities can also be deeply and damagingly impacted by globalization, more particularly by strategic decisions and actions taken by powerful actors with minds focused on promoting international economic cooperation and coordination as in the case of Mercosur. Thus, economic opportunities narrowed significantly in the late 20th century, producing the obvious question at the turn of the millennium: what would the future hold for two cities basing themselves on a slowly declining trading relationship? Undoubtedly, after almost three decades of commercial unrest, the Triple Frontier's golden age has certainly ended. However, we cannot yet forecast the common future of Ciudad del Este and Foz do Iguaçu. Regional trade remained relevant in the early decades of the current century and, while trade flourishes, people will continue making their way back and forth across the Friendship Bridge (*100fronteiras* 2019).

This assumption was confirmed during the writing of this chapter, as it became clear that the devastating consequences of the COVID-19 pandemic were also affecting the Triple Frontier. For the first time since the Friendship Bridge opened, the two cities' lifeline was closed to traffic for an unlimited period. This dramatic action, though necessary, struck a very large part of the region's inhabitants and illustrates how bilateral trade and movement of people and goods on the bridge are the air these twin cities breathe (*El Comercio* 2020). Nevertheless, and despite known challenges and others still to come, it seems that as long as the unique supply on one side of the border continues to complement the other's demand, the historic cities' alliance will overcome any crisis. Thus, the refrain 'you can't have one without the other' is truly fitting.

Notes

1 South America has 13 triple frontiers, 9 involving Brazil, 4 Argentina, and 3 with Paraguay. None enjoy such historical and topical relevance as 'the Triple Frontier', the meeting point between Argentina, Brazil and Paraguay. The term began being popular only in the late 1980s, usually referring to the zone of Brazilian and Paraguayan border cities.

2 A national policy proclaimed during the Estado Novo (1937–1945) to develop and integrate the sparsely populated mid-west and northern regions of Brazil. See De Oliveira Klever (2016).

3 For all agreements between Brazil and Paraguay1872–2018, see Presidência da República, Mensagem (SF) No.27, 17–28.

4 Gaceta oficial del Ministerio del Interior, Asunción, Decree No.24.634, 27.01.1957, 9.

5 Legislative Decree No. 22, *Diário do Congresso*, 16.12.1959, 3165.

6 Gaceta Oficial del Ministerio del Interior, Asunción, Law No. 06/89, *Que denomina "Ciudad del Este" a la ciudad Capital del Departamento de Alto Paraná, que hasta el momento lleva el nombre de "Ciudad Presidente Stroessner,"* 11.08.1989.

7 Dirección General de Estadística, *Atlas Censal del Paraguay*, 146; IBGE, *Censo Demográfico de Brasil 1980*, vol. 1, 4, no. 20 (1982), 72; *Censo Demográfico de Brasil 1991*, No.22, Paraná (1991), 36.

8 Paraphrasing Jeremiah 1/14: "Out of the north the evil shall break forth ..." (New American Standard Bible).

References

Archives

Diário do Congresso Nacional (Brasilia)
Dirección General de Estadística (Asunción), Encuestas y Censos, (Asunción)
Ministry of Interior Archive (Asunción)

Secondary Sources

A *Gazeta do Iguaçu* (1994). A pavimentacao de BR-277 e a construção de ponte da Amizade deslancharam. 10 June, 10.

A *Gazeta do Iguaçu* (2006). Ponte e BR-277 impulsionaram o desenvolmiento. 10–11 June, 14.

Baer, W. (1983). *The Brazilian Economy: Growth and Development*. New York: Praeger Publishers.

Baer, W. and Birch, M. (1987). The International Economic Relations of a Small Country: The Case Study of Paraguay. *Economic Development and Cultural Change*, 35(3), 601–602.

Birch, M. (1992), Pendulum Politics: Paraguay's National Borders, 1940–1975, in Lawrence A. Herzog, ed., *Changing Boundaries in the Americas: New Perspectives on the U.S. Mexican Central American and South American Borders* (San Diego, Center for US-Mexican Studies), 203–228.

Buursink, J. (2001). The Binational Reality of Border-Crossing Cities. *GeoJournal*, 54(1), 7–19.

Clarín (1997). Santuario de impunidad. 1 December. https://www.clarin.com/ediciones-anteriores/santuario-impunidad_0_BJebpkb0tg.html

Da Mota Menezes, A. (1990). *La herencia de Stroessner: Brasil-Paraguay, 1955–1980*. Asunción: Carlos Schauman.

De Brito, J.M. (2005[1938]). *Descoberta de Foz do Iguaçu e a fundação da colônia military.* Curitiba: Travessa dos Editores.

De Oliveira Klever, L. (2016). 'Passado e presente: projeto político e escrita da história na march para u oeste', *Anais do III Encontro de Pesquisas Históricas – PPGH/PUCRS*: 217–228.Dirección General de Estadística, Encuestas y Censos (DGEEC) (November 2004). Atlas Censal del Paraguay, Fernando de la Mora, 146.

Doratioto, F. (2012). *Relações Brasil-Paraguai: afastamento, tensões e reaproximação, 1889–1954.* Brasília: FUNAG.

El Comercio (2020). Paraguay cierra su principal puente de entrada y salida con Brasil por el covid-19. 17 March. https://www.elcomercio.com/actualidad/mundo/paraguay-cierra-puente-brasil-coronavirus.html

El País (1957). Se cumplió ayer la fundación del Puerto Presidente Stroessner. 4 February, 1.

El País (2001). "Comandos" terroristas se refugian en la triple frontera. 9 November. https://elpais.com/diario/2001/11/09/internacional/1005260421_850215.html

Ferradas, C. (2010) Security and Ethnography on the Triple Frontier of the Southern Cone. In: H. Donnan and T.M. Wilson, eds. *Borderlands: Ethnographic Approaches to Security, Power, and Identity.* Lanham: University Press of America, 35–52.

Folch, C. (2018). Ciudad del Este and the Common Market: A Tale of Two Economic Integrations. In: J. Blanc and F. Freitas, eds. *Big Water: The Making of the Borderlands Between Brazil, Argentina, and Paraguay.* Tucson: University of Arizona Press, 267–284.

Garrard, J. and Mikhailova, E. (eds.) (2019). *Twin Cities Urban Communities: Borders and Relationships over Time,* . New York, London: Routledge.

Geographical Magazine (1996). Bridging the Gap. Volume 68(4), 13.

Glick Shiller, N., Basch, L. and Szanton-Blanc, C. (1992). Transnationalism: A New Analytic Framework for Understanding Migration. *Annals of the New York Academy of Sciences*, 645(1), 1–24.

La Nación (1997). El triángulo de los fantasmas. 19 October. https://www.lanacion.com.ar/opinion/el-triangulo-de-los-fantasmas-nid209372/

La Patria (1957). La Marcha hacia el Este. 10 February, 10.

La Tribuna (1965). Testimonio elocuente de hermandad entre dos pueblos constituyó el acto vivido en el puente de la Amistad. 28 March, 3.

New York Times (NYT) (1956). Brazil, Paraguay Start Key Bridge. 7 October, 12.

New York Times (NYT) (1973). Modern Problems Are Visiting Placid Paraguay. 5 August, 20.

New York Times (NYT). (2002). US Expanding Effort Block Terrorist Funds Latin America. 21 December, 12.

Nosso Tempo (1983). Tumulto na Ponte: "Paseros" e policiais trocam palavrões. 30 June–6 July, 6.

Nosso Tempo (1986). Compristas de todo o Brasil invadem Foz do Iguaçu. 20–26 June, 11.

Nosso Tempo (1987). A explosão urbana de Foz do Iguaçu. 27 July, 6.

OECD et al. (2002). *Measuring the Non-Observed Economy: A Handbook*, Paris: OECD Publications Service.

O Estado de São Paulo (1956a). Acordo fronteiriço Brasileiro-Paraguaio. 27 October, 40.

O Estado de São Paulo (1956b). Brasil e Paraguai deverão firmar brevemente um acordo de comércio. 11 November, 20.

O Estado de São Paulo (1969). Brasil e Paraguai de acordo. 28 March, 7.

O Globo (1965). Paraguai e Brasil mais unidos pela Ponte da Amizade. 29 March, 1, 3.

O Globo (1973). Medici e Stroessner presidem a assinatura do Itaipu. 26 April, 1, 8–11.

O Globo (1975). Brasil empresta ao Paraguai. 28 June, 11.

100 fronteiras (2019). Compras no Paraguai: dicas infalíveis para visitar e fazer compras CDE. 22 October. https://100fronteiras.com/brasil/noticia/compras-no-paraguai/

Penner, R. (1998). *Movimiento comercial y financiero de Ciudad del Este: Perspectivas dentro del proceso de integración.* Asunción: Departamento de Economía Internacional/ Gerencia de Estudios Económicos/Banco Central del Paraguay.

Rabossi, F. (2004). *Nas Ruas de Ciudad Del Este: vidas e vendas Núm. mercado de frontera.* Doctoral thesis. Universidade Federal do Rio de Janeiro.

Rabossi, F. (2010). Made in Paraguay: Notas sobre la producción de Ciudad del Este. *Revista Electrónica del IDAES,* 6, 1–22.

Revista ACIFI (2015). As transformações. No. 5, 13 March.

Revista Painel (1980). O difícil tráfego na Ponte da Amizade. April, 6–7.

Revista Painel (1981). Jardim Jupira recebe pavimentação através do plano comunitário. July.

Robinson, J. (2000). *The Merger: The Conglomeration of International Organized Crime.* Woodstock and New York: The Overlook Press.

Silva, F.M. (2018). *Alto Paraná desde una perspectiva histórica.* Ciudad del Este: del autor.

Veja (1987). A hidrelétrica de Itaipu vira um foco turístico e estimula a economia das cidades ao redor do lago. 22 July.

Veja (1995). A fronteira da muamba. 26 July, 74.

Veja (2003). Ele esteve no Brasil. 19 March, 58–61.

Washington Post (2003). Bin Laden Reportedly Spent Time in Brazil in '95. 18 March.

Weisskoff, R. (1980). The Growth and Decline of Import Substitution in Brazil – Revisited. *World Development,* 8(9), 647–675.

Ynsfrán, E.L. (2012). *Un giro geopolítico: El milagro de una ciudad.* Asunción: Fundacion Ymaguare.

Part II

International Twin Cities

E. In North America and the Caribbean

20 Niagara twin cities

'Living Apart Together' on the Canada-US border

Nick Baxter-Moore and Munroe Eagles

As their common names suggest, Niagara Falls, Ontario, Canada, and Niagara Falls, New York, USA, resemble a single conurbation divided by, among other things, a river and a (spectacular) set of waterfalls. The latter makes Niagara Falls one of the world's most popular tourist destinations. However, the Niagara River, separating above the falls and rejoining below, is also an international boundary between the USA and Canada. This border provides an unusually clear and simple setting wherein to assess the impact of an international boundary on the political life of what otherwise might be considered one metropolitan area.

Various geographical, historical, cultural, social, political and economic factors combine to make the Niagara twin cities approximate to what methodologists might consider a 'crucial case' (Eckstein 1975) for cross-border cooperation. Yet, in overviewing binational cities around the world, Buursink argued that, although the two cities 'could be called the perfect border twin cities because of their adjacent location, their size, their age, their common (English) language and even their (rather deteriorated) appearance' (Buursink 2001, 10), in fact they more resemble a couple 'living apart together', coexisting like separated marital partners on different floors of the erstwhile family home. Does this latter description still hold? If so, what might explain the failure of cooperation to emerge in what appear to be highly conducive circumstances?

Garrard and Mikhailova (2019, 9–11) suggest that twin cities frequently exhibit tensions between 'inwardness' (or 'separateness') and 'openness', between competition and cooperation. Our analysis of the two cities of Niagara Falls provides further evidence of how these seemingly contradictory impulses can co-exist. We explore the balance between forces that could conceivably integrate the entire Niagara Falls conurbation ('openness') and the disintegrative forces pushing the two communities apart ('inwardness'). To anticipate our argument in general terms, we find that, even in this most promising setting for cooperative, coordinated binational action involving twin cities, that potential remains largely unrealised.

Our analysis is based partly on interviews conducted with senior politicians and officials in each city. In 2016, we interviewed the two mayors: James (Jim) Diodati, City Councillor (2003–2010) and Mayor (2010–present as this

DOI: 10.4324/9781003102526-26

is written) of Niagara Falls, Ontario; and Paul Dyster, City Councillor (2000–2003) and Mayor (2008–2020) of Niagara Falls, New York. We also interviewed senior planning officials on each side in 2018: John Barnsley, Manager of Policy Planning, City of Niagara Falls, Ontario, who subsequently retired in 2019 after 30 years' service in the city planning department; and Tom DeSantis, Acting Director, Planning and Economic Development, City of Niagara Falls, New York, who had likewise worked for 30 years in city planning.

Binational Niagara as an international borderland

The two Niagara cities lie at the heart of a larger cross-border region. Binational Niagara encompasses the Niagara Region of southern Ontario and the 'Niagara Frontier' (chiefly Erie and Niagara Counties) of western New York State. It includes larger cities like Buffalo, New York and St Catharines, Ontario, alongside smaller tourist centres like Niagara-on-the-Lake, Ontario, home of the Shaw Festival Theatre, and Youngstown, New York, site of historic Fort Niagara (see Figure 20.1). Between the Niagara Peninsula and western New York runs the Niagara River, a 58 km waterway connecting the two lowest Great Lakes – Erie to the south (upstream) and Ontario to the north (downstream). Just over halfway along its length, the river plunges over the Niagara Escarpment to create one of the world's most famous waterfalls, around which grew the Niagara Falls twin cities. Once viewed as a major transportation and communication obstacle,

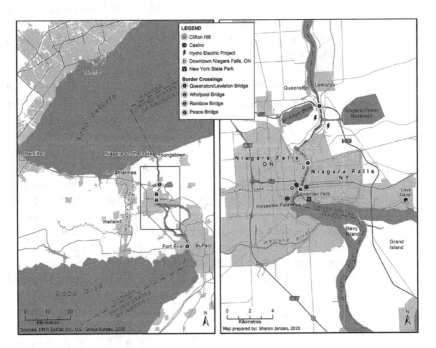

Figure 20.1 The twin cities of Niagara Falls.

Source: Map by Sharon Janzen, by permission.

the Falls became a major 19th-century tourist attraction (see Dubinsky 1999). Contemporaneously, with developing rail travel, the gorge below the Falls became the most direct bridging point for railroads linking southern Ontario's Golden Horseshoe and the growing industrial and port complexes of America's Eastern Seaboard.

The Niagara River had become part of the border between British North America (later Canada) and the newly independent USA under the 1783 Treaty of Paris. That border was seriously contested in the War of 1812–1814 but subsequently confirmed by the Ghent Treaty. Since then, the river has been part of what is often labelled 'the world's longest undefended border,' although that characterisation has recently been challenged, first by responses to the 9/11 terrorist attacks and, more recently, by COVID-19.

After the American Revolution, geographic proximity and continuing cultural similarity along the 'Niagara Frontier' produced extensive cross-border interaction and close cultural ties (see Taylor 2010), as Americans and Canadians crossed the border for family, economic and social reasons. Hence, H.F. Angus, surveying Canadian attitudes toward the USA in the late 1930s, posited that southern Ontario, to which the Niagara peninsula belonged,

> thrust deeply into a highly industrialized and closely populated region of the United States, separated only by rivers or the readily navigated lakes of the St. Lawrence system ... Here is part of Canada obviously accessible by nature to numerus American contacts and influences.
>
> (Angus 1938, 42)

Angus argued further that British immigration to Ontario did not diminish America's influence; instead, Ontario remained 'broadly a frontier of the northern United States; a territory separated by a political boundary, but so contiguous as to be subjected to the varied influences of the great metropolitan areas which grew up across the boundary' (Angus 1938, 43).

The region's cross-border ties remain strong today. The four road bridges crossing the Niagara River are the busiest for personal travel and the second busiest for commercial traffic of all Canada-US border crossings. Only two of these bridges are located within the Niagara Falls conurbation: the Rainbow Bridge, closest to the Falls, which is limited to passenger traffic, and the Whirlpool Bridge, which is further restricted to those enrolled in 'trusted-traveller' programs. Commercial traffic (trucks) most cross at Queenston-Lewiston Bridge to the north, or the Peace Bridge between Fort Erie and Buffalo in the south. Whichever bridge they use, Niagara Region residents cross the border for education, air travel (one-third of passengers leaving Buffalo-Niagara International Airport are Ontario residents), sports (on average, 15,000 Canadians attend the Buffalo Bills NFL home games), medical tourism, recreation and leisure (many western New Yorkers own cottage properties in Canada), and, depending on the US-Canada dollar-exchange rate, cross-border shopping (UB Regional Institute 2007, 2–3). Moreover, for much of the 20th century, residents on one side of the border quite often enlisted in the neighbouring country's military (Siener 2013, 381).

Integrative forces and the twin Niagaras

The twin cities of Niagara Falls appear unusual in several respects. First, they share a name – not unprecedented, but hardly common. Second, they share somewhat similar historical development patterns on either side of a pre-existing border. The City of Niagara Falls, Ontario, was incorporated in 1903, although importantly its predecessor, the Town of Clifton, was first recognised in 1856, a year after the first railway bridge was completed across the Niagara Gorge linking Canada and the United States. On the American side, the City of Niagara Falls, New York, was incorporated in 1892, created from existing regional townships, including Bellevue Village, also known as 'Suspension Bridge,' founded in 1854 and popularly named after the same bridge producing the Canadian settlement (Niagara Falls Info n.d.). Their relative proximity in foundation dates and origins might suggest that these communities are 'twins' in an almost literal sense. Third, the two cities jointly possess a tremendous economic asset – the Falls. In an era of global competition for tourist dollars, this shared custody, alongside shared cultural and historical ties, should encourage cooperative and coordinated efforts to market the binational region.

Nevertheless, the Niagara River is an international border and, as Buursink notes (Buursink 2001, 8), international twin cities, by their nature, are usually on the periphery, both in spatial and political terms. Hence, both halves of the binational Falls are subject to the vagaries of decisions taken many hundreds of miles away in their respective national capitals that make policy on issues like national security, exchange rates, customs and immigration. They are distant even from provincial/state capitals (Toronto, Ontario; Albany, New York) that determine transportation policy, gambling laws, tourism licensing and alcohol regulation (Hartt and Warkentin 2017). Feelings of peripherality and insufficient control over critical aspects of their destiny as border communities are potential sources of a shared sense of grievance believed to foster cross-border cooperation (Konrad and Nicol 2008, 32).

Unsurprisingly, then, Niagara twin-city relations are generally congenial and outwardly harmonious. When asked in interviews how they would characterise the general state of cross-border relationships, both Mayor Jim Diodati of Niagara Falls, Ontario, and Mayor Paul Dyster of Niagara Falls, New York, emphasised cordiality and mutual interest. Both grew up in their respective cities at a time when border crossings were routine and unproblematic, and each knew intimately the city 'across the bridge'. Mayor Dyster recalled riding his bike, carrying no identification whatsoever, to visit his great aunt on the Ontario side. From a Canadian perspective, Mayor Diodati nicely captured the social and cultural integration across the border of their childhood years, ending abruptly with the 9/11 terrorist attacks:

> I always say we're one city divided by a river. And growing up here before 9/11, it was just a way of life … I've got family on both sides …We grew up never having to carry ID. Going to shop over the river, going to dinner, visit, travel … the border was just one more stop along the way.
>
> (Interview)

Sources of differentiation and divergence

Despite extensive similarities and intense cross-border relationships between the Niagara twin cities, there are important differences. First, the two cities' early-20th-century development patterns diverged, partly because of differential exploitation of the Falls' hydroelectric generating power. The American side saw large-scale industrial enterprises emerging, including the petrochemical plants later associated with the Love Canal scandal.[1] On Canada's side, much of the power generated went to major industrial cities like Hamilton and Toronto, while the Falls were exploited as a tourist attraction.

The two cities' relative sizes have changed over the past century, pretty much following their respective economic fortunes. Rapid industrialisation on the American side produced speedy population growth, from c20,000 in 1900 to 50,000 in 1920 and over 100,000 in the 1960 US census, by which time the American side was over double the size of its Canadian counterpart. Since 1960, the two cities' fortunes have reversed. Niagara Falls, New York, has shared the fate of many other de-industrialising 'rust belt' American cities: declining employment in previously well-paid unionised jobs, hollowing out of the downtown core, growing crime and violence, affluent peoples' migration to suburbs and surrounding villages (Thomas 2014). However, its demise was also partly self-inflicted, resulting from disastrous urban-planning decisions following a 1960s mayor's decision, according to critics, 'to bulldoze [the] quaint downtown and replace it with a bunch of modernist follies' (Bloomberg 2010). Hence, while Niagara Falls, Ontario's population has continued growing (from 67,000 in 1971 to 88,071 in Canada's most recent 2016 census), Niagara Falls, New York, has declined by half, from 102,000 in 1960 to 50,193 in 2010, the most recent available US census.

By the late 20th century, the contrast between desolate downtown Niagara Falls, New York, and the booming, bustling tourist zone on the Ontario side inspired Patrick McGreevy to characterise the two cities as 'the end of America' and 'the beginning of Canada'. Reading landscapes on the Canadian and American sides as artefacts of the cultures creating them, McGreevy contrasts the Canadian side's 'orderly and pleasing' appearance with the 'messy and polluted' American side. To use a domestic analogy, the well-manicured gardens along Ontario's Niagara waterfront are Canada's front lawn, while the American petrochemical plants are the equivalent of garbage cans in the alley behind the house (McGreevy 1988, 307, 310). McGreevy ascribes these differences to the River's cultural significance for the two nations. For Canadians, as the site of crucial battles in the War of 1812, the Niagara Peninsula was the crucible wherein an emerging Canadian identity was forged. For Americans, however, what is still called on the US side 'the Niagara Frontier' represented the defeat of (north)westward expansion, and of that particular expression of Manifest Destiny (ibid, 315–316).

As Garrard and Mikhailova suggest (2019, 11), there is often a perception of inequality between twin cities that results in a sense of grievance. The Niagara twins' underlying aesthetic and economic differences have recently created inter-city hostility. Promotional videos produced by Niagara Falls, Ontario, picked up on a 2015 conversation wherein late-night talk-show host Jimmy Fallon and actor Nicole Kidman jokingly ask why the Canadian side of Niagara Falls is so much more attractive as a destination than the American side. The three videos take the form of light-hearted invitations from Mayor 'Jimmy' Diodati to Jimmy Fallon to visit Niagara Falls CANADA for 'the fun,' 'the Vegas' and 'the lights,' each portraying favourable contrast to the allegedly drab, depressed city across the river ('on the other side').[2] Interviewed on television, Mayor Diodati explained: 'The idea was, let's have a fun, creative, unique invitation to Jimmy Fallon, one he wouldn't get every other day,' while admitting he probably took things too far (Wooten 2015). Like many on the US side, Mayor Dyster failed to see the humour: 'Mayor Diodati is a ball-breaker, right? So, they got far down the road with this thing before they realised "wait, we don't hear laughing from the other side", because there was sensitivity on our side' (Interview). Other New York officials were less circumspect: John Percy, president and CEO of Niagara Tourism and Convention Corporation (New York), called the videos disrespectful and unprofessional ('I don't think they needed to take it that far'), while Lisa Vitello, chair of the city's tourism advisory board, characterised Fallon's comments as 'rude' (Scheer 2015; Wooten 2015).

Tourism: competition or cooperation?

This incident highlights a competitive impulse within the cross-border tourism industry, where the time visitors spend on the other side of the Falls represents lost revenue. Yet, increasingly, global competition for tourist dollars should instil a positive, rather than zero-sum, orientation within this sector. According to the UN's World Tourism Organization (UNWTO 2020, 3), international tourism recorded its tenth consecutive year of growth in 2019 as one of the most rapidly expanding and competitive sectors in the global economy. Tourism clearly is the twin cities' leading economic driver; Niagara Falls, Ontario, estimates 12–14 million yearly visitors, while Niagara Falls State Park on the American side claims 9.5 million (Niagara Region 2019; Prohaska 2020). Marketing the Falls' natural wonders should be cooperative for both sides' decision-makers and entrepreneurs. Local mayors seem well-aware of the advantages of joint marketing. Mayor Diodati claimed, 'we don't see it as Niagara Falls, New York versus Niagara Falls, Canada … It is Niagara Falls versus all the other destinations around the world'. However, Mayor Dyster also acknowledged inter-twin competitiveness for tourist dollars: 'In tourism, there is competition, but then there is also cooperation … We're always trying to one-up the other guy, but anybody in the tourism sector would say "We're much stronger together"' (Interviews).

Evidence of competition is apparent in the two tourist authorities' promotional materials, wherein attractions across the border are seldom mentioned. Even each side's tourist maps provide little detail about the other city. When we brought this to the mayors' attention, both thought it regrettable, but perhaps inevitable. Paul Dyster (Niagara Falls, New York) acknowledged 'all our tourism people on both sides of the border recognise we are stronger as a binational destination … when they get up at conferences, they brag about this,' but added 'it is sometimes given only lip service when you look at their promotional materials'.

The simple (and arguably decisive) fact is that tourism agencies have membership drawn exclusively from their respective national sides of the border; this diminishes incentives for actors to cooperate in marketing the entire region.

A related area where competition prevails over cooperation is the casino and gaming industry. Niagara Falls, Ontario, opened its first casino (Casino Niagara) in 1996, enjoying a binational regional monopoly on such entertainment until the Seneca Niagara Casino opened in downtown Niagara Falls, New York, in 2002. Ontario responded with the new Fallsview Casino entertainment complex in 2004. But, unlike other tourism sectors, the American side dominates the cross-border casino market. According to Mayor Diodati in 2016,

> Right now, we're not very competitive. … The Seneca Casino is winning … even with the advantage in currency exchange, less than 5% of the patrons of our casinos are Americans. Yet it's upwards of 40% at all the other [tourism-related] businesses. It's a converse relationship at the Seneca Casino where upwards of 30% [of customers] are Canadians.
>
> (Interview)

Diodati noted there were advantages on both sides. In the Canadian casinos, winnings are untaxed, and the American dollar goes further, but no smoking is allowed. On the US side, smoking is permitted, and gamblers receive free drinks – something Ontario's liquor-licensing laws forbid. In summer 2019, US-based Mohegan Gaming and Entertainment took over management of both Canadian casinos (Spiteri 2019), but it is still too early to tell if the new regime has altered the cross-border competitive balance.

The city planners we interviewed also identified casinos as a source of cross-border competition. When asked whether the relationship was one of benign mutual neglect or zero-sum competition, John Barnsley of Niagara Falls, Ontario, responded:

> It varies between the two. Most of the time I'd say it is sort-of benign … but you see the competitive side rear its head every once in a while, like when the casino was developed on the opposite [New York] side. That's when we said 'Wait a second here'.
>
> (Interview)

Tom DeSantis in Niagara Falls, New York, also saw casinos as a source of competition because 'there is a saturation point for gamblers'. But, he suggested, both cities are 'parochial' and 'see their market as local rather than global'. Instead of complaining about regional competition, he argues, they should be

> marketing to people 5,000 or 10,000 miles away … brand it differently and market it differently … you have the Falls and you're unique. Sell the package in France [or] Norway and, once a week, land a plane – especially in winter when [hotel] occupancy rates are low.
>
> (Interview)

Sometimes, twin-city relationships around tourism are strained by minor irritants. For example, in 2016, the binational Illuminations Board controlling night-time Falls light displays decided to spend $4 million to purchase LED lights. This created disquiet among American Board members who thought the new lights would disproportionately benefit the Canadian side (Law 2017). Hard feelings and installation delays resulted. Once overcome, though, such disagreements appear of limited consequence for the long-term twin-city relationship.

Security issues following the 9/11 terrorist attacks

Long-time Niagara residents often refer to the Niagara River as 'the ditch,' an ironic, slightly dismissive term connoting its former lack of significance in popular consciousness as an obstacle to passage, especially for those living near the border. For much of the 20th century, as mayors recounted remembering their youth, little identification was required to cross, especially if one were local. But 9/11 changed that: increased securitisation and 'thickening' of the border has occurred, due to both unilateral US government decisions and bilateral security agreements. The border is no longer a line, or fence, or bridge, but rather 'a zone' (Longo 2018, esp. 192–194). The consequence has been 'a shifting "emotional geography" of the border that included a new sense of the crossing itself as "less comfortable" and sees the other side as more distant' (Helleiner 2010, 95).

In 2007 the United States announced the 'Western Hemispheric Travel Initiative' (WHTI), making presentation of secure identification required of all border-crossers: a passport, enhanced driver's licence, a NEXUS card, or Permanent Resident card. The WHTI evidently and seriously hurt both Niagara twins by contributing to delay and uncertainty for all potential cross-border travellers. However, it disproportionately harmed the Canadian side by serving as a disincentive to Americans visiting Canada. More Canadians than Americans hold passports: in 2007 when WHTI was announced, 51% of Canadians possessed this document against 27% of Americans; in 2018, 42% of Americans had passports versus 65% of Canadians (McCarthy 2018).

Clearly, increased border securitisation and resulting uncertainties and delays in crossing the Niagara River have hurt the tourism industry, causing frustration on both sides. Meanwhile, the border's increasing prominence as a symbol of national sovereignty and point of separation and control between the two cities gave the mayors common cause and incentives to cooperate. As Mayor Diodati put it: 'Slow-downs don't help tourism, and tourism is the number one industry on both sides of the border' (Interview). Both mayors expressed frustration with the fact that US-Canada border policy often reflects perceived problems or fears arising from the US-Mexico border and failures by border policy makers, largely located in distant capital cities, to take account of borderland residents' interests and perspectives. According to Mayor Diodati,

> So many [border policy] decisions [are] made in vacuum, they're insular, they don't consult with the front-line people. And I think that should be the first thing you do in any business. ... You go to the front-line people and they'll tell you what's going on.

Mayor Dyster echoed this sentiment:

> If you get a bunch of mayors from both sides of the border, mayors sitting around together ... we obviously have a very different perspective on the US side from Washington, which seems a long way away. Ottawa doesn't seem that far away [geographically], but I think our Canadian friends immediately opposite ... feel that they have more in common with us than they do with their federal government.
>
> (Interview)

Political, partisan and jurisdictional impediments to cooperation

Our interviews with twin-city mayors and planners uncovered evidence of more mundane factors impeding cross-border cooperation. Canadian and US mayors and city councillors are directly elected. Both mayors pointed to time pressures and needing to be responsive to their electorates as constraints on pursuing cross-border activities. The natural tendency for elected officials in all jurisdictions on both sides to get caught up in their own small political worlds constitutes a significant impediment to greater cooperation among municipalities. This applies to municipalities on the same side of the border as well as those on opposite sides. According to Mayor Dyster:

> It's a good relationship, a strong relationship. If anything keeps it from being stronger, it's just the scarcity of time. ... We get caught up in our own cycles, and you have to be careful that the interaction with your neighbours across the border doesn't become just an afterthought. That you're not just seeing each other at symbolic events, as opposed to having an opportunity to interact on policy issues.
>
> (Interview)

Mayor Diodati described it similarly:

> at the end of the day, I'm responsible to my taxpayers. He's responsible to his. But we are both able to go high enough up, to take a 10,000-foot view, to realise that we are connected ... We have different purposes and different roles, but we're still always connected and there's no getting away from that.

But Diodati also noted problems in trying to work with a city government across the border divided by partisanship (unlike the US, Canadian municipal elections are generally not contested along party lines):

> Sometimes I'll bring up an issue and they'll say, 'well, the Republican side of the Council won't support it.' ... The Mayor [a Democrat] may think it's a good idea but ideologically the Council may be against him. And I said, 'Even if it's a goodidea?' [to which he responded] 'Even if it's a good idea'. Sometimes you're forced to accept ... that some good ideas won't be chosen because of partisanship. That's frustrating.
>
> (Interview)

Across the river in New York State, rather than blaming partisanship, Mayor Dyster attributed apparent lack of interest in cross-border initiatives among most of his city councillors to their need to focus attention on their district electorates and on issues most important to local voters:

> I think it is not so much partisanship as ... parochialism. The mayor has some ability to see a bigger picture because I'm always going to these things. Even though I'm in the same union contracts, garbage collection, snow-plowing world with them, I also have a chance to glimpse over the wall.
>
> (Interview)

Thus, Dyster continued, even though invited to binational events, most part-time city councillors don't attend.

Planner Tom DeSantis (Niagara Falls, New York) agreed that the vicissitudes of electoral politics militate against long-term cross-border cooperation. There are few political incentives to pursue cross-border initiatives because 'nobody's job description includes international relations'. But DeSantis also saw individual personalities as crucial determinants of whether particular local authorities become involved in cross-border cooperation:

> [One local mayor] never saw the reward in having regular lunches with these guys. He can't see any way this will produce an economic development project that he could stand in front of and cut a ribbon. ... He needs a deliverable. ... It comes down to personalities and the things they value.
>
> (Interview)

The closure of Buffalo's Canadian Consulate in August 2012, victim of Canadian federal budget cuts, further exacerbated problems because there was no longer any single government agency in the region taking responsibility for advancing a binational agenda. In the early 2000s, regular cross-border mayoral meetings took place under the Consulate's aegis. As Dyster noted: 'I think we became too reliant on the Canadian Consulate. ... All the binational stuff we were doing seemed so robust, but if you check carefully where the emails were coming out of ...'. Due to Consulate activities, Dyster explained, binational events would simply appear on municipal officials' agendas. Such events were

> on the calendar. Now *we* have to remember, somebody has to take the initiative. ... There's a big difference between that and the binational mayors [meetings] where you leave the table with a to-do list of things that quickly get pushed into the background when you get back into your single-minded US or Canadian term of reference.
>
> (Interview)

Paradoxically, the closing of the Canadian Consulate coincided with a temporary surge in cross-border cooperation around celebration of the Bicentennial of the War of 1812, also marked as the beginning of 200 years of peace between the

United States and Canada. This was mostly led by NGOs such as the Niagara 1812 Binational Legacy Council, although the Consulate had played a major role in planning the Bicentennial.

Conclusion

Each year, on a Saturday in early May, representatives of law-enforcement agencies from either side of the river (Niagara Regional Police Service in Ontario; Niagara Falls Police Department and Erie County Sheriff's Office in New York State) meet in the middle of the Rainbow Bridge to participate in an annual tug-of-war. According to planning official John Barnsley, this good-natured contest is 'the perfect metaphor' for the contradictory impulses of cooperation and competition between the two cities; they must come together to pull in opposite directions! The tug-of-war might also be a metaphor for the state of inter-city competition around tourism; as of 2019, of 59 contests, the Canadian side had recorded 53 wins (the COVID-19 pandemic closed the Canada-US border, prohibiting a 2020 contest).

In many respects, the two cities of Niagara Falls represent an archetypal model of binational twins. They seemingly have more in common (origins, age, development, size, language, culture, joint possession of one of the world's leading tourist attractions) than most other international twin cities. Yet the border clearly has much more than symbolic import; it is, in many ways, a physical and a psychological boundary separating these two communities. Even though senior officials admit they have much to gain by working together, there are serious impediments – political, economic, cultural, geographic – conspiring to limit cross-border cooperation. These challenges have been compounded by the disruption to Canada-US relations associated with the Trump presidency and the impact of the COVID crisis. Identifying the more general obstacles to cross-border cooperation in a setting wherein such relations should be well-advanced is particularly instructive of the larger challenges facing other international twin cities where local conditions are often less conducive.

Notes

1 In the late 1970s, media revealed that a suburban housing development and a school had been built on the site of a chemical waste dump, producing major health problems for local residents.
2 See: 'Come for the Fun': https://www.youtube.com/watch?v=s3fPr1wWACo; 'Come for the Vegas': https://www.youtube.com/watch?v=ERWhwTzVNMs; 'Come for the Lights': https://www.youtube.com/watch?v=tv1DSlCHeXI.

References

Angus, H F (1938), *Canada and Her Great Neighbour: Sociological Surveys of Opinions and Attitudes Concerning the United States*, Toronto: Ryerson Press.
Bloomberg (2010), 'The Fall of Niagara Falls', *Bloomberg Business*, December 2. https://www.bloomberg.com/news/articles/2010-12-02/the-fall-of-niagara-falls

Buursink, J (2001), 'The Binational Reality of Border-crossing Cities', *GeoJournal* 54 (1), 7–19.

Dubinsky, K (1999), *The Second Greatest Disappointment: Honeymooning and Tourism at Niagara Falls*, Toronto: Between the Lines.

Eckstein, H (1975), 'Case Studies and Theory in Political Science', in Fred Greenstein and Nelson Polsby, eds., *Handbook of Political Science*, Volume 7, Reading, MA: Addison-Wesley, 79–138.

Garrard, J and Mikhailova, E (2019), 'Introduction and Overview', in Garrard and Mikhailova, eds., *Twin Cities: Urban Communities, Borders and Relationships over Time*, London: Routledge, 1–20.

Hartt, M D and Warkentin, J (2017), 'The Development and Revitalisation of Shrinking Cities: A Twin City Comparison', *Town Planning Review* 88 (1), 29–41.

Helleiner, J (2010), 'Canadian Border Resident Experience of the "Smartening" Border at Niagara', *Journal of Borderlands Studies* 25 (3&4), 87–103.

Konrad, V and Nicol, H N (2008), *Beyond Walls: Re-inventing the Canada-United States Borderlands*, Burlington, VT: Ashgate.

Law, J (2017), 'Thomson: New Lights Can Do So Much More', *Niagara Falls Review*, July 14, https://www.niagarafallsreview.ca/news-story/8195849-thomson-new-lights-can-do-so-much-more/

Longo, M (2018), 'A 21st Century Border? Cooperative Border Controls in the US and EU after 9/11', *Journal of Borderlands Studies* 31 (2), 187–202.

McCarthy, N (2018), 'The Share of Americans Holding a Passport Has Increased Dramatically in Recent Years', *Forbes*, January 11, https://www.forbes.com/sites/niall-mccarthy/2018/01/11/the-share-of-americans-holding-a-passport-has-increased-dra-matically-in-recent-years-infographic/#4f9bebfe3c16

McGreevy, P (1988), 'The End of America the Beginning of Canada', *Canadian Geographer* 32 (4), 307–318.

Niagara Falls Info (n.d.), 'Municipal History', http://www.niagarafallsinfo.com/histo-ry-item.php?entry_id=1235&current_category_id=68

Niagara Region (2019), *Niagara Tourism Profile*, Thorold, ON: Niagara Region, https://niagaracanadaadmin.com/investment/wp-content/uploads/sites/7/2019/03/Niagara-Tourism-Profile_FINAL.pdf

Prohaska, T J (2020), 'How Many Visitors Come to Niagara Falls? The Answer Depends on Who's Counting', *Buffalo News*, February 18, https://buffalonews.com/2020/02/18/how-many-tourists-come-to-niagara-falls-the-answer-depends-on-whos-counting/

Scheer, M (2015), 'Local Officials Disappointed in Fallon's Comments on "The Tonight Show"', *Niagara Gazette*, January 5, https://www.niagara-gazette.com/news/local_news/local-officials-disappointed-in-fallons-falls-comments-on-the-tonight-show/arti-cle_5d148d3c-19c3-517c-9fc4-302f05a0a922.html

Siener, W (2013), 'The UN at Niagara: Borderlands Collaboration and Emerging Globalism,' *American Review of Canadian Studies* 43 (3), 377–393.

Spiteri, R (2019), 'Mohegan Takes Over as Service Provider of Falls Casinos,' *Niagara Falls Review*, June 13, https://www.niagarafallsreview.ca/news/niagara-region/2019/06/13/mohegan-takes-over-as-service-provider-of-falls-casinos.html

Taylor, A (2010), *The Civil War of 1812: American Citizens, British Subjects, Irish Rebels, & Indian Allies*, New York: Alfred A. Knopf.

Thomas, G S (2014), 'Niagara Falls and Buffalo Post the State's Highest Crime Rates (Yes, Higher than New York City),' *Buffalo News*, December 12, http://www.bizjournals.com/buffalo/news/2014/12/16/niagara-falls-and-buffalo-post-the-states-highest.html

UB Regional Institute (2007), 'Policy Brief—Defining the Region's Edge,' Buffalo: University at Buffalo Regional Institute, http://ubwp.buffalo.edu/ubri/wp-content/uploads/sites/3/2014/12/Defining-the-Regions-Edge-Policy-Brief.pdf

UNWTO (2020), *International Tourism Highlights, 2020 Edition*, Madrid: United Nations World Tourism Organization, https://www.e-unwto.org/doi/pdf/10.18111/9789284422456

Wooten, M (2015), 'Niagara Falls, Ont. Mayor Apologizes Over Videos,' *WGRZ News*, Buffalo, NY, February 13, http://legacy.wgrz.com/story/news/local/niagara-county/2015/02/13/niagara-falls-ontario-mayor-apologizes-to-wny/23378689/

21 Asylees, removals, returnees

Mexican border cities response and adaptation to mixed migratory flows

Hilda García-Pérez and Francisco Lara-Valencia

Introduction

Cities on the U.S.-Mexican border are at the front line for displaced Latin American and Caribbean asylum seekers moving northwards, and Mexican migrants returning voluntarily or involuntarily from the United States. Historically, these cities have been resettlement points for many. However, they mostly temporarily shelter those seeking refuge, chasing dreams northwards, or returning home southwards. Sweeping changes in U.S. policy to immigration enforcement and refugees during the Trump administration altered migrant flows, creating a border bottleneck and a borderland humanitarian emergency. This suddenly changed the composition and volume of migratory flows, putting pressure on local capacities to handle otherwise transient populations. Mixed migratory flows challenge long-standing beliefs and practices regarding migrants, demanding new ways of addressing migrants' needs and impacts on receiving communities. This chapter explores how local migrant-serving organizations (MSO) responded to the new scenario. MSOs are governmental and non-governmental entities providing services that are critical for migrants' safety and well-being, and they play an important role in articulating collective narratives underscoring the human dimension of migration, and potentially policies enabling the integration of migrants to local communities.

We argue that U.S.-Mexican border cities are developing greater agency and becoming more proactive in responding to challenges from mixed migratory flows. This is rooted in border communities' traditional openness to migration – including cross-border solidarity – and aligned with the growing importance of local-level knowledge and initiatives observable in national migratory regimes globally (Bauder and Gonzalez 2018; Garcia and Bak 2019). From this perspective, border cities are loci for continual negotiation and solidarity due to their unique situations as human-mobility gateways. They are also spaces where local actors repeatedly challenge state-centrist policies, seeking enhanced decentralisation (Ray 2003; Young 2019). As Martinez (1994) observes, border cities are prone to collective scepticism and resistance toward national policies, a practice rooted in years of rejecting centralism, propelled by cross-border interactions, interdependency and remoteness from the centre.

DOI: 10.4324/9781003102526-27

Examining the experience of Ambos Nogales, a conurbation along the U.S.-Mexico border, we describe the local apparatus providing care and protection for migrants and asylum seekers transiting to or from the U.S. Secondly, we examine how this apparatus is regulated and coordinated locally and binationally to meet this population's humanitarian, informational and protection needs. Finally, we explore how the border's humanitarian emergency might be stimulating a 'local turn' in migration governance, inducing reorganization and adaptation of local MSO networks, and the re-ontologization of migrants.

We draw upon mixed-methods research based on content analysis of regional newspapers, social media and policy reports. We also conducted 12 semi-structured interviews with local MSO representatives, exploring organizational activities before and after migration patterns changed, the impact of mixed migration on organizational resources, capabilities, and cooperation, and the organizations' long-term prospects.

New scenario of mixed migratory flows

Cities along the U.S.-Mexico border have long been nodal places for migration between the U.S., Mexico, and Latin America (Figure 21.1). Historically, millions have crossed the region bound for U.S. production and consumption centres, attracted mainly by jobs and wage levels. Recently, however, changing U.S. immigration policies and consequent increased border regulation and delay have sharply shifted the direction, composition and dynamics of migratory flows converging on the region.

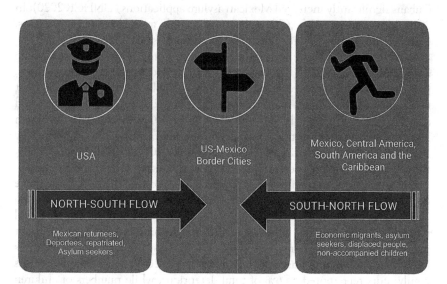

Figure 21.1 New scenario of mixed migratory flows in the U.S.-Mexico borderlands.

Source: Diagram by Hilda García-Pérez.

First, rising Latin American insecurity, violence, poverty and natural disaster in the last decade have uprooted thousands from Mexico, Haiti, Honduras, Guatemala, El Salvador, Venezuela and elsewhere who migrated north amidst extreme precariousness and vulnerability. Driven by pressing survival needs, this flow includes asylum seekers, unaccompanied minors, displaced persons and others moving northwards seeking U.S. shelter and relief. Second, stricter U.S. immigration policies and criminalization of undocumented migrants resulted in massive involuntary and voluntary deportations, generating exceptionally high north-south migratory movements (Coubès 2018). In particular, intensifying U.S. immigration control caused the re-entry/entry into Mexico of millions with varying Mexican familiarity and connection.

South-north flows

Alongside traditional economic migrants, during the mid-2010s, the border saw evermore migrants forced to leave Mexican and Central American towns following violent or catastrophic calamities (O'Connor et al. 2019). Leaving homes, belongings and social networks, displaced people also risked violence during transit toward the border and legal uncertainty around asylum-seeking outcomes due to U.S. discouragement policies. A recent U.N. High Commission for Refugees report suggests numbers of Central Americans receiving its assistance in Mexico rose from 13,300 in 2015 to 140,000 in 2018. Although providing no causal information, the report indicates violence was a contributing factor (UNHCR 2019). Concurrently, Venezuela's deteriorating situation, Central American violence and insecurity, and changed U.S. migratory policies toward Cubans significantly increased Mexican asylum applications (UNHCR 2020). In 2019, Mexico received 70,609 applications, mainly from Hondurans, Haitians, Cubans, Salvadorians, Venezuelans and Guatemalans (COMAR 2020).

In 2016, Haitians arriving at north-Mexican border cities attracted public attention. After the 2010 earthquake, thousands left Haiti for South America, particularly Brazil. However, Brazil's economic and political crisis and consequent unemployment pushed Haitians northwards. Approximately 4500 Haitians and some Africans arrived at border cities (CCINM 2017; París 2018). Although Haitians arrived seeking the temporary protection status (TPS) conferred by President Obama on Haitians (Alarcón and Ortiz 2017; CCINM 2017), many were denied U.S. entry and currently reside in the region (Rivlin-Nadler 2020).

Although Mexican numbers apprehended while crossing the U.S. border have declined substantially, annual apprehensions still averaged nearly 500,000 during 2015–2020, reaching 977,509 in 2019 (USCBP 2020). The U.S. Border Patrol returned many, usually via a procedure allowing deportations to Mexico without judicial or administrative review. Significant amongst these returnees/detainees have been families and unaccompanied minors. In 2019, detainees traveling as family units represented 55.6% of total detentions, while numbers of children reached record levels (Gramlich and Noe-Bustamante 2019).

The last three years have also seen more 'inadmissible' people, mostly Mexicans seeking U.S. humanitarian protection but rejected by U.S. border officers

(USCBP 2019). In 2018, Mexican asylum applications in U.S. courts totalled 10,896, higher than claims filed by Hondurans (8745) and Guatemalans (9214), and only below Salvadorian applications (12,073) (TRAC 2018).

North-South flows

The changing composition and volume of northward migration flows coincided with essential changes in north-to-south flows. During the Obama administration, the U.S. deported nearly 2,800,000 immigrants, significantly above the Bush administration's 1,600,000 (TRAC 2020). In 2018, U.S. immigration authorities ejected 337,300, a 17% increase over 2017 but below the Obama administration's annual average (Gramlich 2020). Most deportations happened in border towns. In 2019, 48.7% of deportees were received by Mexican authorities at the border entry ports of Tijuana, Ciudad Juárez and Nogales, the three cities with the largest increase in deportations between 2015 and 2019 (COLEF 2018).

While border apprehensions declined due to heightened border control, U.S. immigration authorities ramped up apprehension within U.S. non-border states (Schultheis and Ruiz 2017). Arrests in the country's interior totalled 158,581 in 2018, half being Mexicans (DHS 2020). Apprehended under programs like Secure Communities and 287(g), many had no criminal record and possessed limited knowledge of Mexican culture and Spanish (Capps et al. 2011; INM 2019b).

In 2019, approximately 56,000 non-Mexican asylum seekers were returned to Mexico, awaiting asylum-claim verdicts, including 16,000 children (HRW 2020). These migrants come under the Migrant Protection Protocol (MPP), allowing U.S. border offices to return non-Mexican asylum seekers to third safe countries. Most large Mexican border cities received MPP migrants, including Tijuana, Mexicali, Nogales, Ciudad Juarez, Nuevo Laredo and Matamoros.

Contours of an emergency

Although most border towns possess infrastructure providing shelter, food and essential medical services to traditional migrant flows, the new influx tested capacities, particularly in medium-sized and small border towns like the two Nogales (COLEF 2020; París 2018)

Nogales, Sonora, with 255,000 residents, and Nogales, Arizona, with 21,000, were the seventh-largest conurbation on the U.S.-Mexico border in 2015 (Figure 21.2). The two are often referred to as Ambos Nogales, a nomenclature highlighting their adjacency, interconnection and interdependence. The regional economy rests upon export-oriented manufacturing (maquiladoras), international trade and tourism. About 30,000,000 people and 12,000,000 vehicles crossed Ambos Nogales's border entry ports in 2018. Social and economic ties resulting from this interaction are intense, engendering diverse cross-border cooperation and networking.

The confluence in time and space of new migratory flows caused an emergency, altering day-to-day life for local people and institutions in Ambos Nogales. Like

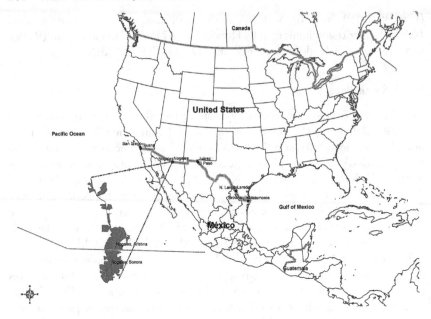

Figure 21.2 Ambos Nogales Location.

Source: Map by Francisco Lara-Valencia.

any emergency, the sudden and unusual nature of events challenged existing resources and operating routines. For local MSOs, the problem lay in addressing extremely vulnerable families and individuals' needs, demanding immediate action while also requiring citizens and authorities to mobilize new resources and adopt new strategies providing humanitarian assistance for newly arriving migrants.

Local governance of migration flows

In 2018, the local system managing migration flows in Ambos Nogales comprised a network of approximately 18 governmental and non-governmental MSOs.

Mexican government organizations are crucial in assisting and protecting migrants. Federal government programs helping deported migrants include the Module for the Care of Repatriated Migrants operating at border crossings. This multi-agency program provides primary health care and orientation to Mexican migrants as part of a collaboration between the National Health Department, Mexican Red Cross (MRC) and National Institute of Migration. An instance of federal-state collaboration is a 2016 program receiving and protecting minors repatriated from the U.S. This program looks after minors referred mainly by the National Institute of Migration. Locally, the Nogales Municipal Council (AYN) collaborates with migrant shelters and soup kitchens, providing food, water and financial support and sometimes facilitating communication and interaction with state and federal agencies. Other entities within the migrant-aiding governmental network are the General Hospital, participating in health campaigns, and the National Commission on Human Rights, investigating alleged human-rights violations.

Contrastingly, Nogales, Arizona, municipal and state agencies are not involved in migrant-protection activities. Governmental activities are circumscribed by regulations and protocols governing how subnational actors interact with migrants and other non-governmental organizations (NGOs), limiting their ability to respond to sudden migration-flow changes.

However, NGOs on both sides of the border are more adaptable, responding faster to changing conditions. Overall, NGOs on the network mainly provide free temporary aid to migrants – shelter, food and local transportation. Humanitarian organizations like San Juan Bosco (SJB), the Home of Hope and Peace (HHP) and the House of Love and Mercy (HLM) provide temporary accommodation. Kino Border Initiative (KBI) mainly organises a migrant's dining room, runs a small shelter and offers legal assistance, including legally accompanying persons seeking U.S. asylum. Such organizations mostly predate the emergency. Previously, they mainly assisted repatriated Mexicans or domestic and international migrants transiting to the U.S.

The emergency also pushed new players into the local MSO network. For example, the Rotary Club (RTY) operated a temporary shelter in Nogales, Mexico, and provided humanitarian aid to Haitians. Similarly, a private transportation and rental car company, Taxi Amigo, offered migrants free local transport when other network organizations asked them.

Essential for local MSO network functioning are organizations playing intermediary roles, either liaising between NGOs or connecting NGOs with governmental entities, foundations and universities on both sides of the border. They can also advocate and lobby with legislators/political leaders to influence policy.

Intermediary organizations enjoy high prestige and considerable influence, as evidenced by their ability to moderate resource flows, promote partnerships and affect overall network activity and effectiveness. Interviews identified San Juan Bosco Migrant Shelter (SJB), the Kino Border Initiative (KBI), the Grupo Beta for Protection of Migrants (GBP) and the Fundación del Empresariado Sonorense, A.C. (FESAC) as intermediary organizations.

SJB is Nogales's oldest and most important migrant shelter, established in 1982 by the Loureiro family. Over three decades, it has sheltered thousands of migrants from Mexico and elsewhere (Flores 2014), attending around 230,000 migrants annually (Lara 2019).

U.S. and Mexican Jesuits established KBI in 2008, providing humanitarian aid to arriving migrants (KBI 2014). Legally present in both countries, KBI's binational structure allows it to function as a bridge attracting resources from humanitarian and solidarity organizations and cross-border groups. In 2018, KBI provided nearly 56,000 meals in its dining area and hosted 839 women and children in its dormitory (KBI 2019). Alongside this work, KBI provides legal advice to asylum seekers and other migrants and communicates with legislators and political leaders on both sides to influence national and sub-national policies (KBI 2019).

Established in 1994 in Nogales, GBP is a unit of the National Institute of Migration, a Mexican federal governmental agency. GBP provides humanitarian aid and human-rights guidance to transiting migrants, rescues migrants and provides first-aid services regardless of migratory status or nationality (INM 2019a). They coordinate with other local MSOs, but particularly KBI and SJB. GBP also maintains a close collaboration with the U.S. Border Patrol, coordinating border rescue operations. In 2018, Nogales GBP undertook over 145,000 actions supporting migrants, including search and support to displaced migrants, first aid, hospital transfers, water and food provision, and legal guidance (INM 2019a).

FESAC is an independent non-profit foundation operating in several Mexican cities. FESAC's network activities comprise developing financial resources to support non-governmental projects, promote philanthropic spirit among local businesses and support NGO professionalization. In Nogales, FESAC has actively connected local and international financial sources and migrant-serving NGOs (FESAC 2020).

The emergence of mixed flows

For Ambos Nogales MSOs, the flow's changed composition rather than its volume was what generated an emergency. They felt that demand for services and care remained just like previous years: the arrival of Haitians, Central Americans and migrants of other nationalities compensated for decreasing numbers of Mexicans deported by U.S. immigration. As one informant observed, 'the decline in volume [of southbound migrants] has been very noticeable for the last four years … [in 2014] we were used to receive about 600 a day, but now [2018], we received only 50' (GBP). However, all agreed the new migrant flow was 'different' from receiving only southbound deportees/returnees or traditional flows from the south.

Bodies in motion

For decades, local organizations attended migrants in movement. This included Mexicans from the U.S. or those transiting northwards. They needed shelter, food and other services while returning to pre-established southward destinations or while crossing to the U.S. Most Nogales aid centres were designed to facilitate mobility.

The mobility principle is tangible in routines an MSO characterized as the 'three-days of pillow and food' policy: 'three days are enough for migrants to communicate with relatives, and receive information about the city's resources' (SJB). This assumes migrants' relatives can provide information and finance to resume journeys north or south quite quickly. Additionally, municipal agencies and bus companies offer returnees free/low-cost transportation to anywhere in Mexico. Allegedly, the three-day policy discouraged long stays and reduced risks of 'creating an atmosphere of too much comfort in shelters, while encouraging migrants to find quick solutions to their new situation' (SJB). The policy was also seen as a way to reduce migrants' risk of 'being kidnapped by organized crime, start using illegal drugs, or experience harassment' (AYN). However, not all organizations supported the policy, one informant suggesting 'it is becoming increasingly unfit' (KBI).

Migration-flow changes complicated the system for two main reasons. First, returnees include many adults who have been U.S.-resident most of their lives. One informant explained, they 'no longer have support networks in Mexico … they now have strong ties in the U.S.' Many 'are deciding to stay at the border' (KBI). Second, migrants also include people coming from the south, like Haitians, Central Americans and displaced Mexicans arriving to apply for U.S. asylum. Fluctuating U.S. asylum policy and uncertainty around the application process has prolonged their borderland stay. One informant said, 'Haitians came to stay for months … to solve a problem … if they needed to stay a year, they would' (AYN).

Changes in flow created a mismatch between migrant's needs and Nogales's shelters' routines and practices. The resultant emergency challenged support systems' effectiveness. Referring to Haitians' arrival in 2016, one informant explained, 'Haitians … gradually arrived and exceeded the shelters' capacity' (SJB). Furthermore, 'migrants are mobile by nature; they stay one, two, three days and continue their travel. But [Haitians] stayed for months … depleting shelter capacity' (AYN). An unforeseen and unique phenomenon now alters previous practices and routines. For one informant, the new flows were qualitatively distinct: it 'was not like receiving Salvadorans or Mexicans. Arriving migrants now stayed until the "gringo" government received them'. This generated problems in local shelters being 'full of Haitians'. Meanwhile, 'Mexican returnees and migrants from Central America were arriving, and there was no capacity to provide care for all' (AYN).

Families, not individuals

Changing flow composition pressurized other aspects of the local support network. Shelters traditionally service mostly male adults. Many women, mothers with children and adolescents, even unaccompanied children, added new elements.

Thus, for privacy and security reasons, SJB subdivides sleeping areas by sex, with a private area for women and children, another for adult men. For some families, particularly women, separating husbands, if only for sleep, created tension between migrants and shelter providers. An informant explained, 'you could say, it is a culture-matters situation … it was very difficult to separate [Central American women] from their husbands … they wanted to sleep together … here [in the shelter], we have regulations' (SJB).

Children also generated constant concern due to lack of parental supervision: risk of accidents being a recurring concern. One informant noted, 'We don't offer childcare'. Shelter staff could not supervise children inside or outside the facility, so parental expectations and providers' capacities were often mismatched. While shelters expected mothers to take full responsibility for children's behaviours, some requested childcare support. Childcare caused friction in a shelter, producing displacement: 'Some ladies were upset ... eight families are gone' (SJB). Lack of parental supervision also wasted resources like water and food, reprehensible to one interviewee (SJB).

Rice, beans and Wi-Fi

Cultural differences between Mexicans and other migrants also caused confusion and friction. For example, Haitians were uninterested in eating pizzas, tortillas and other flour products locals commonly consumed, their food being heavily rice-based. The community was very supportive of the Haitians, with large food donations. 'Citizens brought a lot of food. It was food and more food. Every day food was spoiled ... It was shocking to see ... much food going to waste!' (GBP). This caused bewilderment, some people characterizing Haitians as 'problematic people', needing treatment with 'energy' (RTY). However, as another inform- ant explained, this misunderstanding changed with time and dialogue. So, after moving Haitians to a temporary shelter, 'I asked them, "what do you want to eat[?]" ... and they told me ... "chicken, rice, beans, and wifi"' (RTY). This helped Rotarians start requesting different food donations, assisting trust and cooperation.

Alongside rice and beans, Haitians also requested Wi-Fi. A participant stated, 'at first, I thought it was a different meal, but then, they made it clear they wanted ... internet access (to) ... make contact with relatives in Brazil and Haiti' (RTY). This addressed basic communication needs, enabling migrants to make decisions about their journey and maintain affective connections with family and friends, essential for emotional stability.

For Ambos Nogales, sudden exposure to cultures with different lifestyles and attitudes caused confusion, testing the local MSOs' ability to deal with the emer- gency. Locals tended to allocate migrant groups to categories, simplifying indi- vidual experiences in purely normative terms. Some informants saw Haitians as 'conflictive' but also more positively as people 'with different behaviours than the traditional migrant', 'well-dressed' and 'well-organized'. Contrastingly, some Central American migrants, like Hondurans, were identified as 'always com- plaining', 'unsatisfied' and 'violent'. However, some Guatemalans were charac- terized as 'very noble' or some women as 'careless mothers'.

Local responses

Ambos Nogales has long maintained support and care systems effectively respon- sive to migrants' basic needs. Comprising mainly voluntary organizations moti- vated by humanitarian spirit, the system's adaptability and resilience is evident in their growth and diversification. Facing the emergency, and viewed retrospec- tively, the system's dynamic and orientation went through three functional key moments or periods.

The first is the pre-emergency period. Here, the system emphasised providing temporary services and care. Organizations primarily focused on humanitarian actions to improve well-being. SBJ is typical: in 1982 the Loureiro family estab- lished the shelter after observing deported and northbound migrants sleeping on streets and sidewalks, sometimes in sub-zero temperatures. Previously, most migrants spent most of the day in public squares awaiting the best times to cross to the U.S. (Flores 2014). Initially, SJB provided only daytime food, although it soon began providing beds for migrants passing through the city. Assistance was temporary, assuming migrants were moving bodies temporarily detained and needing help to continue. Therefore, immobility was undesirable and should be avoided.

The pre-emergency period saw strong coordination between all local MSOs. One participant described the prevailing environment: 'in the past, we (had) excellent communication ... monthly meetings, where migrants' issues were addressed jointly, but now, that is not happening' (KBI). Mixed migrations caused disagreements: some MSOs resisted ending the 'three-days' policy, while others started implementing new routines accommodating longer stays, and actions supporting new migration groups. The impasse was mediated by an intermediary MSO. As a participant observed: 'Most agencies have excellent communication with *Grupo Beta* ... if a migrant is facing an issue, we could call *Grupo Beta,* and they will support us taking them to the hospital or bus station' (KBI).

The second key moment occurred during the emergency. It was characterized by incursions from new actors and embryonic inter-organizational networks being formed. During the emergency, some social clubs and impresarios decided to intervene, focusing their action on a segment of the migratory flows, bringing new resources and perspectives. A shelter was improvised on the facilities of the RTY site. During 2016–2017, it provided shelter and other services to almost 200 Haitian and African migrants. After the local system's near-collapse under stresses from massive numbers of Haitians arriving, the Rotarians assumed responsibility for equipping a social saloon with showers, heating and internet service.

Alongside other agencies, RTY coordinated a donation campaign to obtain beds, clothing, and food. Local merchants, business organizations linked to the club, alongside individual citizens supported the move. The contribution of 'people from Tucson, Phoenix…and Nogales, Arizona' was crucial. Participants said the emergency was addressed using human-rights approaches rather than assistential perspectives. Guidance and rights protection were increasingly emphasised, alongside legal accompaniment during asylum applications (RTY). Information exchange and management networks were established with U.S. immigration authorities, resulting in creating a waitlist for asylum seekers interviews. Rotarians were contacted by U.S. immigration authorities when the latter was ready to receive asylum seekers. Alongside Taxi Amigo, RTY volunteers could drive Haitians to border stations at any time.

The third key period sees humanitarian efforts under a new light and starts envisaging more space in the city for migrant contributions to its future. In Ambos Nogales, some organizational discourses highlighted migrants' economic and social *potential*, noting the need to influence state and federal migration policies, and questioning discriminatory and xenophobic positions and practices. In particular, some business organisations began articulating new narratives about migrant potentialities. As some noted, 'if more migrants want to come [to Nogales], it would be better' (MRC). For businesses and local government, migrants might help re-activate the local economy and diversify the demographic landscape. Economically, newcomers might fill the city's labour-force gaps, addressing high turnover affecting maquiladoras (RTY). Meanwhile, having people of different nationalities could diversify border demographics (i.e., interracial marriages) and the local economy (i.e., entrepreneurialism). For some, a more cosmopolitan hospitality industry could emerge, potentially impacting local tourist and service sectors (AYN).

This period offered glimpses of more local and autonomous ways of managing migratory dynamics, opening possibilities of multi-level governance. Several MSOs identified multi-level coordination and cross-cutting programs as pre-conditions for more humanitarian and effective migration management. To such actors, a significant obstacle to migrant economic integration was federal and state governments' narrow vision. Some expressed regret that governments could not see migrant economic potential for Ambos Nogales (MRC) when other border cities already had Haitians and other migrants working in their local industry (Anonymous 2018; Martinez 2018).

Final thoughts

International experience suggests that how a city defines its relationship with asylum seekers, refugees, displaced persons and other migrants largely determines how much these populations get integrated into urban life and the extent of their contribution to the overall city's stability and well-being (DPDAM/UNDESA 2017; GAUC 2016). Therefore, the spectrum of local responses to immigrants is varied, with cities playing increasingly prominent roles in migration policy and producing what some see as a local turn in migration regimes (Zapata-Barrero et al. 2017). Such a local turn involves creatively linking migrant/refugee integration with local growth and well-being, thereby quelling fear and conflict. As Poppelaars and Scholten (2008) suggest, this is possible because cities tend to be considerably more pragmatic and solution-oriented than national governments.

Ambos Nogales has long solidarity traditions towards migrants, evolving a support system showing remarkable resilience and adaptability. For decades, volumes of migrants transiting north or deported from the U.S. have received help and relief in local shelters and soup kitchens (COLEF 2020). Sustained by a network of MSOs, these shelters facilitated mobility when migrant workers moved back and forth across the border. The three-day food and pillow scheme, telephone access, temporary employment, legal advice and transport aid was the right humanitarian solution considering the motivations, orientation and composition of migratory flows until recent years. However, the sudden radical change in migratory flows, galvanized by border securitisation, exposed limitations in the prevailing local system, stimulating adjustments in functioning and orientation. In particular, mixed migration flows produced significant accommodations in the following areas:

1. *Understanding new vulnerabilities and needs.* Local MSOs responded to the unique vulnerability of returnees, displaced persons, asylum seekers and other migrants. Changing from the three-day meal and pillow scheme to an

indefinite time in shelters was a first step. Other steps were experimenting with new aid approaches, including legal and administrative accompaniment of asylum seekers, education for migrant minors, day centres, job guidance and healthcare. In implementing these innovations, MSOs engaged in networked activities that reinvigorated cooperation.

2. *Re-conceptualizing migrants as resources with knowledge and skills that can diversify and enhance local human capital.* Narratives were generated promoting local action facilitating migrants' and asylum seekers' integration within a strategy benefitting this population and ensuring it contributed to the region's prosperity, security and stability.

3. *Constructing a more inclusive and diverse network.* Mixed migration created opportunities for more community participation, particularly actors interested in contributing to migrants' social and economic integration. This expands resources and capacities available locally to deal more effectively with future migratory challenges. It may also contribute to a collective sense of agency and local affirmation.

Ultimately, the foregoing changes are expressions of a possible transformation towards forms of migration governance hitherto unseen on the U.S.-Mexico border. The complete metamorphosis of Ambos Nogales's local migratory system depends largely on the sustained mobilization of local resources and strategies seeking to integrate migrants through policies and programs benefitting both migrants and local communities.

References

Alarcón, A. R. & E. C. Ortiz. 2017. Los haitianos solicitantes de asilo a Estados Unidos en su paso por Tijuana. *Frontera Norte*, 29, 171–179.

Anonymous. 2018. Contrata maquiladora a un grupo de haitianos. *El Heraldo de Chihuahua*. 05.09.2018.

Bauder, H. & D. Gonzalez. 2018. Municipal responses to 'Illegality': Urban sanctuary across national contexts. *Social Inclusion*, 6, 11.

Capps, R., M. Rosenblum, M. Chishti & C. Rodríguez. 2011. *Delegation and Divergence: 287(g) State and Local Immigration Enforcement.* 67. Washington, DC: Migration Policy Institute.

CCINM. 2017. *Caso: Haitianos y Africanos en la Frontera Norte de México.* Ciudad de México: Consejo Ciudadano del Instituto Nacional de Migración.

COLEF. 2018. *Encuesta sobre Migración en la Frontera Norte de México.* Tijuana, Mexico: El Colegio de la Frontera Norte.

COLEF. 2020. Migrantes en albergues en las ciudades fronterizas del norte de México. In *Documentos de Contingencia 2: Poblaciones vulnerables ante COVID-19.* Tijuana: El Colegio de la Frontera Norte.

COMAR. 2020. *Solicitantes de la Condición de Refugiados en México.* Ciudad de México: Comisión Mexicana de Ayuda a Refugiados.

Coubès, M. L. (2018) Deportaciones de mexicanos desde Estados Unidos: ¿qué está cambiando con el nuevo gobierno de Estados Unidos? *Coyuntura demográfica*, 87–95.

DHS. 2020. *U.S. Immigration and Customs Enforcement Fiscal Year 2019 Enforcement and Removal Operations Report.* 32. Washington, DC: Department of Homeland Security.

DPDAM/UNDESA. 2017. Migrants and cities: A public administration perspective on local governance and service delivery. In *United Nations Expert Group Meeting on Sustainable Cities, Human Mobility and International Migration*. New York: United Nations Secretariat – Population Division.

FESAC. 2020. *Quiénes Somos*. Nogales, Sonora: Fundación del Empresario Sonorense.

Flores, M. 2014. Cumple albergue 'San Juan Bosco' 32 años y más de un millón 800 mil migrantes atendidos. *Infonogales*. 30.01.2014.

Garcia, A. Ó. & J. M. Bak 2019. Institutional solidarity: Barcelona as refuge city. In *Solidarity and the 'Refugee Crisis' in Europe*, eds. Ó. Garcia Agustín & M. Bak Jørgensen, 97–117. Cham: Springer International Publishing.

GAUC. 2016. Forced displacement in urban areas: What needs to be done?, 9. Global Alliance for Urban Crises.

Gramlich, J. 2020. How border apprehensions, ICE arrests and deportations have changed under Trump. In *FactTank: News in the Numbers*. Pew Research Center.

Gramlich, J. & L. Noe-Bustamante (2019) What's happening at the U.S.-Mexico border in 5 charts. *FactTank*. 15.12.2019.

HRW. 2020. *Q&A: Trump Administration's "Remain in Mexico" Program*. New York: Human Rights Watch.

INM. 2019a. *Grupos Beta de Protección a Migrantes*. Ciudad de México: Instituto Nacional de Migración.

INM. 2019b. *Registro e Identidad de Personas*. Ciudad de México: Instituto Nacional de Migración.

KBI. 2014. *Annual Report 2013*. 16. Nogales, AZ: Kino Border Initiative.

KBI. 2019. *Annual Report 2018*. 24. Nogales, AZ: Kino Border Initiative.

Lara, G. 2019. Migrant shelter 'saturated' by influx of Mexican asylum-seeker. *Nogales International* 2/21/2019.

Martinez, G. 2018. Maquilas en Baja California contratan a haitianos. *El Economista* 2/16/2018.

Martinez, O. 1994. The dynamics of border interactions. In *Global Boundaries-World Boundaries*, ed. C. Shofield, 1–15. London: Routledge.

O'Connor, A., J. Batalova & J. Bolter. 2019. Central American Immigrants in the United States. In *Spotlight*. Washington, DC: Migration Policy Institute.

París, P. M. D. 2018. *Migrantes Haitianos y Centroamericanos en Tijuana, Baja California, 2016–2017. Políticas Gubernamentales y Acciones de la Sociedad Civil*. CNDH-COLEF.

Poppelaars, C. & P. Scholten (2008) Two worlds apart: The divergence of national and local immigrant integration policies in the Netherlands. *Administration and Society*, 40, 335–357.

Ray, B. (2003) The Role of Cities in Immigrant Integration. https://www.migrationpolicy.org/article/role-cities-immigrant-integration (accessed 11.11.2018).

Rivlin-Nadler, M. 2020. *Haitians in Tijuana Look Back at a Decade of Displacement Following 2010 Earthquake*. San Diego, CA: KPBS. https://www.kpbs.org/news/2020/jan/14/haitians-tijuana-look-back-decade-2010-earthquake/.

Schultheis, R. & S. A. Ruiz. 2017. *A Revolving Door No More? A Statistical Profile of Mexican Adults Repatriated from the United State*. Washington, DC: Migration Policy Institute.

TRAC. 2018. *Asylum Decisions and Denials Jump in 2018*. Syracuse, NY: Syracuse University.

TRAC. 2020. *Transactional Records Access Clearinghouse*. Syracuse, New York: Syracuse University.

UNHCR. 2019. *North of Central America Situation*. 3. Geneva, Switzerland: United Nations.

UNHCR. 2020. *Global Trends, Forced Displacement in 2019*. Copenhagen, Denmark: United Nations High Commissioner for Refugees.

USCBP. 2019. *Southwest Border Migration FY2019*. Washington, DC: Department of Homeland Security.

USCBP. 2020. *Southwest Border Migration FY2020*. Washington, DC: Department of Homeland Security.

Young, J. E. (2019) Seeing like a border city: Refugee politics at the borders of city and nation-state. *Environment and Planning C: Politics and Space*, 37, 407–423.

Zapata-Barrero, R., T. Caponio & P. Scholten (2017) Theorizing the 'local turn' in a multi-level governance framework of analysis: A case study in immigrant policies. *International Review of Administrative Sciences*, 83, 241–246.

22 Ouanaminthe and Dajabón

Two unequal cities on the Haitian-Dominican border

Lena Poschet El Moudden

Introduction

Few borders exist that separate two countries as dramatically disparate as the Dominican Republic and Haiti. It is as if each country is located on a separate island. Their ethnic, historical, linguistic, social, economic and religious differences are striking.

The equilibrium of forces between the two parts of the island of Hispaniola has changed greatly over time and undergone a profound process of differentiation rooted in its colonial division between France and Spain. It has continued to characterise relations between both modern nations (Théodat 1989, 2003). Nevertheless, all prejudices notwithstanding, the people living in the border regions have established cross-border relations, which, even if principally based on economic interests, have been much more frequent and important to both nations than officially acknowledged (De Jesus Cedano and Dilla 2005; Silie and Segura 2002).

The border towns of Ouanaminthe (Haiti) and Dajabón (Dominican Republic) located on either side of the Massacre River in the northern part of the island, share the characteristics of other border twin cities. They can to some extent be compared to cities on the border between Mexico and the USA, both countries with dramatic differences of income and wealth. Much of their interdependence has been due to economic cross-border exchange and is the motor for their development. The two countries are profoundly unequal economically: GNI/capita in 2018 in Haiti is USD 1360; in the Dominican Republic, it is USD 7760. The ambiguous relations between their populations, alongside unstable interstate relations at the central-government level, have made them a rather shaky pair of twins. Contacts between the local population have been frequent; for some, determining survival. But the experience of a common urban space has been mostly restricted to the market in Dajabón.

The economic interdependence has been palpable when twice a week, people from almost the entire northern region of both countries have travelled to these towns, to buy and sell goods or offer services in Dajabón's busy binational market. On other days of the week, tons of cargo have been transported to Haiti, to the extent that the once-small borderland settlements have become an important junction in the northern part of the island's economic subsystem. Moreover,

DOI: 10.4324/9781003102526-28

the 2002 construction of a textile factory in a duty-free zone in Ouanaminthe, producing textiles for the U.S. market, has connected the borderland towns to the international economy.

My research explores the influence of the intermediation function of the border twin cities on the development of their urban space. From the beginning, I was fascinated to see how much the historic development of both parts of the island shaped the spatial organisation and appearance of both towns. In Ouanaminthe I found decayed colonial-style warehouses and shops, evidence of a certain ancient splendour and importance of commercial exchange. By contrast, Dajabón's townscape is marked by the development of public buildings around 1950, demonstrating by their monumental style power and control, testimony to the epoque when the Dominican regime aimed to reinforce the frontier's Dominican identity. My first impressions confirmed that understanding the history would be central; without this perspective, we cannot comprehend the relation that both cities have with the border and within their own broader national and cross-border network of cities.

Most of what follows is based primarily on data collected during four periods of fieldwork from 2002 to 2004, mainly using qualitative methods: observing and mapping the occupation and use of urban space, the cartography of new urban extensions, plus interviews with (1) local stakeholders (30 politicians, representatives of civil society, entrepreneurs, landowners and educators on both sides of the border), and (2) residents of deprived neighbourhoods (15 residents of Ouanaminthe and 15 from Dajabón). My approach focused on the practices of local stakeholders in politics, economics and civil society. Although the chapter concentrates on a period up to the end of 2004, a brief and necessarily fleeting update at the end suggests the overall situation has changed only a little.

The island's twofold identity – a brief incursion into Hispaniola's history

To understand the issues currently characterising this island's division, we must go back to the 17th century, when the then colonial powers, France and Spain, negotiated its division. The Spanish arrived first when in 1492 Christopher Columbus dropped anchor near the island, establishing the first settlements in its eastern part. French settlers arrived around 1625 on the island's western part, proclaiming it a French colony in 1665, a status officially confirmed 32 years later by the Rijkswik Treaty. The determination of the demarcation line between the colonial powers occurred in 1777 through the Aranjuez Treaty. With few exceptions, this delimitation still matches the current border. Before and even after this treaty, frequent territorial disputes in the borderlands resulted in military excursions organised by the central governments. At the end of the 18th century a permanent demographic imbalance emerged between the two parts of the island. The French side contained 500,000 inhabitants including 30,000 whites (mostly Creoles), while the Spanish side comprised 125,000 Hispano-Dominicans – including 15,000 slaves (Moreau de Saint-Méry 1796). This unilateral demographic pressure generated stories denouncing the invasion of Spanish territory by French colons, supposed to take possession of its land.

The border region became both the focus of military interventions and the location for contact and exchange, vital to the local population that lived through trade.

The foundation of the two cities of Ouanaminthe and Dajabón reflected this paradoxical situation. Ouanaminthe, on the French side of the river, became an independent parish in 1758. It was described by Moreau de Saint-Méry (1796) as a pleasant small town. Its relative prosperity continued after Haiti's independence in 1804. Dajabón, its Spanish counterpart, was founded sometime between 1771 and 1776 as a military outpost. In comparison to Ouanaminthe, the same author described it as a rather modest, uninteresting town. Its inhabitants were heavily involved in – then illegal – cross-border trade. By the end of the 18th century, the parish of Ouanaminthe contained 7500 inhabitants, of whom 208, according to Moreau de Saint-Méry (1796), were white, 270 freed slaves and about 7000 actual slaves. Moreover, 308 men – half white and half people of colour – were armed. Meanwhile, Dajabón parish comprised 4000 inhabitants, mostly poor people (ibid.). These were rather significant numbers at this time, indicating cross-border trade's importance for both colonies.

Haiti's independence, the brief Haitian occupation of the entire island (1822–1844) and finally Dominican independence in 1844 did not dramatically alter cross-border relations: production in each part of the island did not change significantly, and cross-border trade continued. Although much of the Dominican population welcomed the Haitians as liberators from the Spanish administration, this short period of territorial unification ironically even emphasised the Spanish colons' nationalism and the emergence of anti-Haitian resentment. These were encouraged by the elites, fostering the independence movement of the Dominican Republic (Muñoz 2003).

During American occupation of the island (1915–1924 in the Dominican Republic, 1916–1935 in Haiti), the occupying administration installed strict control of cross-border trade and collected taxes in order to reimburse the credits that the Dominican administration contracted with several European nations. This foreign takeover of binational commerce triggered resistance amongst the local elites. Especially in the island's northern part, resistance to the occupying force fostered a binational solidarity movement and an emerging borderland identity (Baud 2000). The two sets of local elites were able to reinforce binational contraband networks. Meanwhile, the Americans altered established relations between the two agricultural production systems by starting intensive exploitation of sugar-cane plantations on the Dominican side. Since the latter was sparsely settled and thus lacked a local workforce, the new plantations hired mainly Haitian labourers, initiating Haitian immigration that continues to this day (Muñoz 2003). Dominican settlers, unable to survive by cross-border trade, left the border for elsewhere on the island. Many Haitians arriving in the Dominican Republic revived anti-Haitian and racist resentments, culminating some years later under Rafael Trujillo. He came to power shortly after the American occupation ended and installed a military dictatorship from 1930.

Very systematically, Trujillo set going a policy of Dominicanising the border-lands. One early move was the 1937 massacre of Haitians living on Dominican territory (estimations vary between 15,000 and 30,000). The Dajabón region was strongly affected. This cruel and unforeseeable action created strong resentment in the Haitian population, thereby establishing strong anti-Dominican sentiment in Haiti. The Trujillo regime also militarily strengthened the border, reinforced border settlements through new infrastructure (schools, administrative buildings, churches, military posts, roads, monuments) and repopulated the region with mainly European settlers. Due to their influence, the 'Dominican' border landscape physically distinguishes itself from the appearance of the settlements on the Haitian border through its demonstrative and monumental architecture (Augelli 1980).

After the period of dictatorships on the island (Trujillo, 1930–1961 in the Dominican Republic; François and Jean-Claude Duvalier, 1957–1986 in Haiti), the border's status changed, progressively opening to economic exchange. Following the military coup against President Jean Bertrand Aristide's election in Haiti in 1991, there was a brief but major turnaround in cross-border trade, with the United Nations imposing a trade embargo on Haiti. However, even during this time, the frontier had a particular status, since contraband was one of the only ways to supply the Haitian market. The UN embargo emphasised the unequal relations between the border twin cities but also perversely facilitated trading across the border. As one of my interviewees noted, 'The embargo has impoverished Haiti, but it was also the time when the people of Dajabón began to handle money' (Interview with a Dajabón inhabitant, September 2003).

Binational trade has been an important income source for the Dominican Republic. By absorbing low-quality goods produced in the Republic and unsuitable for export to other countries, Haiti has been the Dominican Republic's second export destination after the USA. According to official sources, between 1996 and 2005 the value of these goods (mainly construction materials and processed foods) grew from USD 24,000,000 to USD 96,000,000. In the opposite direction, exports grew from virtually zero to USD 9,000,000 (CEDOPEX 2006).

For Haiti, the Dominican Republic has not been an important export destination for goods, but it is very important for the export of labour. Haitian migration to the Republic, starting under the American occupation of the island in the 1920s has been estimated in 2004 at 500,000 people working in agriculture and construction, mostly living permanently in the Dominican Republic (OIM/FLACSO 2004).

Unequal urban development

The Figure 22.1 map shows how both towns have grown in different ways. Dajabón's major growth occurred before 1990, whereas Ouanaminthe's area more than doubled between 1990 and 2004 (from approximately 90 to 230 hectares). This period coincides with the embargo on Haiti, when hardship drove migrants to the cities and borderlands, contraband being a way of assuring survival. It also

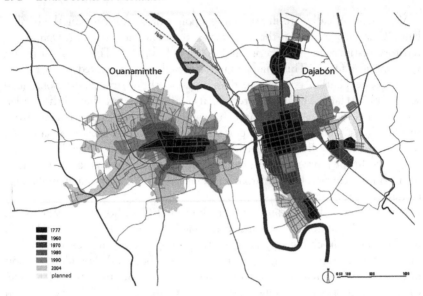

Figure 22.1 Growth of Ouanaminthe and Dajabón, 1777–2004.

Source: Elaboration by Lena Poschet El Moudden (Poschet El Moudden 2006)

reflects the countries' different economic systems: during the 1990s, the Dominican Republic extensively developed free zones close to large agglomerations, making the emerging border-region economy less attractive.

Both towns lack urban planning and even adequate application of building regulations. Thus, spatial growth has been entirely governed by landowning structure and land-occupation practices. Land around Dajabón has belonged mainly to two landowners, whose strategy of enhancing land value by developing well-equipped middle-class neighbourhoods has restricted urban sprawl. In Ouanaminthe, land has belonged mostly to small farmers. The selling and illegal occupation of quite small plots has produced sprawling urban growth of poor housing along rural footpaths, with entire areas inaccessible to motor vehicles. Furthermore, population growth has been absorbed by densification of existing neighbourhoods.

Land has become a conflict point. Over the 20-year period up to 2005, Dajabón land prices rose from very low value up to USD 40 per square metre in better locations. The town centre, where the market takes place, has been progressively transformed into a commercial area, with warehouses in former dwellings and improvised food stands open for just two market days per week. In both towns, a process of internal segregation has begun, with evermore businesses occupying the town centre and the emerging middle class gathering in specific neighbourhoods. In the eyes of Dajabón's mayor, this development was very rapid and upgraded Dajabón's status at the national level: 'Ten years ago, I never thought, that Dajabón would be called "the capital of the borderlands". This development is incredible' (Interview, July 2004).

Both towns have lacked infrastructure, but Dajabón could at least provide a minimum standard, whereas Ouanaminthe has lacked nearly all necessary services and infrastructure. In Ouanaminthe the yearly budget was about USD 30,000 in 2004, while Dajabón municipality received a yearly sum from central government equivalent to USD 360,000, and collected taxes from the binational market equal to about a third of this amount (2004 values).

Central states versus local society

The two countries have very different governments. On the Dominican side, the government, a republic with an elected president, shows the stability of a rather well-functioning liberal democracy. The Haitian side evolved very differently: although possessing a democratic regime, corruption, poverty and a very unstable social and economic situation made it very difficult, if not impossible, to form stable governments over several decades.

Nevertheless, the responses of both governments concerning the towns' unequal relationship barely registered: neither addressing the striking disbalance of the economic exchange nor the discrepancy of resources and infrastructures each side of the border. Hardly any mutual agreements exist at the central-government level, and those that do have not been operational. The Dominican government's main concern regarding the border has seemingly been around migration and maintaining its enduringly strong military presence, regulating and administering the border. Meanwhile, Haitian central government border regulation has been non-existent; in 2004, only a few international UN soldiers could be seen during daytime in Ouanaminthe. The absence of central-state support left local governments confronted with a complex situation.

> We have neither electricity, nor clean water, but we have all the problems of the world and we don't know when and how we can resolve them. The municipality operates in order to operate, that's all – we are by no means helped by the central government.
>
> (Interview with Mayor of Ouanaminthe, August 2004)

Existing cross-border relations mainly resulted from individual practices and the two sets of civil society that gradually organised cross-border networks aimed at reacting to the arbitrary administration of transit at customs, human-rights infractions, and environmental issues. Organisations like the Groupe d'Appui aux Rapatriés et Réfugiés (GARR) on the Haitian side and Solidaridad Fronteriza or Centro Puente in Dajabón have largely underpinned local cooperation efforts. These organisations maintained relations with international solidarity networks like the Convergence des Mouvements des Peuples d'Amérique (COMPA) and in 2004 played important roles in Dajabón and Ouanaminthe's local governance: they have been the only organisations having much credibility regarding cross-border relations. Meanwhile, international development agencies in the borderland have initiated projects of binational interest, but they are rather disconnected from local institutions while being strongly influenced by their direct relationships with the central governments. For example, the International Monetary Fund financially supported Ouanaminthe's free-trade zone in 2002, and the European Community financed the construction of Dajabón's covered market (opening in 2012) – both projects without involvement from the local governments.

Disconnection from central government is common in the borderlands, where governance models frequently are invented to regulate and supply services in binational regions in general. They have often anticipated more formalised treaties (Sparrow 2001). For instance, Dajabón's mayor decided of his own accord in 2003 to help Ouanaminthe's waste elimination, and Haitians can get medical care in the hospital of Dajabón.

Economy reinforces inequalities

Over 30% of Haitian-Dominican trade registered by Dominican customs passes through Ouanaminthe and Dajabón (the remaining 70% cross at one of the other three crossing points). The binational market in Dajabón has been the most important in the island: according to my interviewees, during a single market day about USD 1,000,000 of goods were being normally exchanged in 2004. The market represents an important income source for both cities' inhabitants.

The export of goods from the Dominican Republic to Haiti led some entrepreneurs to establish warehouses on both sides of the border. These entrepreneurs also introduced a transnational dimension to trading as they re-exported merchandise imported from third countries (for example, rice, imported from the USA to Haiti has been re-exported from Haiti to the Dominican Republic). Even if the population of both towns can take advantage of this growing trade, most profits have been reaped by companies located in urban centres larger than Dajabón or Ouanaminthe: notably in Cap Haitien, 64 km away in Haiti, and Santiago de Caballeros 145 km away in the Dominican Republic.

Regularly, the Dominican Republic government has expelled Haitian migrant workers to Ouanaminthe, where some then settle. Occasional revelations of illegal networks organising cross-border migrant transit in Ouanaminthe or Dajabón show how these cities have played particular roles in the system that supplies the Dominican Republic with its labour force.

The booming border economy has had contradictory influences on service development. While public infrastructure has lagged far behind what is needed, each city had four banks in 2004 that were branch offices of the two countries' most important financial institutions (before 2000, there was one branch office in each town). The major national and international telecommunications companies have been present in Dajabón. Ouanaminthe's population has frequently used the Dominican mobile phone network, also available in the Haitian border region. Despite the absence of tourist attractions (although the market regularly attracts a few foreigners), hotel numbers in both towns have been rising. This private development, initiated by economic players at the national levels or by local investors, clearly shows both cities consolidating their insertion as transit points for goods and people in each national city network. This statement by the owner of Magazine Commercial, one of the most important export firms trading in both towns, confirms this point:

> We started with a small warehouse in Ouanaminthe and today we have one big warehouse in the town centre and a small depot in front of the market. We also own 50,000 square meters of land located on the main road for further business development. We have as well bought warehouses in Dajabón. ... Our commercial relation is not only a relation between Ouanaminthe and Dajabón, but concerns the trade between the Dominican Republic, Haiti and other countries that can satisfy the Haitian market.
>
> (Interview Dajabón, August 2004)

This venture also concerns the development of the then declining seaport of Manzanillo in 2005, located about 30 km northwards.

Haitians often criticised the capitalist, even colonialist, relationship imposed by their neighbours. They disliked not only the fact that the binational market takes place in Dajabón, but also that Dominican businesses have developed subsidiaries in Ouanaminthe, taking advantage of cheap labour and the absence of local production:

> The Dominicans need us for the market, that's all. There is no possibility to live together on an equal base. But also we [Haitians] have to develop our infrastructure; in the present situation we cannot share with them – first we have to grow ... stronger and then we can start sharing.
>
> (Interview with Ouanaminthe resident, November 2003).

The binational market has seemed the only space where Haitian and Dominicans intermingle: in 2005, about 10,000 people actively participated either as vendors or buyers (Figure 22.2). However, this mingling is equal only in appearance. Haitian participation has depended on the arbitrary attitude of customs and the military's discriminatory regulation of the market making permanent installation of market stalls difficult, such as the regulation that merchandise imported from Haiti (mainly used clothing) may remain in Dajabón only for three consecutive markets. They have then had to be taken back to Haiti. It is unsurprising that Haitians have less attractive displays and lower prices than Dominican vendors. Haitian products have also been considered lower quality. A Haitian merchant described her situation thus:

> I used to go to the market in Dajabón every week, but for the moment I stopped going there as I don't find any profit there. There are assassins, and some people steal merchandises – no use to make a declaration to the police.
>
> (Interview with Ouanaminthe resident, August 2004.)

Dajabón residents interviewed, on the other hand, have not needed to cross the river to make a living. Half have never been to Ouanaminthe but consider it insalubrious, dangerous and out of control: 'I am scared over there, I have

Figure 22.2 Massacre River on a market day, 2003.

Source: Photo by Lena Poschet El Moudden.

been one or two times to Ouanaminthe, I know the park and the church in the centre of the city , I stayed one hour, not more' (Interview with a Dajabón resident, August 2004). Simply crossing the river has been considered dangerous, the general perception being that only people involved in illegal activities or with low morals (e.g. seeing prostitutes, participating in voodoo rituals) visit Ouanaminthe.

An interesting example of the ambiguous inter-city relationship is the free zone installed in 2002. It is located on Haitian territory, but on the Dominican side of the Massacre River. Consequently, controlling this space has been easy for the Dominican military, which has frequently violated the Haitian border, notably to intervene in Haitian workers' protests for better working conditions.

On the one hand, the zone's presence created new tensions between the two populations, as the workforce is exclusively Haitian but the supervisors and technical personnel are Dominican. On the other hand, the fact that this zone has been imposed by the central governments, without consulting the local population and governments, produced a common struggle against the military mobilizing civil society in both towns. This mobilization has continued with some success in the battle for workers' rights. Civil society commitment in the two cities has also rallied international networks and labour organizations.

> The story of the free zone is somehow confusing and surrounded by secrets. … The consultation of the local population has been done in an arbitrary way. Thus, when the government decided its construction in 2002,

everyone hurried to denunciate this on local, national and international level … But the government pursued its enterprise, destroying the culti-vated area without compensating the farmers. And now the factory is run-ning, the wages are much too low: 432 gourds (about 11 USD) weekly, in difficult working conditions.

(Interview with the coordinator of the *Comité de défense Pitobé*, Ouanaminthe, October 2003)

Perceptions of the free zone nevertheless differ generally according to which side of the border one is on. In Dajabón, economic players and some of the popula-tion could see an opportunity to develop business and attract higher-income res-idents, but in Ouanaminthe it has been perceived as a problem, overburdening existing infrastructure with additional, mostly unskilled migrants.

They say the free zone will create employment, but on our side [Haiti] the government has taken no initiative to help us cope with this situation. Peo-ple come from all over the country, Port au Prince, Jacmel, Gonaives, and they do not necessarily find work here.

(Interview with Ouanaminthe's Mayor, October 2003)

Two opposite cities without binational space

Interest in developing further border permeability must clearly be located at local level. For the central governments – especially in the Dominican Republic – the border has remained a source of conflict, principally due to illegal immigration. Furthermore, the lack of transparency and impartiality in customs administra-tion profited a minority of influential entrepreneurs allied to central adminis-trations and governments. Small local traders did not benefit, and probably still have not benefitted, from such favouritism.

Local governments and local populations have been confronted with prob-lems going beyond borderlines, like discrimination and labour rights conflicts, population growth and environmental deterioration. The aims of active local civil society involved with cross-border relations has reflected these concerns. In both towns there has been awareness that increasing trade could only con-tribute to their positive development if the disequilibrium in the established relations was remedied. But there have nevertheless been divergences about how to achieve a more equal relation and who must take action.

The major common problem for local stakeholders has been that, for the cen-tral governments and transnational companies, the partial permeability of the border, restricted to economic exchange, remained the best option. Important issues, like customs regulations, military intervention, installation of transna-tional companies and, last but not least, most illegal cross-border activities, have not been within border-population control.

The ambivalence the local population has felt about the border fostered the aspiration of actors in both towns to develop activities independent of the border economy by revitalizing links with the rural environment, persuading the two towns to take their role as regional centres more seriously. Private initiatives in Ouanaminthe and Dajabón reinvest in the development of agriculture and in educational programs, e.g. for professional qualification.

Dajabón's dominance can be attributed to a more stable national environment and superior infrastructure, giving the Dominican town the leading role in border relations from the start. Opening the border enhanced this unequal relationship, as good infrastructures play primary roles in the economy. Ouanaminthe could become the poor neighbourhood of Dajabón.

However, as of 2006, the border's partial permeability and the instability of the cross-border relations was not fostering any converging development between the towns. Their role as complementary interfaces was not creating common urban space characterised by common social and spatial organisation that would give to the local population the sense of belonging together. The only space where both populations mingle is the binational market – under severe control of the Dominican police and military and limited to the duration of the market. Otherwise, cross-border social relations remain mainly confined to the actions of civil-society organisations or local leaders – like the mayor of Dajabón in 2003 – whose awareness that the problems of Ouanaminthe are also the problems of Dajabón made him help the municipality of Ouanaminthe with waste collection.

At the end of my research, I left the island with the impression of two opposite border towns, twins by destiny, but unable to construct through their own initiative a solid relationship enhancing stability and equal exchange. Resources and support from the two national governments were lacking, and most of the wealth generated by the binational exchange disappeared into the pockets of a minority.

Epilogue

As the uncertain tense in the foregoing pages indicates, although my research ended in 2004, little has probably changed since then. There has been development but mainly benefitting the Dominican side. Since 2006, the year I finished my thesis (Poschet El Moudden 2006), the situation seems to have evolved towards greater inequality. In 2019 exports from the Dominican Republic to Haiti were 522,000,000 USD – over five times higher than in 2004, while Haiti's exports to the Dominican Republic declined dramatically to USD 700,000 (ONE 2020).

With international help, the infrastructure of both towns has been improved and now supports the logistics for economic exchange in a more national and international perspective. In 2010 the reconstruction of the road (financed by the EU) connecting Ouanaminthe to Cap Haitian has been finished so that heavy trucks no longer get stuck in the mud when it rains. In quest of improving this route, a new bridge links the towns. On the Dominican side, in front of the new bridge, a covered market opened in 2012, offering space to Dominican and Haitian vendors. The traditional market still stands and operates in the streets of Dajabón. According to the web-based journal *Haiti Libre* (11 March 2019), the market now generates about 1,200,000 to 1,500,000 USD per week through the exchange of merchandise. The same source suggests it is now the arbitrary behaviour of Haitian customs that makes it difficult for Haitians to make their living from the market.

In 2021 information on Ouanaminthe and Dajabón can easily be found on the internet. Along with the development of the road and the new market, private infrastructure of both towns also evolved. In Ouanaminthe three new, quite luxurious hotels opened recently and in Dajabón I have seen, one new hotel residence. Nevertheless, most information I could discover relates to discrimination against Haitians, arbitrary treatment and injustice.

In the light of what I could gather, it seems that the path for Ouanaminthe and Dajabón to become a pair of more equal twins remains a long one.

References

Augelli, J. P. (1980). Nationalization of Dominican Borderlands. *Geographical Review*, 70(1), 19–35.

Baud, M. (2000). State-Building and Borderlands. *Cedla Latin America Studies*, 87 (Towards a Borderless Latin America), 41–79.

CEDOPEX. (2006). *Reportes Estatisticos: CEDOPEX*.

De Jesus Cedano, S. & Dilla, H. (2005). *De problemas y oportunidades: intermediación urbana fronteriza en República Dominicana*. Revista Mexicana de Sociología, no. 201.

Muñoz, M. E. (2003). *Apuntes para una Interpretacion Historica de las Relaciones Domenico-Haitianas. Paper presented at the La Frontera: Prioridad en la Agenda Nacional del Siglo XXI*, Santo Domingo.

OIM/FLACSO (2004). *Encuesta sobre inmigrantes haitianos en la Republica Dominicana.* Santo Domingo: FLACSO/OIM.

ONE (2020) Sector externo, Available: https://www.one.gob.do/economicas/sector-externo, Accessed 01.03.2021

Poschet El Moudden, L. (2006) *Villes à la Frontière et transformation de l'espace, le cas de Haïti et la République Dominicaine. Thèse 3655 (2006)*. Lausanne: Ecole Polytechnique Fédérale de Lausanne.

Saint-Méry, L. E. M. d. (1796). *Description topographique et politique de la partie espagnole de l'isle de Saint Domingue, Tome premier* Philadelphie.

Silie, R., & Segura, C. ed. (2002). *Hacia una nueva visión de la frontera y de las relaciones fronterizas*, FLACSO, Santo Domingo.

Théodat, J-M, (1989). Haïti – Quisqueya: une double insularité. *Mappemonde 51* (1998.3)

Théodat, J-M, (2003). *Haïti – République Dominicaine. Une île pour deux 1804–1916.* Karthala, Paris.

Part III

Twin Cities in Fiction and Editors' Dreams

23 Twin cities in China Miéville's fiction

Cathrine Olea Johansen, Ruben Moi, Ekaterina Mikhailova and John Garrard

Real-world twin cities keep multiplying. They have also existed for several decades in art, notably literature and film. Artistic representations of twin cities vary from both being fictional (like Gotham City and Metropolis from DC Comics – home cities respectively of Batman and Superman) to mixed fictional/real twins (like 'London Above' and 'London Below' in Gaiman's *Neverwhere*) to real twin cities brought into the fictional world (like Copenhagen and Malmö in the TV crime series *The Bridge*).

In the DC comics, Gotham and Metropolis are divided by Delaware Bay, but their distance changes from one comics issue to another – sometimes adjacent, sometimes far apart. Brayson (2016) suggests the two cities represent New York City: Metropolis is New York City by day, and Gotham New York City by night. In Snyder's 2016 film *Batman v Superman: Dawn of Justice*, Metropolis and Gotham are indeed portrayed as adjacent twins, separated by Delaware Bay. The title explicitly refers to the border between night and day and between the two superheroes.

'London Above' and 'London Below' in *Neverwhere* – a fantasy television series and its companion novelisation (1996) by Neil Gaiman – are twin cities, with the former being familiar London and the latter its magical double in and beneath the sewers. London Below is replete with dangers and fragmented into different baronies that occasionally go to war. Its inhabitants include characters from different historical epochs (e.g. Roman legionaries, Black friars and medieval Earls) and with varying degrees of supernatural power (from homeless and rats with few supernatural abilities to vampires and wizards). London Below is a place 'where history diversifies and comes alive' contrasting with 'ossifying and capitalist-driven' London Above (Elber-Aviram 2013). Familiar London toponyms acquire new meaning in London Below: Angel, Islington from a locality, becomes a character – an actual angel; the Knightsbridge becomes a Night's Bridge which must be traversed in complete darkness and confronting whatever a bridge crosser personally fears the most.

The TV crime series *The Bridge* (2011–2018), set in Malmö and Copenhagen, portrays real-life Swedish-Danish twins classified as 'engineered twins' by Garrard and Mikhailova (2019, 9). The series focuses on twin cities' darker side – murder and mystery – raising questions on how twin cities (both cross-national, and even sometimes internal, twins, and their border areas) cope with

DOI: 10.4324/9781003102526-30

problematic cross-border criminal and other issues that can upset more amicable commercial cooperation and cultural exchange stimulated, indeed engineered, by the opening of the Øresund Bridge in 2000.

This chapter focuses primarily on yet another example of fictional twin cities – those created by China Miéville in his *The City & the City* – a unique crime-fiction novel, filled with surprise, wonder and suspense. In its genre-breaking novelty and hermeneutic ambiguity, *The City & the City* offers insight into the concept, history and current ideas around borders and twin cities. By examining this novel, we attempt to showcase how studying fictional borders can contribute to understanding real-world twins by elucidating and 'interrogat(ing)' how certain types of borders or border practices remain visible, or legitimate, or acceptable' (Rosello and Wolfe 2017, 6); and are invariably interesting, as the chapters in this volume have successfully established.

***

China Miéville is a contemporary British fantasy-fiction writer. His works range from essays and comic books to novels often described as part of *weird* fiction. This genre frequently develops from supernatural stories within the gothic tradition, mixing horror, tension and science-fiction tropes with an otherwise ordinary world. Here is Miéville describing the genre:

> I don't think you can distinguish science fiction, fantasy and horror with any rigor, as the writers around the magazine *Weird Tales* early in the last century … illustrated most sharply. So I use the term 'weird fiction' for all fantastic literature – fantasy, SF, horror and all the stuff that won't fit neatly into slots.
>
> (*The Guardian* 2002)

The City & the City is a crime-fiction novel published in 2009 and a prominent example of border-conscious literature where twin-city relationships are imagined in enigmatic ways. Much of the novel's unnerving element rests upon the frequent haziness of real-world borders: where exactly they are, how sustained and how policed, and how the 'other' side can be 'unseen' while those so doing remain acutely aware of its existence.

Readers of *The City & the City* follow detective Tyador Borlú's search for a young woman's killer in Besźel and Ul Qoma, the two city states that occupy the same physical space or, using the novel's fantasy terminology, are 'grosstopical'. The investigation leads Borlú across the border from Besźel into Ul Qoma, to attempt cooperation with the latter's police, showing readers the cities' weird relationship – the source of the novel's mystery and tension. Different groups emerge as the crime investigation progresses: neo-nationalist groups and politicians (exploited by the ultimate villains of the story for their own purposes), and city-unification activists. There is a possible third city (Orciny) which, while ultimately proving mythical, nevertheless is central to the story, plus an American hi-tech company.

Miéville's novel 'reifies our everyday practices of ignoring certain things around us, using a science fictional novum: the institutionalised practice of "unseeing'" (Schimanski 2016, 106). 'Unseeing' is central to the novel since seeing, hearing or sensing the opposite city is illegal. The practice of unseeing depends on citizens' abilities to follow the rules and stick to their side of the border. Twin-citizens, from childhood, are trained and told the various social and symbolic differences between Besźel and Ul Qoma, enhancing the reality of there being two cities, not one: 'The early years of a Besź (and presumably an Ul Qoman) child are intense learnings of cues. We pick up styles of clothing, permissible colours, ways of walking and holding oneself, very fast' (Miéville 2009, 80). Tourists and researchers must undergo courses and tests before entering either city, as must travellers from one city to the other who have to train themselves 'to unsee what they were previously allowed to see' (Schimanski 2016, 110). Here, things get complicated. A concrete border like a fence or a big wall would have been accepted and comprehended easier. However, the two cities' borderline is invisible and incorporeal. It is a key element, building on the idea of the two cities 'turning their backs to each other' – a conscious act of choosing to ignore 'the other side', while remaining intensely aware of it in the frequent manner of real-life twins: both internal and external.

The inter-city space and the cities' invisible border are guarded by 'Breach' – a mysterious and powerful policing squad that catches you if you 'breach' between the cities, a reference to Kafka's *The Trial* (ibid, 108). Breach might be seen as a neutral zone, vitally important for maintaining inter-city balance and relationship: 'The powers of the Breach are ... highly circumstantially specific. The insistence that those circumstances be rigorously policed is a necessary precaution for the cities' (Miéville 2009, 83). Without Breach, the current inter-city situation and relationship could disintegrate. Breach then represents the conditions upon which the twin cities depend and the fifth characteristics of real-life twin cities – persistence, even though the case for their continued separation, especially for internal twins, may seem mysterious to outsiders. Without Breach, one city could take over the other, or they could eventually merge, or more likely simply go to war.

The invisible but evident border between Besźel and Ul Qoma suggests hybridity, a potential for subversion and co-dependence. Without Breach, the possibility for a twin-city coalition exists. Such change could be beneficial for either city, but unification without borders could also cause more change than either city could sustain. Besźel and Ul Qoma's implicit agreement to uphold the current relationship illustrates the stability of their co-existence.

Yet, however extensive, the powers of Breach are ultimately limited to preventing crossing from one city to the other by any other route other than the single authorized crossing point – Copula Hall. Much of the early plot hinges upon how the story's villains manage to evade discovery by transporting the murder victim from Ul Qoma, where the murder occurs, into a dumping place in Besźel through this crossing point in a legally authorized vehicle, leaving Breach unable to touch them. Similarly, the ultimate villain nearly escapes both Breach and the two police forces by walking along the border zone and out of both cities without crossing into one or the other and breaching.

As implied, the only legal border checkpoint between Miéville's cities is the formidably large and impressively built ('like a coliseum') Copula Hall, which functions as the only portal where inter-city crossing is possible without fear of capture by Breach. Copula Hall is one of a very few places with the same name in both cities: 'because it is not a crosshatched building, precisely, nor one of staccato totality-alterity, one floor or room in Besźel and the next in Ul Qoma: externally it is in both cities; internally, much of it is in both or neither' (Miéville 2009, 72).

As befits its size and architecture, this is a multi-functioned institution: not just a crossing point but also customs control, and the formal site for inter-city negotiations via the Oversight Committee drawn from delegates/politicians from both cities. It also provides the site and staffing for the extended mental adjustment process (unseeing the city one has left and seeing one's destination) that citizens crossing for any extended period, like Tyador Borlú, must undergo before emerging into the neighbouring city.

***

The narration begins in the middle of a scene, giving readers no chance to catch up with Miéville's world. Gradually, they gather the cities are situated in post-communist Central/Eastern Europe. Schimanski (2016, 108) describes the twins as 'caught between globalisation and neo-nationalism, each with violent aspects in the novel'.

Nationalism appears in various forms. Readers learn about Balkan Muslim refugees settling in Besźel's previously Jewish ghettos and about Jewish and Muslim minorities that Besźel formerly had. The two minorities were said to get along quite well, having 'a centuries-old humorous dialogue ... about the intemperance of the Besźel Orthodox Church' (Miéville 2009, 22). Later, readers learn about Besźel's little Ul Qoma-town – a 'small community of Ul Qoman expatriates' who could be bullied by locals or even beaten in the streets (ibid., 53), but also where one could nervously sample a little of the other city's cuisine and even culture. The Ul Qoma-town residents had migrated for different reasons, one being political prosecution, a hint on Ul Qoma's totalitarian past.

At one point, the novel mentions Besźel, perhaps even Ul Qoma, being created from ruins (one factor stimulating interest in their archaeology and thus the site in Ul Qoma around which quite a lot of the plot revolves). The novel implies that the cities existed as one until the frequently mentioned 'Cleavage'. The novel contains suggestions that inter-city relations have improved in recent years, thus aligning it with the experience of many twin cities in our current world.

Due to sharing the same spatio-temporal location, Besźel and Ul Qoma share weather conditions, as a Besź proverb goes 'Rain and woodsmoke live in both cities'. Popular alternatives of this proverb feature fog, rubbish, sewage, pigeons and wolves (Miéville 2009, 54).

Their spatio-temporal identity notwithstanding, Besźel and Ul Qoma are quite distinguishable, having different but interdependent structures. Both have their own governance, police forces, regulations, culture, food and language written using different alphabets. Both treat the invisible border with respect and fear, with the two police forces maintaining vigilance towards crime within each city and cross-border. Other separating features are currency and economics, fashion and outlook, religion and language. Schimanski (2016, 108) describes the languages as faux-Hungarian/Slavic/German in Besźel and faux-Arabic/Urdu in Ul Qoma.

The consequences of separation are highly problematic for both sets of citizens:

> If someone needed to go to a house physically next-door ... but in the neighbouring city, it was in a different road in an unfriendly power ... foreigners rarely understand. A Besź dweller cannot walk a few paces next door ... without breach. But pass through Copula Hall and she or he might leave Besźel, and at the end of the hall come back to exactly (corporeally) where they had just been, but in another country, a tourist, a marveling visitor, to a street that shared the latitude-longitude of their own address, a street they had never visited before, whose architecture they had always unseen ... a whole city away from their own building, unvisible now they had come through, all the way across the Breach, back home. Copula Hall (is) like the waist of an hourglass
>
> (Miéville 2009, 86).

***

The emergence of any novel today, to properly be a novel, also relates to the borders and theories of its own medium, literature. *The City & the City*, as the title's doubleness indicates, places it on the border between literature and the real world, and on the margins of literary theory and literature. Fiction often locates ambiguously on the many borders of arts and society; Miéville's novel certainly does.

A border poetics reading uses border planes as a conceptualising tool. Border planes often are layered and 'juxtaposed upon one another, enabling possible 'allegorical transfers of meaning from one to the other' (Border Poetics 2020a). Using the border-poetics reading of *The City & the City*, one could identify several different border planes.

The topographical plane typically comprises everything from physical borders between nations to the human body's own boundaries. In Miéville's novel, the topographical plane with Breach, the liminal border zone at its core, is particularly important because of how close the cities are, but also how very separated. It is difficult to imagine two cities so close, yet so actively working to maintain separation, yet again so intensely mutually aware. In both the novel and with real-world twins, borders and borderlands are often crucial to their relationships.

The novel's temporal plane plays out the borders and representation of time – a plane with a central role, not just because it inevitably refers to the order of events, but also because the temporal border is represented within the novel's different parts: 'Every border-crossing is a temporal border, dividing time into a period before the crossing and a period after the crossing' (Border Poetics 2020c). The crossing between the novel's first two parts also entails the crossing from Besźel to Ul Qoma, showing how temporal and topographical border planes mutually interact. In *The City & the City*, the hero's investigation forces him to cross the border from one twin to the other. The fact that he can and must drag his investigation across borders illustrates how the cities (as in the real external twin-city world) depend on cross-border interaction, indeed cooperation, even though Borlú crosses into Ul Qoma with significant strain and difficulty:

> They sat me in what they called an Ul Qoma simulator, a booth with screens for inside walls, on which they projected images and videos of Besźel with the Besź buildings highlighted and their Ul Qoman neighbours minimised with lighting and focus. Over long seconds, again and again, they would reverse the visual stress, so that for the same vista Besźel would recede and Ul Qoma shine.
>
> (Miéville 2009, 160)

Unseeing his own city is difficult. Inspector Borlú must not only unsee all the familiar places; he must for the first time start seeing a city he has spent most of his life mostly unable to see, or illicitly glimpsing. The days he spent learning how to unsee Besźel illustrates how the temporal and topographical border planes are so intertwined. The complex combination of the tempo-topographical border plane enables readers to understand the difficulty and complexity of this cross-border interaction. Although the transformation process from Besź citizen to Ul Qoman visitor may not fully resemble any real twin city, it may be a metaphor for political and economic refugees crossing a border, trying to forget what they left behind.

The novel works thoroughly on the epistemological border plane of knowledge about how blocking basic human senses automatically creates difference between knowing and unknowing (Border Poetics 2020b). From Besźel, Ul Qoma is perceived as unknown, and vice versa. Each city's citizens can cross the inter-city border at Copula Hall and become aware of both the cross-hatched border space itself and the other city. Yet, the twins are supposedly unknown to each other, suggesting knowledge is limited and not ultimately available to all. An important distinction here is that knowledge of what lies on the other side is not unattainable to citizens, but it is *limited to one city at a time*. Unknown elements across borders exist everywhere, also between real border twins. We do not have access to every private home, still less to the other city. Similarly, with novels, readers' knowledge of location and characters is limited to the unfolding plots, narrative constructions and viewpoints. They must accept piecemeal and limited information; maybe entailing excruciating cliff-hangers at the end.

Breach itself functions as the epistemological border: 'the border between the known and the unknown' (Border Poetics 2020b). The invisible border, wherein Breach resides, contains the ultimate truth between the twin cities, functioning as someone all-knowing and all-seeing, even though, crucially, not all-powerful. The difference between Besźel, Ul Qoma and Breach becomes very clear because of the different figurations of the novel's epistemological borders, and because both cities are greatly inferior to Breach in knowledge.

The symbolic border plane introduces the binary opposition between Besźel and Ul Qoma. Although similar and interdependent (just like many real-life internal and external twins), the two cities function as binary oppositions because being in one city automatically excludes the other. Twin-city citizens are taught from childhood to unsee and unhear each other. The symbolic border plane interconnects epistemological, temporal and topographical border planes, demonstrating how closely they interlink. It also shows how Miéville's cities may be perceived and compared to real-life twin cities.

***

The City & the City raises numerous theoretical and principal questions around the twin-city phenomenon. Its plot and themes place themselves upon the many borders of twin cities that appear similar yet different, here yet there. Thereby, the novel joins some twin-city scholarship in questioning the accuracy of the 'twin' metaphor (Arreola 1996). It poses the questions: What are the borders? Where? How constructed, maintained and surveilled? By whom? Exploring these questions is highly important as international borders vary intensely from hazily defined and little surveilled to closely drawn and tightly policed. Equally, the 'hardness' of internal twin city borders have varied greatly over time, generally becoming steadily more permeable in recent decades due to ongoing conurbanisation/metropolitanisation, even while the cities themselves determinedly persist, true to their twin-city breed – like the age-old twin-city relationship of Manchester and Salford, now engulfed into Greater Manchester. There, as in many internal and external twin cities until recent decades, unseeing has been central to their existences: for example, Salford citizens remembering their childhood in the 1960s when Manchester 'was a foreign place you did not go into' (Garrard and Kidd 2019, 49).

Miéville's novel reflects and refracts many real-world, border-fraught communities. On the one hand, Besźel and Ul Qoma draw a comparison with Minneapolis and St. Paul in that they have long seen themselves as twins, sharing a common border. But they have a long history of intense mutual antagonism and, although relations in recent decades have become far more peaceful and cooperative, they have never been, and seem unlikely to become, identical or formally merged. The same seems true of several other internal twins featuring in this volume and in Garrard and Mikhailova (2019). Despite close proximity, both Besźel/Ul Qoma and Minneapolis/St. Paul keep their individual identities and characteristics. Recent years have seen Minneapolis/St. Paul increasingly recognising mutual interdependence, with considerable inter-city negotiation and over-arching governmental structures spanning the broader twin-city metropolitan region (Garrard and Mikhailova 2019, chapter 1). Yet, the multiple layers of history, culture, language and religion of Besźel and Ul Qoma resemble Narva and Ivangorod, external twin cities on the Estonian-Russian border with a five-century story of numerous divisions and unifications (Lundén 2019).

The reference to the common root, the subsequent Cleavage and gradually improving Besźel/Ul Qoma relationships alludes to the integration-disintegration-reintegration phases of the Hungarian-Slovak twin cities on the Danube described in chapter 12 of Garrard and Mikhailova (2019). There, Esztergom-Štrurovo and Komarom-Komarno were forced into separate existences from 1920 to 1990 when their residents mostly experienced economic degradation and depopulation, their residents witnessing deportations and forced migrations.

Besź Ul Qoma-town with 'cramped kitsch' buildings' design and 'mongrel' street names described by the novel's hero as 'a provocation ... at the universe in some way' (Miéville 2009, 53) reminds of Russia-themed parks, squares, museums and restaurants in Manzhouli where buildings, statues and inscriptions combine both traditional Russian and distinct Chinese features (Chapter 14 here). This 'squat self-parody' (Miéville 2009, 53) created by Ul Qoma expatriates in the novel and Chinese entrepreneurs in Manzhouli partly serves the same purpose of entertaining visitors and providing them with an unusual cultural experience within their comfortingly habitual area without the need for stressful border crossing.

The invisible border separating Besźel and Ul Qoma may resemble borders in international twin cities located within a supranational union – e.g. twin cities on EU member states' borders like Happaranda-Tornio on the Swedish-Finnish border (Garrard and Mikhailova 2019, chapter 16) or 'eurocities' on the Spanish-Portuguese border (Chapter 8 here). In some eurocities, crossing the border takes no longer than crossing the street. However, in some adjacent border twins where there is no such supranational arrangement, crossing the invisible border wherever you like is considered illegal – as in Miéville's novel. Crossing legally takes longer, being only possible through defined border checkpoints. A similar situation existed in 1991–2007 in Valga and Valka, contiguous twin cities on the Estonian-Latvian border, – from the USSR's collapse until the Schengen regime's implementation. It took authorities some time to agree upon exceptional border-crossing procedures for those living at Savienība, a street bisected by a border (Lundén 2019, 242). Another real-world example is Ambos Nogales, US-Mexican twin cities on the border that is heavily guarded despite being extensively crossed (Chapter 21 here).

As in the novel, policing across internal and external twin-city borders has long been particularly problematic, at least until recent decades. Manchester and Salford showed this vibrantly. Parliament had united them in 1797, but the two sets of Police Commissioners rapidly began 'unseeing' each other by meeting at opposite ends of their building, staying that way until Parliament recognised the inevitable and separated them in 1830. Cross-border crime notwithstanding, they remained separate until the emergence first of the Manchester and Salford Police in 1968 and the Greater Manchester Police in 1974 (Garrard and Kidd 2019). So even more did Minneapolis/ St Paul where there were two quite different policing regimes, and attitudes towards certain sorts of criminality (Garrard and Mikhailova 2019, Chapter 1). In external twin cities typical crimes are drug trafficking as in US-Mexican twin cities, smuggling as in Myawaddy-Mae Sot (see ibid., Chapters 10 and 21) and document forgery and money laundering as in Ciudad del East and Foz do Iguaçu (Chapter 19 here).

Fiction frequently attends to the dark, hidden and forbidden, as Miéville's novel often does. In accenting murder, difference and surveillance more than friendship, similarity and open society, the novel may point towards initiating points for twin cities in post-conflict environments like Europe's following the Second World War: as in Gubin/Guben (Garrard and Mikhailova 2019, chapter 12). The reference to the transition from 'a totalitarian communist past' to 'a form of limited turbocapitalism' in Ul Qoma (Schimanski 2016, 108) echoes twin cities pertaining to the two confronting ideological blocks after the Cold War such as Gorizia-Nova Gorica on the Italy-Slovenia border, Bad Radkersburg-Gornja Radgona on the Austria-Slovenia border and Česke Velenice-Gmund on the Czechia-Austria border (Garrard and Mikhailova 2019, chapter 14). In both sets of examples town twinning united former enemies and fostered peace and reconciliation. However, as Arieli (Chapter 10 here) suggests with the case study from the Jordan-Israeli border, twin-city initiatives in post-conflict environments are fragile, often having to keep a low profile to avoid risky publicity.

In view of the murder investigation, and the (in)visibility, surveillance and state organisation in a place where two cities appear contemporary and coterminous while unseeing each other, the novel reflects several aspects of divided cities – like Vukovar (see Chapter 9 here) and Belfast in the decades following the 1968 Troubles.

The novel also resonates with realities of twin cities on extremely 'hard' international borders like Nikel and Kirkenes on the Soviet Union-Norway border, or Tallinn and Helsinki on the Soviet Estonia-Finland border, with the Iron Curtain in between. There, 'unseeing' each other was less about inwardness as a cornerstone of twin citizens' local identity, and much more about the state's demand for loyalty and its severe punishment for 'breaching', be it tuning into Radio Liberty or foreign TV from across the border.

The fact that Besźel and Ul Qoma are 'spatially intermeshed to an extreme degree' (Schimanski 2016, 110) reminds of Baarle-Nasau and Baarle-Hertog – the complex urban area of enclaves and counter-enclaves on the Dutch-Belgian border. Billé (2017) describes these cities as 'an intricate jigsaw puzzle' where 'an individual may cross the border several times on her way to the corner store'.

Most fictional twins in the film and literature examples discussed in this chapter – in *Superman* and *Batman*, *Neverwhere* and *The City & the City* – are mutually antagonistic just like many internal twin cities, vividly exemplifying the interdependence and unequal relationship typical of real-world twin cities. Gotham and Metropolis, London Above and London Below and Besźel and Ul Qoma are all like Manchester and Salford until recently: 'too near neighbours to be good friends' (Garrard and Kidd 2019). Fictional twins invite speculation upon values and life questions uniting and separating two parallel places, and concerning the immaterial, constructed and subjective often escaping historical archives, public documents and social statistics, particularly in divided communities and border areas where common ground for dialogue is non-existent, disputed or complex.

Writers and directors, as seen in *The Bridge*, *Superman* and *Batman*, not only imaginatively present twin-city and border-area complexities, but they frequently draw upon such conditions for developing their fiction and broadening their arts, widening how writers can develop fiction as an imaginative art. It is exactly the creative and fanciful aspect of art that creates openings for disclosing dark pasts, revealing current malfunctions, imagining future possibilities and asking new questions of documentary and oral primary sources. Frequently, such imaginings, weird and wry as much as brilliant and bright, retain wider fields of interest the less connected their settings and plots to identifiable locations.

Overall, we might think there are two questions worth answering about what Miéville's fiction can tell us about real-world twins. First, does the novel reveal real insights into what primary-source material has shown internal and external twins are like? Here, the answer is fairly clearly affirmative. As shown in this chapter, Miéville's twins resemble two specific subgroups of twin cities – internal twins until a few decades ago, rather than now, and external twins straddling hard rather than permeable international borders.

Second, does the novel provide insights into real twin cities that might not otherwise occur to us? Here, the answer is more problematic. Certainly, Miéville probably has much to say imaginatively about what it may feel like being an honestly intended middle-level police officer trying to resolve a serious crime straddling twin-city borders, especially external twin cities where the border is hard. And this is central to his novel. Against this, there are two real-world features missing from the novel due to Besźel and Ul Qoma being imagined as discrete city states without external hinterlands and influences coming from afar – like state and central governments. One force is conurbanisation. As many chapters in this volume have shown, conurbanisation has greatly conditioned the experience and self-identity of both internal and external twin cities over many recent decades, their undoubted persistence and continued emergence notwithstanding. The other force is ongoing city's empowerment within larger governmental and regulatory frameworks producing increased multi-layered territorial governance with blooming formal and informal negotiations.

References

Arreola, D. (1996) Border-City Idée Fixe. *Geographical Review*, 86 (3), 356–369.

Billé, F. (2017). Sectional. Theorizing the Contemporary, *Fieldsights*, October 24. Available: https://culanth.org/fieldsights/sectional (accessed 15.02.2021)

Border Poetics (2020a) *Border Planes*. Available: http://borderpoetics.wikidot.com/border-planes (accessed 22.03.2020)

Border Poetics (2020b) *Epistemological Border*. Available: http://borderpoetics.wikidot.com/epistemological-border (accessed 22.03.2020)

Border Poetics (2020c) *Temporal Border*. Available: http://borderpoetics.wikidot.com/temporal-border (accessed 22.03.2020)

Brayson, J. (2016). Have Metropolis & Gotham Always Been Neighbours? Available: https://www.bustle.com/articles/150553-metropolis-gotham-are-twin-cities-in-batman-v-superman-but-has-that-always-been-the (accessed 22.07.2020)

Elber-Aviram, H. (2013). 'The Past Is Below Us': Urban Fantasy, Urban Archaeology, and the Recovery of Suppressed History. *Papers from the Institute of Archaeology*, 23(1).

Garrard, J., and Kidd, A. (2019) 'Too Near Neighbours to Be Good Friends': Manchester and Salford, in Garrard, J. and Mikhailova, E. *Twin Cities: Urban Communities, Borders and Relationships over Time*, Abingdon: Routledge, 37–50.

Garrard, J. and Mikhailova E. (eds.) (2019) *Twin Cities. Urban Communities, Borders and Relationships over Time*. New York and Abingdon: Routledge.

Lundén, T. (2019) Border Twin Cities in the Baltic Area – Anomalies or Nexuses of Mutual Benefit? in: Garrard, J., and Mikhailova, E. eds., *Twin Cities: Urban Communities, Borders and Relationships over Time*, Abingdon: Routledge, 232–245.

Miéville, C. (2009) *The City & the City*. London: Pan Books.

Rosello, M. and Wolfe, S.F. (2017) Introduction, in Schimanski, J. and Wolfe, S. F. eds., *Border Aesthetics – Concepts and Intersections*, New York: Berghahn Books, 1–24.

Schimanski, J. (2016) Seeing Disorientation: China Miéville's *The City & the City*. *Culture, Theory and Critique*, 57(1), 106–120.

The Guardian (2002, 16 May) China Miéville's Top Ten Weird Fiction Books. Available: https://www.theguardian.com/books/2002/may/16/fiction.bestbooks (accessed 15.08.2020)

24 Conclusion

The twins that got away

John Garrard

In this volume and its predecessor *Twin Cities: Urban Communities, Borders and Relationships over Time* (Garrard and Mikhailova 2019), we have become acquainted with well over 120 twin cities of all sorts and sizes across all continents and straddling many internal and external borders. We hope we have established their distinctive characteristics, and above all their considerable interest for anyone interested in the modern urban and international world.

Nevertheless, we are aware of significant numbers of twins, both internal and external, who would have warranted their own chapters, including one *type* of internal twin not featured in either volume: cities nesting inside cities. However, they all 'got away', our best efforts to find authors notwithstanding. I will try to review some of them now. Here, even though I have used a number of sources in composing this conclusion, I must acknowledge considerable indebtedness to Wikipedia. Its entries are of uneven quality. However, at its frequent, well-evidenced and referenced best, Wikipedia can provide excellent jumping-off points for deeper reading and serious research – as it has done here, and as it did when I starting thinking about twin cities as a breed in what turned out to be the lead-up to the Manchester Metropolitan University Conference where the two editors met.

Before casting a wistful eye at twins that escaped, we should admit that, while twin cities are resilient, some have disappeared. We would have liked a chapter on those that succumbed, and tried to formulate reasons/variables why some died when the vast majority have not – if only to throw light on the many survivors.

One unsurprising reason for twin-city demise is the passage of extended time. Thus, most of Anna Anisimova's mediaeval English twin towns (Chapter 2) have now merged, and/or become part of wider entities, though, as she notes, they took a long time over it. Weymouth and Melcombe Regis became a double borough called Weymouth in 1571, though they ceased being separate civil parishes only in 1920, suggesting, as with Brighton and Hove covered here, that there can be life after death. East and West Looe continued as separate parliamentary boroughs (albeit decayed and 'rotten'), until the UK's 1832 'Great' Franchise Reform Act'. Even now, they continue as separate electoral districts of Cornwall Council (originally Cornwall County Council). And Portsmouth and Gosport continue as separate and thriving twinned entities with independent metropolitan and non-metropolitan councils respectively. Meanwhile, most of

DOI: 10.4324/9781003102526-31

Chapter 7's Prussian/German cross-border twins have also disappeared even as internal twins; though we note the author's point that, surprisingly given the amount of change and territorial expansion involved, the process did not produce many twins of either sort to begin with.

On an equally modest scale, but many years later, and a whole continent away, we have the two very deprived and majority-black cities of Helena and West Helena (Arkansas, USA) merging by mutual referendum-endorsed consent in 2006 because doing so would enhance their bargaining power for funds in the wider world.

On a larger scale, there is the merger in 1997 of the popular, some said vulgar, UK seaside resort of Brighton (long-beloved of London day-trippers) with the smaller, visually indistinguishable but altogether more upmarket town of Hove just along the coast. This resulted from a more general local government reorganisation by central government, and occurred despite ferocious and near-unanimous opposition (77% according to an opinion poll) from Hove. Central-government determination was clearly one factor here; another was Brighton's oft-expressed desire for merger over very many decades, eying up Hove's wealthy tax base (Middleton 2019).

This is dramatically and revealingly different from, say, Manchester and Salford, where Salford's long-standing determination to resist amalgamation was matched by Manchester's equally long-standing superior indifference (Garrard and Mikhailova 2019, chapter 2). It also contrasts with Minneapolis and St. Paul (ibid., Chapter 1), where constant conflict was matched by mutual indifference, which was eventually rendered irrelevant by the Twin Cities Metropolitan Council's formation which achieved conurban region-wide coordination in the areas where needed while leaving the twins mostly self-governing. Furthermore, we could deem Brighton and Hove as only semi-amalgamated even now: witness the retention of both names in their municipal (though not constituency) title, and the frequent reply of 'Hove, actually' when the latter's residents are asked whether they 'come from Brighton?' (Middleton 2019). Moreover, they still possess a Premier League Division football team entitled *Brighton and Hove Albion*.

Another very different British example can be found further eastwards around the coast in the Medway Towns, formerly Rochester, Chatham and Gillingham who underwent their final of several stages of merger in 1998. Again, central government was important in several of these stages, including the last. So too perhaps was growing conurbanisation – their joint and increasingly joined population comprising 278,556 in 2019. However, equally important is the fact that the three towns had long since lost much of whatever twinned, triple or individual identity they may once have had. Not all, since there was intermittent resistance from at least two towns on grounds of endangered 'equal' status. However, their overall loss was because they jointly possessed a strong naval identity centring on the Medway River, dating back to the Royal Naval Dockyard built in 1567, and involving great naval events, heroism and considerable suffering. This explains why merger had been under intermittent consideration, with active moves from 1903 and with the first formal request lodged by the three councils' Medway Towns Joint Amalgamation Committee in 1944 as the Second World

War wound towards its end. If we deem resilience as a key characteristics of twin cities, then maybe twin or tri-city-ness was never very evident from the start.

Some much larger and later internal twin cities have also merged. Our first volume (Garrard and Mikhailova 2019, chapter 4) contains the most famous example – Buda, Pest and Óbuda, which succumbed to overwhelming national priority and prestige to become the joint capital of newly created Austria-Hungary in 1867. However, and significant of twin resilience, is the fact that it took radical city-wide town replanning to efface the former twin identities, and even now, the bus service still runs with a logo separating the two names (ibid., fig. 4.2).

Lagos and Ikeja in Nigeria also eventually amalgamated (Garrard and Mikhailova 2019, chapter 5). However, amalgamation took a long time, as did Fort William and Port Arthur (Chapter 3 in this volume) who very hesitantly became Thunder Bay in 1970, leaving much dysfunctional twin-city detritus behind them.

Beyond our two books, we have the far larger example of Guangzhou, great trading port, capital of Guangdong Province, and China's third largest city (12,780,000: 2010), apparently 'integrating' with the almost equally venerable Foshan (7,200,000: 2010) 16 km southwest and separated by the Pearl River. Whether they count as amalgamating is uncertain; though, if they do, it may be part of the mindset behind a much broader Beijing-led plan reported as early as 2011 in *Forbes*, called 'Turn the Pearl River Delta into one', to merge all nine cities along the Pearl River into a mega-city then calculated at 260,000,000 people (Chang 2011). The same authoritarian mentality, of course, has also impacted Hong Kong/Shenzhen reviewed later in this chapter. If this mindset is crucial, then one thing it takes to merge twin cities is central-government authoritarian diktat, alongside massive conurbanisation.

....

In the examples presented, it is notable that most twin cities retained traces of their former twinned selves long after formal disappearance. This confirms our central contention that twin cities are resilient, whether within national borders or straddling them. Conurbanisation may change them but only rarely produces their disappearance; even then, only when other factors are present, like determined central government. Indeed, as with some distanced US twins (e.g., Dallas-Fort Worth and Hartford-Springfield), it may even reinforce inter-linkages. Unsurprising, therefore, there are many still-existing twins we would have like to have discovered more about.

Let us begin with Australia, represented by just one twin city: Gold Coast–Tweed Heads (Chapter 4 in this volume). We would have liked at least one other, most notably Perth and Fremantle. Of comparable size during Western Australia's 19th-century Convict era, the two are in a classic dominant-subordinate relationship today: Fremantle with its 29,000 inhabitants serves as port for Perth, and its 2,000,000 citizens. While Perth has been Western Australia's administrative centre since its 1829 foundation, Fremantle was its conflictual rival. A local lobby group had tried making it the capital of the Swan River Colony (Jones 2007, 169). And today, despite the two cities functioning within one metropolitan region and reaching greater harmony, both dedicatedly project their own identities.

Given its many twin cities, the United States is also somewhat underrepresented in the two volumes, its southern border with Mexico aside. We would certainly have liked a couple more to set alongside Minneapolis-St. Paul, perhaps testing their typicality. Many share the characteristic of being part of much broader metropolitan/conurban areas.

At the relatively modest end of the scale, 3 km apart in Illinois, are Urbana (population 2020: 41,684) and Champaign (90,739) surrounded by farming communities but also intervening suburban areas, creating a Champaign-Urbana metropolitan area containing 226,033 people.

Far more massively scaled are Dallas and Fort Worth, 50 km apart and core cities of the huge northern Texas Metroplex. While considerably distanced, these twins share an airport and host multiple jointly titled civil associations: including Executives, Realtors, Black Journalists, Black Psychologists, an Apartment Association, a Hospital Association, a Limousine Association, several non-profit organisations, a Japanese-American Society, a Herpetological Society and the Dallas-Fort Worth Tourism Council. If civil society conceives its remit in such explicitly twinned ways, this clearly indicates strong twin-city self-perception: particularly given that Texas is America's second most populous state.

Almost equally impressive are Hartford (Connecticut's capital), sporting its own flag, and Springfield, Massachusetts. They are 42 km apart but share Bradley International Airport located halfway between. Typical of their kind, they owe their separate existence to a bitter dispute between two of the founders of the original Connecticut colony in 1635, leading one to migrate, alongside his followers, across the Connecticut River into the Massachusetts neighbouring colony. Since then, relationships have been rivalrous, each competing for business even as economic times grew harder from the mid-20th century and into the 1990s, with cooperation also handicapped by intervening state jurisdictions – rather like Australian Gold Coast-Tweed Heads.

Oddly, there was one conspicuous exception to their 20th century non-cooperation: the Hartford Springfield Street Railway starting in 1901, and the product of merging a few smaller rail companies and connecting two slightly earlier railways designated as Hartford and Springfield. Its trams proudly labelled 'Hartford and Springfield ST RY Co.' It went bankrupt, was foreclosed, and replaced by The Hartford and Springfield Coach Company (Connecticut Trolley Museum n.d.).

Only in the 21st century, again like many of their species, have the two started more extensive cooperation: first as the key centres of the 'Knowledge Corridor' embracing 32 universities and colleges and home to 160,000 students. This seems actually to have been part of a much broader Hartford-Springfield Economic Partnership, embracing business, municipalities and cultural organisations. Furthermore, the twins have become two of the three centres in the New Haven, Hartford and Springfield high-speed intercity rail link. They also cooperate in a professional hockey partnership, and, internet searches suggest, an accumulation of jointly labelled associations: the Hartford and Springfield-designated Alumni

Association (unsurprising given the 'Knowledge Corridor'), a Dating Group, an Association for Systems Management, an Actuaries Club, an Underwriting Association, a Collie Club, Geriatric and Adult Psychiatrists, an Auction Co-op and the Airport Admirals Club. The two cities' respective populations in 2019 look modest: Springfield, 153,606; and Hartford, 122,105. However, they are twin centres of a much larger intervening metropolis of urban/suburban settlements totalling 2,180,000.[1]

Any of these US twin cities would have warranted chapters. So too would another US-Canadian cross-border city. Here the prize would be Detroit (Michigan) and Windsor (Ontario). This is North America's largest cross-border conurbation, spanning the Detroit River, with a 2021 population totalling 5,750,000. Partly because of its trading-centre role, partly due to commonly hosting a car industry, it hosts many partnerships and agreements – including the cross-border tunnel, the Ambassador Bridge and the new Gordie Howe Bridge (completing in 2024), cross-border security, legal cooperation, commuting and healthcare. Internet searches also suggest significant numbers of Detroit-Windsor civil-associational links, remarkable for a cross-border twin.

While the Middle East is quite well represented in the two volumes (Garrard and Mikhailova 2019; Chapters 10 and 22 here), one important pairing certainly escaped: Tel Aviv and Jaffa. These might be seen as a sort of test-bed for Israeli-Muslim relations, whose emergence and relationship has been deeply studded by inter-ethnic violence, alongside much international intervention like so much Middle Eastern history, including the League of Nations and the UN.

Jaffa dates back to 1800 BC. Its modern history hesitantly emerged under the Ottomans in the 16th century. It has been mostly Muslim but was becoming slowly more Jewish and somewhat Christian through the 19th century. In 1909 Jewish residents, attempting improved living conditions, established Tel Aviv as Jaffa's garden suburb within Ottoman-ruled Palestine (Golan 1995, 385). Tel Aviv grew rapidly under the League of Nations British-administered mandate 1917–1948. In 1934 it was formally separated from Jaffa's municipal area (ibid.). Jaffa also grew after 1917, reaching 110,000 inhabitants, including 35,000 Jews, in 1947.

When Israel began emerging in 1945–1948, Jaffa was initially to be part of a UN-mandated Palestinian state – an Arab exclave in the Jewish territory (allowing a brief glimpse of a cross-border twin city). However, eventually Jaffa was incorporated entirely into Israel and taken over by Tel Aviv, notwithstanding initial reluctance due to the costs of Jaffa's rehabilitation, necessitated by its destruction during the 1948 Arab-Israeli war (LeVine 2001, 243). Following Israeli-government commitment to Jaffa and Tel Aviv's unification in 1950, the Tel Aviv-Jafo municipality emerged. In Israeli post-war 1950s planning, Jaffa was considered slumland (ibid., 250). Today, at 460,613 (2019), Tel Aviv Jafo is Israel's largest city after Jerusalem and its economic and technological hub. Jaffa is Tel Aviv's semi-incorporated outlier, ten-fold smaller, with 16,000 Muslim inhabitants. As LeVine notes (2001, 252), 'Jaffa has served as the historical "other" of Tel Aviv'. After multiple renewals Jaffa was 'liberated from its Arab identity' and became a tourist attraction and site of elite development (ibid.).

Nevertheless, it seems implausible to include Tel Aviv-Jafo under our list of properly amalgamated twin cities.

The two volumes have featured many cross-border places. One of the more spectacular are Lomé and Aflao (Chapter 11 here) – Togo's capital city jammed alongside much smaller Aflao just over the Ghana border. This prompts editorial longing for one well-known example of twin capitals:[2] Kinshasa (capital, Democratic Republic of Congo – hereafter DRC) and Brazaville (the much smaller Republic of Congo's capital: RC), together forming Africa's largest urban conglomeration.

Both their individual characteristics and their twinned relationship were set around 150 years ago by competing colonial legacies, respectively Belgian- and French-donated; legacies that, after independence were enhanced by Cold War rivalries and strategic calculations led by the USA and USSR, with Zaire's (now DRC) Mobuto regime American-backed and the RC's left-leaning Marien Ngouabi's government supported by Communist Russia (Burke 2017).

Kinshasa and Brazaville's individual characters and inter-relationships also stem from their geographical positions just 4 km apart interfacing across and at the western end of the Pool Malebo on the massive Congo River. The two are located just upstream of the mighty Livingstone Falls rendering the route to the Atlantic unnavigable, forcing both colonial administrations to construct entirely separate rival railways to the coast over 300 miles away to carry expanding multiples of goods from the massive and extensively navigable Congo River Basin to the East, and offloaded in the two respective cities.

Characters and relationships are further set by their evermore dominant positions within the two Republics. Both have massively rising populations: Kinshasa Metro Area at 14,970,000 in 2021 has risen by 53% since 2011 (9,788,000); Brazaville Metro Area at 2,470,000 in 2021 has risen by 50% since 2011 (1,646,000).[3] In population terms, and even more in economic terms, both dominate their countries. Both spectacularly suck population from their rural areas (in Kinshasa's case well beyond), with Kinshasa pulling 390,000 yearly. Unsurprisingly, both cities and city administrations are beset by massive problems around even the most basic infrastructure like drainage and sewage disposal, electricity supply, etc., all worsened by massive inequality, with slums of the Indian 'gunthawari' sort and very little urban planning: Kinshasa's being described as resembling 'an overgrown village' (Adejuwon et al. 2020).

Given these legacies and geographical positions, enhanced by the perceived political necessities of those leading the extensively corrupt and authoritarian regimes, the two cities' relationships unsurprisingly have been at best mutually indifferent, more often rivalrous, even warring ('within easy shelling distance of each other') to both city populations' considerable cost (Burke 2017). This in turn has been worsened because the two cities, like their countries and regimes, have had a classically dominant-subordinate relationship since their respective foundation. As already evident, Kinshasa vastly outranks Brazaville in population and area; so does the DRC compared with the CR. Their Malebo-Pool border is hard, with no way across other than either frequent but crowded ferry boats or by 'the shortest international flight in the world' between their airports.

However, it is hardly impermeable, with many on both sides over the decades seeing the border as a source of profit, opportunity and sometimes escape when inter-regime conflict has become military. A bridge has been intermittently mooted, and in 2020 there was agreement to actually build one, partly financed by the African Development Bank (PIDA Virtual Information Centre n.d.). This might indicate something more cooperative emerging, partly fuelled by the possible emergence of enhanced democracy on the DRC side since 2019. If so, this alongside the 2018 Brazaville Declaration to conserve marshland (UNEP 2018), may indicate the two cities are embarking on enhanced conciliation, as seen in many other twin-city locations.

However, cooperation will not shift the relationship's inherent inequality. Kinshasa is not just far bigger than Brazaville; it is also massively more important in terms of soft power. This is not just due to its far larger higher-education base but also because, partly in consequence, its cultural artistic and particularly musical base is influential not just within the DRC, but across much of Africa, dramatically illustrated by its UNESCO City of Music designation since 2015. This survives despite frequent conflict and disruption, and partly because of the multi-ethnic character of many amongst the millions drawn to Kinshasa. As sometimes suggested, the relationship of these rival and hitherto overly adjacent capitals may just be 'work in progress', but with little doubt about the winner if the work is at all competitive.

One feature particularly of cross-border twin city relationships noted in both books has been the tension between, on the one hand, the often-open and implicitly or explicitly contractual relationships developing between populations near international borders and, on the other, the impact of developments initiated far away by national governments and even international agreements. This takes us into another realm of twin cities and twin-city developments we would have like to have discovered more about.

Thus, the two volumes have hosted chapters on towns straddling the US-Mexican border – one on 'the resiliency of Los Dos Laredos' as points of mutually advantageous interaction and crossover for both Mexican and US citizens; the other here in Chapter 21 on migratory flows northwards and southwards and how these have been dealt with and adapted to by civil society on both sides while also being deeply impacted by changing US migrant policies.

However, even brief investigation reveals the unsurprising fact that these centres interact with other border settlements further south, most notably on the Mexican-Guatemalan border. Two cross-border urban pairs impacted not just by enhanced migrant restriction under Donald Trump but also by reactive Mexican policies towards northward migrant flows stand out: (1) Frontera Corozal (Mexico) and Bethel (Guatemala), and (2) Tapachula (Mexico) and Tecun Uman (Guatemala). During 2018 just before Trumpian policies took hold and while Mexico began responding, large and small 'migrant caravans' moving northwards along Highway 307: 'the Grand Pacific Corridor of the migrant' and hosting 5000 migrants monthly (Time Magazine 2018). All were fleeing crime, poverty and living costs in Honduras. Their passage was being facilitated on both sides of the Usumacinta River crossing by people smugglers (taking time off from ferrying people to nearby Mayan ruins) and helping make Bethel a 'one-industry town'. Both Frontera Corozal and Bethel were prospering from the 'booming migration industry', based on a multi-faceted array of networks of 'coyotes' (people smugglers), 'corrupt officials, crooks and concerned citizens' with Mexico's President Obrador, now engaged in implementing enhanced restriction, complaining that 'everyone gets paid' (Lakhani 2019).

This implementation becomes visible if we turn to the other border twins Tapachula/Tecun Uman, who were seemingly impacted rather earlier than Frontera Corozal/Bethel (Cervantes 2019), suggesting the enforcement of Mexican restrictions was disrupting not just northward migrant flows (which had reportedly almost stopped) but also traditional trade across the Suchiate River, affecting not just traditional boat-traders but also many small and dependent commercial concerns. Meanwhile, 20 miles north of the border, Tapachula found itself hosting many stranded Latin American, African and Indian migrants, very much like Ambos Nogales (Chapter 21). By 2021, Guatemalan security forces, presumably pressurised by neighbouring Mexico, were reinforcing border-security against travellers from Honduras: 'hundreds of migrants' were arriving in Tecun Uman 'to regroup and circumvent the authorities of Guatemala and Mexico through rafts with which to cross the river' and then' filter through the blind spots they have along the 700 km of border that both countries share' (Agencia EFE 2021).

We have so far caught brief glimpses of *particular* twin cities we would have liked chapters about to supplement those in our two volumes. However, there is another internal twin-city *type* without representation in either, and warranting its own chapter: cities nesting *inside* other cities. There are three examples: first, Taiwan's vast capital metropolis of Taipei located wholly inside even-larger New Taipei; second, the two substantially autonomous Cities of Westminster and especially London (the capital's one-square mile financial quarter) inside the vast metropolitan area presided over by the Greater London Authority; the third, the self-governing city state of Vatican City located within the Italian capital of Rome. Of these we have room for just two: the first because it is gigantic, and the third because it is wholly unique.

Taipei is Taiwan's densely populated capital, with 2,602,418 people in 2020, occupying 272 sq km. The city rose rapidly from the Republic of China's establishment in 1949 by the fleeing Kuomintang forces after defeat by the Communists in the Chinese Civil War. It acquired the top 'special municipality' status in 1967, adding to its seat-of-national-government function. However, its overcrowding meant its population levelled and then began falling with people fleeing outwards. Taipei sits adjacent to Taiwan's northern coast. However, it is surrounded on all four sides by the now demographically and territorially dominant New Taipei. In 2020, its rapidly rising population reached 4,030,054, occupying an urban, rural and even mountainous area of 2053 sq km. New Taipei dominates Taiwan's northern coastline. In 2010, it too acquired special municipality status, replacing its former 'county' designation.

Since the two cities share central-governmental institutions and have similar economies, they are hardly rivals. However, their vast conurban condition and deeply inter-related existence has not sapped their determined separation. And they show customary twin-city resilience. There have been intermittent plans to amalgamate them, alongside the neighbouring very large port city of Keelung, on the grounds that unity would render them a 'global megacity'. These resurfaced most recently in 2013–2014. However, though apparently enticing for some, and notwithstanding a major public enquiry, the plan was dropped after extensive public consultation exercises. These, themselves a tribute to Taiwan's vibrant democratic culture, showed amalgamation lacked significant support in any of the three cities' populations (Taipei Times 2014).

We now turn to the world's oddest twin city: Vatican City State located within the city of Rome, a pairing that arguably has become both internal and cross-border. Rome, as generally known, dates from its foundation in 753 BC, centre of the Roman Empire until the latter's fall in 476 AD, subsequently becoming capital of the Papal States and, since 1870, of Italy. In 2020, the city's Commune population was 2,808,293 with the broader Metropolitan City weighing in at 4,253,414. These two respective areas (1508 and 5352 sq km) are governed by an elected City Council alongside 15 municipalities and the Metropolitan City Council and elected Mayor alongside 121 municipalities. Alongside its municipalities, of course, Rome also houses Italy's legislative and governmental machinery.

In Rome's centre, spanning all of 0.48 square kilometres, a 2019 population of 825, and governed by what is essentially an absolute monarchy and elective theocracy, lies Vatican City. After Rome's fall, its Popes (who began residing in Rome only from 1583) governed the Papal States until their absorption into independent Italy in 1871, after which they lived in uneasy co-existence with Rome and Italy until the Lateran Treaty with Mussolini in 1929 (modified 1984) created the independent Vatican City state. This has its own Swiss Guards and Gendarmerie Corps, and the Holy See conducts Vatican diplomatic relations with overseas embassies scattered through Rome city. It issues its own coins and stamps, manages an eight-team football championship, and is a central feature of Rome's vast tourist industry. Arguably far more important is the fact that the Pope and Vatican City are the centre of vast, if declining, soft power via the world's 1.2 billion Roman Catholics across the world, including many of these two volumes' twin cities.

Another vivid example of the interdependence between Vatican and Rome, a common twin-city feature, is the Vatican's only pharmacy, arguably one of the world's busiest, serving about 2000 customers daily and handling over 40,000 prescriptions annually (Glatz 2008; Rome Reports 2020). Over 50% of its customers come from outside Vatican City – from Rome and beyond. Customers frequent the pharmacy both to acquire 'foreign, hard-to-find medicines' and to 'fill prescriptions at cut-rate prices': in 2008, the Vatican pharmacy was selling products at prices 12% to 25% below Italian drugstores generally (Glatz 2008). Admitting its high demand, the Vatican pharmacy started selling cosmetics and beauty-care products in mid-2000, both quickly becoming an important source of its revenue (ibid.).

Finally, and reinforcing our enduring sense that twin-city relationships are affected by developments both nearby and faraway, we would like to know more about how relationships between the twins in both volumes have been affected by developments since our authors researched and wrote about them. Updating is necessarily hazardous, partly because it can verge on prediction and partly since it is always likely to be undermined by the ever-changing present: embracing, for example, not just Donald Trump but now Joseph Biden. However, let us survey the preceding factors as best we can.

Trumpian hostility to northward US-bound migration, and towards migrants established in the US for greater or lesser periods, has already evidenced its impact on the two Nogales in this volume, and from the glimpse this conclusion has allowed into the Mexican-Guatemalan border situation. It has probably also impacted the deeply transactional relationship between the two Laredos revealed in Garrard and Mikhailova 2019, chapter 11). So too has COVID-19: when we asked one of the chapter's authors John Kilburn back in October 2020 for comment, he told us COVID's impact could 'only be described as "horrible"'. Kilburn referred particularly to 'significant travel restrictions from Mexico to the US, significantly impacts on tourism, shopping, day-labor and family visits for those travelling from Nuevo Laredo to Laredo as part of their everyday life'. As of late 2020, the restrictions were due for renewal, and probably still continue, though north-to-south border movement in the Laredos has been seemingly less affected because, for example, US citizens' medical visits (to consult cheap Mexican doctors) were exempted.

Meanwhile, we have already seen from this volume's Chapter 21 and from earlier in this conclusion how Trumpian migration policies impacted the situation on the US-Mexican and Mexican-Guatemalan borders and the Mexican towns along the 'migrant highway'. Chapter 21 also explored the impact of southward flows from the US caused by accelerating ejections of long-standing US residents of Latin American origins and without secure legal status.

The Biden Presidency has now changed policy again, particularly by pausing ejections and ceasing to repatriate unaccompanied children and teenagers crossing into the US from Guatemala, El Salvador and Honduras. Both changes will have impacted all these towns. Unsurprisingly, it is the latter that has drawn international-media attention, particularly because there has been a huge increase in migrants, especially unaccompanied children, arriving at the US border (100,441 migrants arriving in February 2021; Greeve 2021). In the short term at least, this has seemingly induced chaos in US reception centres, while children await distribution to US-established relatives able and willing to house. Such centres have been beset by massive overcrowding: for example, 4500 housed in facilities designed for single men, and 3000 teenage boys located in a commandeered Dallas convention centre (ibid). Moreover, it is not just real policy changes but also *perceptions* of them amongst people fleeing poverty, crime, violence and hurricanes particularly once enhanced by people smugglers eager to profit.

Meanwhile, the impact of COVID-19, Hong Kong protests and resulting Chinese crackdowns has certainly affected the Hong Kong-Shenzhen relationship, though precisely how and how much is hard to say right now. As noted earlier, it needs setting within the context of China's declared determination to integrate Hong Kong into the Greater Bay Area, including the Shenzhen conurbation.

The wide-ranging report on the relationship in the light of Hong Kong protests from Bloomberg News (2019) presents 'competing visions of China's destiny' in the two cities, noting Hong Kong's 'waning allure' for ambitious young Chinese in Shenzhen which was 'young, hopeful and looks optimistically to a future where it can help push China's drive to dominate the next century through an innovating economy that sidesteps political freedoms.' This was being fuelled by Shenzhen's 7.5% economic growth against its neighbour's relative stagnation. Meanwhile, many Hong Kong young people no longer much fancied Shenzhen either, feeling it materialistic. And China's authorities were actively trying to prevent news of protests reaching the neighbouring city via smart-phone border checks, something perhaps understandable because some Shenzhen people were crossing to join the protests.

This may underpin a later report in *Prospect* (Mitter 2021) suggesting a possible dilemma for the Chinese authorities. On the one hand, enfranchising the rather numerous Shenzhen residents crossing daily into Hong Kong to work might help dilute the radicalism of the latter's elective legislature; on the other, it might also politicise some in Shenzhen. It also noted an *Economist* 2019 poll suggesting over 60% of Hong Kong's young people identified themselves as Hong Kongers, not Chinese.

Understandably, COVID-19 has also impacted the relationship, possibly in connected ways. Various sites suggested that, after the initial flight of many in Hong Kong across the border fearing infection, both city authorities began restricting border travel to just two entry points – at Shenzhen Bay and the Hong Kong-Zhuhai and Macau Bridge. The Security firm Garda World (2021) suggested things might ease from China's end, but Hong Kong's restrictions seemed indefinite.

Finally, we come to the situation along Brazil's triple frontier (featured in Chapter 19) created by the chaotic handling of the COVID-19 pandemic by President Bolsonaro's populist regime, producing 314,000 deaths and 100,000 new daily cases at the point of writing, alongside increasing signs of regime instability following resignation by the heads of the three armed forces. One veteran Brazilian diplomat was quoted as saying, 'The other day I saw a pretty strong article saying Brazil was starting to be seen by its neighbours as a sort of leper colony … and it's probably true'. The situation is seemingly causing increasing consternation amongst those neighbours, with increasing pressure on relevant governments to close borders against Brazilian incomers: pressures now reverberating into border areas. For example, Uruguay's authorities were reportedly 'racing' to vaccinate residents in its border region with Brazil: 'the idea is to create an epidemiological shield', according to a Uruguayan governmental advisor (Phillips et al. 2021).

Overall, the situation here as elsewhere is ongoing. Given how susceptible both internal and particularly external border twins are to changing situations nearby and also originating faraway, the editors can only long for a continuous river of updating information flowing into both volumes.

Notes

1 Here and in what follows, population data has been retrieved from citypopulation.de, unless other source is provided.
2 The other we know about is Tallinn and Helsinki (see Chapter 1).
3 UN Population Projections for Kinshasa [https://www.macrotrends.net/cities/20853/kinshasa/population] and Brazaville [https://www.macrotrends.net/cities/20848/brazzaville/population]. Accessed 01.04.2021.

References

Adejuwon, J.O., MacGaffey, W., Cordell, D., and MacGaffey, J. (2020). Kinshasa. *Encyclopedia Britannica*. Available: https://www.britannica.com/place/Kinshasa (accessed 29.03.2021)

Agencia EFE (2021) Migrantes hondureños llegan a cuentagotas a la frontera de México-Guatemala. Available: https://www.efe.com/efe/usa/inmigracion/migrantes-hondurenos-llegan-a-cuentagotas-la-frontera-de-mexico-guatemala/50000098-4444359 (accessed 29.03.2021)

Bloomberg News (2019) Hong Kong versus Shenzhen: Two competing visions of China's future. Available: https://www.bloomberg.com/features/2019-hk-v-shenzhen/ (accessed 29.03.2021)

Burke, J. (2017) Face-off over the Congo: The long rivalry of Kinshasa and Brazzaville, *Guardian*, Available: https://www.theguardian.com/cities/2017/jan/17/congo-rivalry-kinshasa-brazzaville-river-drc (accessed 28.03.21)

Cervantes, R. (2019) Recent migration policies deepen the divide at the Mexico-Guatemala border. *Arizona Public Media*. Available: https://www.azpm.org/p/home-articles-news/2019/9/11/157956-recent-migration-policies-deepen-the-divide-at-the-mexico-guatemala-border/ (accessed 29.03.2021)

Chang, G.G. (2011), 'A city of 260 million: Where else but China?', *Forbes*. Available: https://www.forbes.com/sites/megacities/2011/04/11/a-city-of-260-million-where-else-but-china/?sh=4a9b1a5755a9 (accessed 29.03.2021)

Connecticut Trolley Museum (n.d.) Local street railway history. Available: https://www.ct-trolley.org/about/hartford-and-springfield-street-railway/ (accessed 29.03.2021)

Garda World (2021) China: Authorities to ease COVID-19 measures in Hong Kong from Feb.18. Available: https://www.garda.com/crisis24/news-alerts/444056/china-authorities-to-ease-covid-19-measures-in-hong-kong-from-feb-18-update-52 (accessed 29.03.2021)

Garrard, J., and Mikhailova, E. (Eds) (2019) *Twin Cities: Urban Communities, Borders and Relationships over Time*. New York: Routledge.

Glatz, C. (2008) World's busiest pharmacy? Vatican drugstore offers cut-rate prices. *Catholic News Service*. Available http://webarchive.loc.gov/all/20080611005258/http://www.catholicnews.com/data/stories/cns/0802820.htm (accessed 13.03.21):

Golan, A. (1995). The demarcation of Tel Aviv-Jaffa's municipal boundaries following the 1948 war: Political conflicts and spatial outcome. *Planning Perspectives*, 10(4), 383–398.

Greeve, J. (2021) Surge in migrants seeking to cross Mexico border poses challenge for Biden. *Guardian*. Available: https://www.theguardian.com/us-news/2021/mar/10/migrants-surge-us-mexico-border-biden (accessed 30.03.21)

Jones, R. (2007). Port, sport and heritage: Fremantle's unholy trinity. *Geographies of Australian Heritages: Loving a Sunburnt Country*, Routledge, London, 169–185.

Lakhani, N. (2019) 'Everyone gets paid': Mexico's migration networks are thriving despite crackdown, *Guardian*. Available: https://www.theguardian.com/world/2019/sep/12/mexico-people-smugglers-migrants-coyotes-guatemala (accessed 28.03.21)

LeVine, M. (2001). The 'new-old Jaffa': Tourism, gentrification and the battle for Tel Aviv's Arab neighbourhood'. *Consuming Tradition, Manufacturing Heritage: Global Norms and Urban Forms in the Age of Tourism*, Routledge, London, 240–272.

Middleton, J. (2019) Amalgamation and Hove. Available: http://hovehistory.blogspot.com/2019/03/amalgamation-and-hove.html (accessed 28.03.21)

Mitter, R. (2021) China is tightening its grip on Hong Kong. *Prospect*. Available: https://www.prospectmagazine.co.uk/magazine/china-is-tightening-its-grip-on-hong-kong-protests-free-speech (accessed 28.03.21)

Phillips, T., Goñi, U., and Daniels, J.P. (2021) 'The heart of darkness': Neighbors shun Brazil over Covid response. *Guardian*. Available: https://www.theguardian.com/global-development/2021/mar/30/neighbors-shun-brazil-covid-response-bolsonaro (accessed 13.03.21)

PIDA Virtual Information Centre (n.d.) Brazzaville-Kinshasa road/rail bridge. Available: https://www.au-pida.org/view-project/2014/ (accessed 28.03.21)

Rome Reports (2020) Vatican pharmacy transitions into robotics age. Available: https://www.youtube.com/watch?v=Nfmyf7QdUSY (accessed 13.03.21)

Taipei Times (2014) Three-city merger debated again. Available: https://www.taipeitimes.com/News/taiwan/archives/2014/07/07/2003594539 (accessed 30.03.2021)

Time Magazine (2018) Caravans help migrants travel safely. Available: https://time.com/longform/caravan-guatemala-mexico-border/ (accessed 13.03.21)

UNEP (2018) Historic agreement signed to protect the world's largest tropical peatland. Available: https://www.unep.org/news-and-stories/press-release/historic-agreement-signed-protect-worlds-largest-tropical-peatland (accessed 30.03.2021)

Index

Page numbers followed by 'n' refers to notes numbers

Printed in the United States
by Baker & Taylor Publisher Services

Printed in the United States
by Baker & Taylor Publisher Services